Lecture Notes in Earth Sciences

Lecture Notes in Earth Sciences

Edited by Somdev Bhattacharji, Gerald M. Friedman,
Horst J. Neugebauer and Adolf Seilacher

22

Ivan I. Mueller S. Zerbini (Eds.)

The Interdisciplinary Role of Space Geodesy

Proceedings of an International Workshop held at
"Ettore Majorana" Center for Scientific Culture,
International School of Geodesy – Director, Enzo Boschi –.
Erice, Sicily, Italy, July 23–29, 1988

Springer-Verlag
Berlin Heidelberg GmbH

Editors

Ivan I. Mueller
Dept. of Geodetic Science and Surveying
The Ohio State University
1958 Neil Avenue, Columbus, Ohio 43210, USA

S. Zerbini
Dipartimento di Fisica, Settore Geofisica
Università degli Studi
Via Irnerio 28, Bologna 40126, Italy

ISBN 978-3-540-51161-8   ISBN 978-3-540-46173-9 (eBook)
DOI 10.1007/978-3-540-46173-9

2132/3140-543210 – Printed on acid-free paper

PREFACE

Our planet is evolving and changing; its surface is capable of unleashing great violence as its crust is created and destroyed. Quite remarkably, it has been only recently that the fundamental elements of this evolution were fully appreciated, and only within the last decade have there been technologies capable of directly measuring the global motions of the Earth's crust which are one of the most visible manifestations of these processes.

Before the advent of space technologies, the nature of contemporary global plate motions went largely unobserved. These motions were understood from the geological records, and plate rates for million year averages were established. Fortunately, the revolution in geophysics brought about by the general acceptance of plate tectonic theory has been paralleled by significant advances in space geodesy oceanography and geophysics. New space technologies have rapidly matured, yielding new insights and capabilities for more completely understanding the dynamical properties of the Earth, its oceans and atmosphere. Likewise, the evolving earth sciences capabilities from space are fostering new questions and goals made possible through the creative exploitation of satellite missions.

A workshop entitled "The Interdisciplinary Role of Space Geodesy" was held in Erice, Italy, on the island of Sicily on July 23-29, 1988, to discuss the directions and challenges of space geodeys for the decades to come. This international gathering was made possible by the E. Majorana Centre for Scientific Culture int he framework of tis International School of Geodesy. The workshop was sponsored by the Italian Ministry of education, the Italian Ministry of Scientific and Technological Research, the Sicilian Regional Government, the Italian National Institute of Geophysics, and the National Aeronautics and Space Administration of the United States.

This volume is the result of the dedicated effort undertaken by an international group of scientists and administrators who have contemplated the challenge of the future of space-based earth science for the next decade. Recognizing the need for defining new milestones both in science and technology, they have developed a detailed report of what could be achieved and what challenges remain after twenty fertile years of space exploration.

This workshop was based upon a similar conference which was held in 1969 in Williamstown, Massachusetts, USA. The so-called "Williamstown Report" was a remarkable document which designed programs, detailed challenges, and gave a focus to the application of space and astronomic techniques in the study of solid earth and ocean physics. This Williamstown report was a valuable guide which was instrumental in laying the foundation for the evolution of space geodesy ring the last two decades. Much of what was proposed in 1969 has now been realized. However, the questions which confronted those at Williamstown are equally important today: Where is the technology going and what science can be accomplished as the technology is improved? What space flight missions are most desirable? What are the major problems in further understanding the dynamical Earth? And, most importantly, how can space geodesy uniquely contribute to the solution and edification of these goals?

During the Erice workshop the issues of Williamstown have been revisited. The international Earth science community has addressed the role of space geodesy in the Earth sciences of the 1990's with the same openness, breadth, and enthusiasm which was undertaken in 1969. The availability of highly accurate and transportable systems has made it increasingly possible to detect movements and deformations of the major tectonic plates. This has spurred an interest for new campaigns leading to improved monitoring of regional and highly temporally resolved crustal motions especially in zones of high tectonic activity. These activities require the broad collaboration of the international Earth science community who have joined together to cooperate in major programs. This international cooperation is necessary and must continue as global and regional projects develop and require sharing of data, instruments and science. Major advances in the solution of problems affecting our planet, such as improving our understanding the seismic hazard and abating its risks, are promises of the future. International cooperation in a multitude of other areas such as planetary space exploration and improved understanding of the consequences of man in our planetary ecosystem can also be fostered in the spirit of Erice.

The workshop was organized in panels corresponding approximately to the chapters in this book. The panel membership in indicated in the List of Participants at the end of the book. Those who contributed directly to the writing are listed under the individual chapter/section headings. Professor William M. Kaula who led the Williamstown conference provided the Introduction and coordinated the thankless, but all important, effort to synthesize the workshop's key conclusions and recommendations, which are presented in the Overview.

We would like to thank all the participants in the workshop for their contributions to the realization of this volume. In particular, we acknowledge the panel co-chairmen for their work in structuring the panels and in leading the discussions which produced the final panel reports. Sincere thanks are due to Mrs. Maria Chiara Jannuzzi who helped organize the workshop and to Dr. Pinola Savalli, Dr. Alberto Gabriele and Mr. Jerry Pilarski for their support and excellent organization at the Ettore Majorana Centre. Special thanks are tue to Dr. Edward Flinn of the Geodynamics branch at NASA Headquarters for his continuous and enthusiastic support of this initiative. Let us express our gratitude to Professor Antonino Zichichi, director of the Ettore Majorana Centre for his attention and support in the development of Earth sciences. Lastly, our sincere gratitude to Professor Enzo Boschi, director of the School of Geodesy, is warranted, not only for his support in making the Erice workshop possible, but also for his longstanding interest and continued support of geodesy at all levels.

31 December 1988
Columbus, Ohio   Ivan I. Mueller
Bologna, Italy   Susanna Zerbini

CONTENTS

CHAPTER 4   INTERACTION WITH OTHER DISCIPLINES AND PROGRAMS

CHAPTER 5   INSTRUMENTATION

CHAPTER 6   DATA ANALYSIS

CHAPTER 7   REFERENCE COORDINATE SYSTEMS

CHAPTER 8   EDUCATION

APPENDICES

OVERVIEW

The purpose of the workshop was to recommend geodetic and geomagnetic programs and missions, and the development of methods and instrumentation for their implementation. The scope of the conference included all mechanical and magnetic aspects of the solid earth, oceans, and core measurable by space geodetic and geomagnetic techniques.

The conference at Erice in 1988 most closely resembled that at Williamstown in 1969, not only in focusing on geodynamics, but also in bringing together those knowledgeable about the scientific problems with those expert in the relevant techniques. Several other studies in recent years have examined the application of space techniques to the solid earth and oceans, but have generally been broader in scientific scope while not including as detailed consideration of technical feasibility.

Four main themes emerged from the conference.

• The accuracy and mobility of observing systems *now* enable frequent and spatially dense measurements of great value to the study of the oceans circulations and the tectonics of deformation zones. Hence there should be an increased emphasis on field observation programs with currently available technology. This is a *change* of circumstances evolving since about 1978.

• Significant potential exists for further improvement in instrumentation of value to oceanic and tectonic physics. The effective applications of these improvements will require more intense attention to environmental effects.

• The principal areas where space techniques alone do not appear capable of getting the resolution and accuracy needed to solve tectonic problems are: the gravity field over the land; submarine topography; and precise positioning on the sea floor. However, significant improvement in measurement of the gravity field to support both tectonic and oceanic studies appears feasible.

• Space geodetic techniques can contribute significantly to the solution of environmental problems of great contemporary concern; most notably, to the study of:

(1) global warming and other aspects of climate by measurement of ocean dynamic heights, the height of sea level relative to the land, and the topography of the ice sheets; and

(2) earthquake hazard alleviation, through measurement of tectonic deformations precursory to earthquake occurrence.

The eleven recommendations gathered in this summary are endorsed by the workshop participants as *both* scientifically urgent *and* technically feasible. Further recommendations, some not satisfying both these criteria, appear in the detailed chapters. The first five recommendations are intended to be implementable with current technology; the sixth, partially so, while the next five require appreciable R&D. The ordering is not intended to imply priority other than the urgency to get going with programs of frequent and abundant measurements by currently available techniques.

I. Pursue a vigorous oceanographic research program using state-of-the-art altimetry from Topex/Poseidon and ERS-1 in conjunction with ERS-1 and NSCAT scatterometry, and surface observations, and lay the basis for continuing monitoring.

Oceanographers have never before been able adequately to sample the global ocean with surface techniques. Seasat and Geosat altimetry have clearly sampled the mesoscale component of oceanic variability with unprecedented coverage. Building on both these and advances in surface instrumentation, oceanographers have defined a fully optimized altimetric mission:

Topex/Poseidon. This mission, as well as ERS-1, will provide valuable data to the Tropical Oceans Global Atmospheric (TOGA) program and will be part of the World Oceans Circulation Experiment (WOCE). In this context, the objectives of Topex/Poseidon are to measure the variable circulation and model its dependence on wind forcing, to determine the ocean tides, and to lay the basis for continuing monitoring. These programs of ocean altimetry and the associated scatterometry will contribute importantly to the improved understanding of climate problems such as the global warming.

II. The major effort in positioning techniques in the next 20 years should be sustained, repeated measurements of dense networks at centimeter-level accuracy to determine the time dependence and spatial distribution of deformation within and across zones of intense tectonic activity. Measurement frequencies should range from daily to annually over a decade or more, with station spacing from 3 to 30 km and network dimensions from 10 to 1000 km, depending on region.

It is now well established that the relative motions between major tectonic plates on a decade time scale agree with the motions on a million-year time scale within one cm/yr. Hence attention is now concentrated on the zones of deformation between the plates, which may be several hundred kilometers wide. As indicated by geologic field work and terrestrial geodetic metworks, the relative motion across these zones is distributed among many faults. These faults are the sites of earthquakes, but some also are characterized by gradual motion: "creep." Accuracy of one mm/yr would be valuable; hence centimeter-accuracy baselines will give appreciable information in a decade. It is important that these baselines be measured in many places, with configurations and time intervals varying in accord with indications from geology, terrestrial geodesy, seismicity, and other surface techniques. Tectonic zones also have a great variety in character, and hence these centimeter-accuracy networks should be observed in several regions, preferably those where other data and research furnish guidance.

The applications recommended here are far from the ultimate system desirable for the monitoring of possible earthquake precursors (see Recommendation IX), but will make major contributions to the understanding of the underlying tectonics essential to interpretation of phenomena related to earthquakes. This development of understanding will be a slow, difficult process; hence it is important to facilitate it with more data as soon as possible.

III. Measure the vector magnetic field to a one nanotesla accuracy globally at two altitudes: (1) 200 km, in a mission lasting some months; and (2) above 600 km, in a mission lasting several years.

The MAGSAT mission of 1979-80 revealed many interesting patterns correlated with geologic features of the crust. These correlations could be multiplied and sharpened many-fold by a lower altitude more accurate system as is now feasible. But such a low altitude system is unsuitable for study of the long wavelength magnetic field, which has temporal variation because it is generated by motions in the fluid iron core. A magnetometry mission lasting a decade or more, requiring a higher altitude, would be the greatest possible advance in unraveling the nature of the motions constituting this geodynamo.

IV. Measure the variations in topographic height of the land and the ice to an accuracy of one meter with a horizontal resolution of 100 meters.

While much of the Earth's surface has been covered by airborne photogrammetric techniques, the resulting topographic maps vary greatly in quality and are far from a coherent data base. Improvement in this database is essential to the interpretation of the gravity field over the land (VI below), as well as supporting a variety of tectonic, geomorphological, hydrological and ecological studies. Most urgent is monitoring of Alpine, Antarctic, and Greenland ice sheets, whose response to the global warming is quite uncertain. but which could greatly affect sea-level and climatic evolution.

The design of the mission will require some study, since the best technique— radar, laser, microwave interferometer, or optical stereo— is not clear at this point. It is also highly desirable, of course, to improve bathymetry — submarine topography — for interpretation of sea surface altimetry and other purposes. However, this is a problem of marine, rather than space, technology.

V. Install microwave transponders on all altimetry missions and other spacecraft in precisely determined orbits sensitive to variations of the gravity field, so that they can be tracked by GPS or GLONASS or other systems enabling continuous coverage.

The greatest limitation of dynamic satellite geodesy is the lack of continuous tracking, because of dependence on ground stations. Radio tracking of close satellites from navigation satellites, such as NAVSTAR, would solve this problem. In particular, the installation of a GPS receiver on the Topex/ Poseidon altimetry spacecraft should reduce the radial errors of its orbit down to an acceptable level of a few centimeters. GPS receivers on Topex/Poseidon and other spacecraft sensitive to the gravity field, such as ARISTOTELES and Gravity Probe B, should also obtain the geoid to sufficient accuracy to infer the temporally invariant oceanic dynamic heights on a basin scale with a few centimeters error.

VI. Develop and implement spaceborne techniques to determine the spatial variations of the gravity field to a resolution of 100-km or better, plus airborne techniques to determine variations in selected area to a resolution of 10 km.

An improved gravity field globally is needed to obtain variations (1) over the land to support tectonophysical studies and other applications of gravity; and (2) the oceans, to support ocean dynamical studies using centimeter-accuracy altimetry (Recommendation VIII below). The 100 km resolution proposed should be feasible with a satellite-borne gravity gradiometer (or satellite-to-satellite range-rate) at 160 km altitude, sustainable for a mission duration of six months. The GRADIO gradiometer on the ARISTOTELES mission (altitude 200 km, 5 mgal accuracy goal) will contribute significant insights to solving problems of gradiometry. To assure the 100 km resolution, it should be followed by a superconducting gravity gradiometer mission of higher performance (altitude 160 km, 1 mgal accuracy goal).

The 100 km resolution will still be inadequate for tectonophysical studies in areas of intense deformation, as described in Recommendation II above. For areas lacking adequate surface gravimetry (e.g., the Andes and parts of the Alpide belt), a demonstrated technique that should attain the 10-km resolution is airborne gravimetry with altitude control by differential radiointerferometry using GPS signals.

VII. Develop and implement techniques to monitor changes in the rotation vector of the solid earth to 0.0001" (0.1 mas, equivalent to 0.3 cm on the surface) to the highest possible frequency (at least several cycles per day).

The variations of the Earth's rotation are valuable integral constraints on global-scale shifts in the fluid parts of the earth that impose torques on the solid earth. Currently the rotation angle and polar direction are monitored to 0.0015" by Very Long Base Line Interferometry (VLBI) at five-day intervals and Satellite Laser Ranging (SLR) at three-day intervals. The feasible improvement in accuracy should pick up much higher frequency atmospheric effects known to exist. It will also greatly increase the probability of detecting a change in the pole path arising from an earthquake, plus determining whether there is significant aseismic motion associated therewith. Complementary geophysical, oceanographic, and atmospheric data should be taken.

The evident technique for the high frequency monitoring is VLBI, in which appreciable improvement of instrumentation appears feasible. Also required are improvements in the reference frames— particularly location of radio sources— and in corrections for environmental effects (see Recommendation X). GPS techniques may provide a valuable supplement in polar monitoring.

VIII. *Develop and implement a satellite altimetry capability of one centimeter, to measure the ocean height every week with a 25 km resolution.*

The 25 km spacing would entail some combination of multiple satellites and side-scanning capabilities. Such resolution is necessary to sample the full range of mesoscale variability of the oceans at the 1-cm level. It would also be valuable for tracking other transient ocean dynamic phenomena, such as Kelvin waves in the equatorial zone. The 1 cm altimetry at intervals of 10 km or finer would also give a greatly improved geoid for tectonophysical purposes, allowing inference of departures from standard models of cooling, spreading lithosphere.

Full application of the one-centimeter altimetry to ocean dynamics probably requires an improved separate determination of the gravity field (Recommendation VI), while its full application to tectonophysics requires improvement in bathymetry (Recommendation IV).

IX. *Develop a capability to measure baselines to one millimeter or better, vertically as well as horizontally, over distances from 1.0 to 100 kilometers, with the flexibility to measure evolving events within an hour and the stability to monitor motions in the solid Earth over years.*

Appreciable improvements over the system discussed in Recommendation II are desirable to study transient phenomena in zones of intense tectonic activity, as well as to extend monitoring networks to regions of lesser, but non-zero, activity, such as eastern North America and northwest Europe. Emphasis on the geological context and on other date types in site selection would continue indefinitely.

A millimeter accuracy system would also yield much new information about the temporal and spatial variations of the solid earth in response to other forces, such as tides, oceanic and atmospheric loading, and post-glacial rebound.

Currently, the evident technique for this future system would be laser ranging from a spacecraft to reflectors on the ground, such as the GLRS proposed for the EOS spacecraft. But such a system depends on good weather, and it may not be able to respond to an evolving event without being installed on multiple spacecraft. Hence consideration should be given to an improved CW microwave system, much more optimized for geodetic application than GPS or GLONASS. In addition, any tracking system, whether laser or microwave, should be flown on higher orbiters optimized for geodetic purposes.

The nature and extent of the application of space techniques to temporal change in the solid earth should be closely coordinated with developments in *in situ* measurements by strainmeters, tiltmeters, dilatometers, gravimeters, and tide gauges, all of which are expected to be improved in both instrumentation and correction for environmental noise.

The development of accurate baseline measurements on the ocean floor is also desirable to measure the nature of spreading at ocean rises and to obtain the deformation pattern on the oceanward side of subduction zones. But even if an accuracy of one centimeter is obtainable, it is not clear whether diversion of resources to this expensive activity would be the optimal allocation until deformation on land is much better understood.

X. *Develop techniques to determine environmental effects as necessary to make positional measurements truly accurate to one millimeter over many years' duration.*

To achieve meaningful one millimeter accuracy over baselines for the study of tectonic activity and other geophysically significant phenomena (Recommendation IX), a great variety of environmental "noise" must be removed. This noise can be divided into two categories: (1) that affecting the transmission of signals; and (2) that affecting the reference marks.

In category (1), the tropospheric delay, both wet and dry, is most critical. Effective calibration thereof will depend on development of remote sensing instrumentation such as multi-channel water vapor radiometers, Raman scattering lidars, multi-color laser ranging and coherently combined laser/microwave systems. In addition to instrumentation, there need to be developed improved modeling and mapping algorithms including observation strategies and ancillary systems to measure meteorological parameters.

In category (2), these are a variety of effects ranging from frost heave of reference marks to regional oscillations caused by ocean and atmospheric loading and ground water motion. Again, a combination of measurement and modeling of effects is indicated in a development program aimed at specifications for site selection, monumentation, supplemental marks, environmental measurements, etc.

XI. *Improve education in space geodesy for both geodesists and other scientists and engineers.*

The advances in space instrumentation and technology, as well as in the interpretation of geodetic data, require appreciable raising of the levels of technical expertise and interdisciplinary interaction. The techniques on which the greater accuracy and detail depend come from a variety of expertise outside the traditional scopes of geodesy, solid earth physics, and physical oceanography. The space geodesist needs to be educated in meteorological effects, orbital dynamics, optical and electronic techniques, and the implications of geodetic measurements of models of ocean circulation and tectonophysical activity. The earth and ocean scientists need to become aware of the capabilities and limitations of the space geodetic techniques as well as to adapt their models to take better advantage of the constraints afforded by geodetic techniques. Although space geodetic techniques have been evolving for three decades, there still are lags on both the geodetic and geophysical sides in integrating them into curricula. There also persists the need to inform policy makers about the applicability of space geodesy to problems of the Earth and environment.

The global nature of the problems addressed by some geodetic techniques makes it desirable to enlist the efforts of scientists and engineers from many countries. But most developing countries are much more concerned about economic development than environmental quality. To obtain this assistance, attention should be paid to the economic benefits of space techniques to surveying and mapping programs, an aspect not touched on in the scientifically oriented parts of this report.

Chapter 1

INTRODUCTION

William M. Kaula

The Overview is a synthesis of the recommendations generated in the final stage of the workshop. Successive drafts were revised in response to comments by members of the editorial board and others. Hence the Overview is a consensus document.

This Introduction is more idiosyncratic. It is based on the opening presentation at the workshop in Erice, July 24, 1988, an attempt to review the leading scientific issues to which space geodesy applies. While the Overview is essentially an update of the principal recommendations of the "Williamstown Report" (MIT, 1970), this Introduction is more similar to the summary of the GEOP conferences (Kaula, 1979). It has been revised by taking advantage of the detailed chapters, but it is more selective in subject matter. The scientific topic areas selected are:

1. the Earth's rotation and core-mantle interaction;
2. mantle convection;
3. regional tectonics and earthquakes;
4. ocean dynamics;
5. Venus-Earth differences.

The emphasis is mainly on the scientific questions, except in problem area 3, tectonics and earthquakes, where there are questions as to whether space techniques can achieve the resolutions and precisions needed.

1. THE EARTH'S ROTATION AND CORE-MANTLE INTERACTION

"Yet that things go round and again go round
Has a rather classical sound"

Wallace Stevens
The Pleasures of Merely Circulating

As detailed in Chapter Two, there has been significant advance in the description of the Earth's rotation (a vector quantity, involving direction of the pole as well as rate) by laser and radiointerferometric techniques. Many of the conjectures of Munk and MacDonald (1960) have been answered, but some remain, such as the mechanism of dissipation of tidal and wobble energy in shallow seas, and the nature of core-mantle coupling evidenced by the discrepancy of the nutation from that for a homogeneous Earth. This advance in understanding of global behaviour also depends significantly on improvement in other global data, such as atmospheric angular momentum (Newell, 1974); sea level; secular change of the zonal harmonics of gravity (Yoder et al., 1983; Sabadini et al., 1988); and change in glacial volume (Meier, 1984; Yoder and Ivins, 1985).

But the secular change of the rate of rotation has a special fascination, as an inexorably unidirectional phenomenon. Hence there is a sort of symbolism in its now being measured precisely enough that a possibly anthropogenic effect— the global warming— is now edging above the noise level. The budget (in $10^{-11}/yr$ for ω/ω) now runs roughly:

Total observed:	-19.
Tidal friction:	-26.
Alpine Glacial melting:	- 2.
Ocean warming:	- 1.

Leaving for post-glacial rebound: +10. (Peltier, 1988).

The oddity is that the sea level rise of 1.0-1.5 mm/yr is essentially accounted for by Alpine glacial melting (Meier, 1984) plus ocean warming. One wonders why the polar ice caps are now changing so little, since Antarctica apparently declined in recent geologic time (Nakeda and Lambeck, 1988). Will the ice caps remain stable if there is further increase in atmospheric CO_2? At present, climatic theorists cannot say whether further warming would lead to a waning of the glaciers from melting, or to a waxing from the increased evaporation causing more snow. Surely this is a problem to which satellite-borne altimeters apply, to monitor the ice sheets of the world to get as early warning as possible of changes in their heights and extents.

It can be expected that continued progress will be made on those aspects of rotation dependent on interaction with the accessible fluid and icy layers. The closer altimetric monitoring of the ocean and the cryosphere should be a significant contribution to coping with the global warming that is inevitable in the coming decades.

But at periods longer than the annual and Chandler, elucidation will be more difficult. While the atmosphere dominates the short-term oscillations, its limited mass and low viscosity make it a poor dissipator and angular momentum exchanger on these time scales. The ocean is also limited as an angular momentum source and sink, but appears to have the right degree of sloshiness to make it the dominant dissipator, as shown by integrating the work function over the tide heights (Lambeck, 1980, p. 329). Hence we are forced to turn to the fluid core. It is clearly massive enough to account for the observed angular momentum swaps on decade and longer time scales. But the nature of this transfer (probably by variations in dynamic pressure, rather than viscous or electromagnetic drag) is not adequately understood, and hence whether it can be jerky as well as smooth. And it is quite obscure why there is not perceptible tidal dissipation —electromagnetic or mechanical— at the core-mantle interface as well. At present, there is quite a long list of questions regarding core-mantle interaction:

1) How to reconcile the small departure from hydrostatic ellipticity (about 0.5 km) of the core-mantle boundary inferred from nutation (Gwinn et al., 1986) with other indicators of the other undulations in this boundary?

2) How sharp and real is the "geomagnetic jerk" of 1969?

3) What decade-scale rotational variations should be expected complementary to the geomagnetic indicators?

4) How does mantle convection determine the undulations of the core-mantle boundary?

5) What is the heat flow out of the core?

6) Do core-mantle boundary interactions contribute to the damping of the polar wobble? (Even though they can't to tidal dissipation.)

Anent item 1), the seismological inference of undulations of some kilometers can probably be dismissed as weakly determined, since it accounts for a very small part of the variance in seismic velocities. More significant are models of mantle convection using the gravity field (Richards and Hager, 1984), which agree with the nutational value *if* the mantle is uniform viscosity— but uniform viscosity does *not* agree with $d\omega/dt$ vs. dJ_2/dt, which requires an increase in viscosity with depth (Yoder et al., 1983).

To those concerned about the fundamentals of Earth behavior, the Earth's rotation inspires a love-hate feeling. On the one hand, it is an extraordinarily sensitive integrator of a variety of effects that are intriguing to unravel (and thus attractive to talent); on the other, is it often remote from basic causes. But (thanks to the absoluteness of angular momentum conservation) the constraints it provides, while few, are quite rigid. The end-of-the-line in refinement of VLBI and other techniques is not yet in sight, so we can expect more such constraints from changes in rotation and other global properties.

2. MANTLE CONVECTION

"... the eternal rocks beneath: a source of little visible delight, but necessary."
Emily Bronte
Wuthering Heights

It is now twenty years since the geoscientists' favorite paradigm, plate tectonics, split the problem of long term solid earth dynamics into two parts:
• the underlying processes of mantle convection, explaining what drives plate tectonics. and further inferring therefrom the overall thermal and compostion evolution;
• the snapping, cracking, and melting in the marginal zones (which may be 100's of kilometers wide) of the plates, which lead to mountain building, volcanism, earthquakes, etc.
This rather oversimplified division has a fair match with the two distinct contributions made by space geodesy. Mantle convection is constrained by the gravity field, while marginal zone tectonics is constrained much more by measurements of positional differences.
The measurements of the gravity field to the resolution feasible from space —currently no better than 400 km over land— clearly relate mainly to mantle convection, to varying degrees to such questions as:
1) How compartmentized is mantle convection— what are the major regimes of flow and the rates of interchange among them?
2) What is the effective "molecule" of mantle convection: the size of the element that retains its identity (as defined isotopically) over long durations?
3) To what extent are there trade-offs between thermal and chemical heterogeneities?
4) What is the effective rheology for convection among its varied thermal and compositional regimes?
5) How heterogeneous is the lower mantle?
6) Are the subducted slabs the main pattern drivers, or are there other boundary-layer instabilities, or source inhomogeneities, approaching them in importance?
From the elegant representations of the geoid by Marsh et al. (1988), Tapley et al. (1989), and their predecessors, some signficant constraints have been placed by Hager (1984) on some of the above questions, particularly the last two. But being able to do so depended to a large degree on the developments of seismic tomography (Dziewonski and Woodhouse, 1987). Thus gravity is still in the classic circumstance of its interpretation being heavily dependent on other data. It also is model dependent; the inferences as to the size of the density anomaly constituted by the subducted slabs and the identification of lower mantle density variations with seismic velocity variations depend on flow models with assumed rheologies, in which the ratio of lower to upper mantle viscosity is virtually the only adjustable parameter inferrable from the data. It is a correct application of Occam's razor for the fluid dynamicist to say that whole mantle flow is the simplest model fitting the gravity field, plate velocity pattern, and oceanic topography. But next it should be pointed out that our current ignorance of the rheology of the mantle beyond a few 100 km deep allows a variety of alternative models. Even the requirement of the subduction zone mass anomaly extending well below the 670-km discontinuity (Hager, 1984) might be mitigated somewhat by better accounting for phase transitions (Anderson, 1988), and explained away by a thermal, rather than mechanical, coupling of upper and lower mantle flow. The majority of those who consider the geochemistry of the mantle favor a distinct layering (Wyllie, 1988). Even those who endorse whole mantle convection propose a rather scruffy lateral heterogeneity and invoke an appreciable increase in viscosity with depth to obtain the requisite separation times well in excess of 1.0 Gy (Hofmann and White, 1982). Such a model is my inclination, but I am not one who spends much time staring at isotope ratios.
It is difficult to see how improvement in the gravity field will help answer the above grand questions about mantle convection. The pacing elements are much more compositional, rheological, and fluid dynamical. But there is a great need for better resolution of the gravity

field for problems of continental tectonics, particularly in active areas, such as the Alpide and Andean belts. Oceanic tectonics has been significantly advanced in recent years by the altimetric measurement of the geoid; a comparable resolution of the gravity field over the continents should yield a comparable, or greater, benefit in attacking such problems as the nature of continental rifting —passive or active— and the physics underlying the complexes of underthrusting and folding in compressive belts.

3. REGIONAL TECTONICS AND EARTHQUAKES

"But the solid earth argues against us."

William Golding
The Spire

The processes in marginal zones described in Section 2 as "snapping, cracking, and melting" constitute the prime area of solid earth science to which space geodesy applies. The sizes of the main elements resisting deformation —kilometers to tens of kilometers— is comparable to the distances that can be measured most effectively by space techniques, most notably GPS. The time should be *long past* that the main interest is in confirming steady plate motion rates over distances of 100's of kilometers. To those knowledgeable about the nature of mantle convection, a sluggish thing (high Prandtl number in the jargon of the trade), it has been irksome to see emphasis on the measurement of long baselines to the neglect of the short. Rocks know each other mainly through their nearest neighbors, and to understand how the plate tectonics driven by mantle convection leads to observed lithospheric structure, composition, and temporal behavior requires a dense network of repeated geodetic measurements. Since budgets are finite, space geodetic efforts should be concentrated on areas where problems have been defined by other data: geological, seismological, petrological, geothermal, hydrofracture, etc. This concentration is necessary because of the complexity of behavior. This complexity is adumbrated by Figs. 1 and 2, which are pictures of the tectonic evolution of southern California according to the two most abundant data types, terrestrial geodesy over the last half century (Snay et al., 1986) and geological fault displacement estimates over the last few million years (Bird and Rosenstock, 1984). These crazy-quilt pictures are undoubtedly incomplete, as symptomized by their *not* including the faults of the three most recent earthquakes wreaking more than $100 million of damage (San Fernando 1971, Coalinga 1983, Whittier 1987). The problem area of regional tectonics has, of course, important practical implications, in that it provides the setting essential to solving the problem of earthquakes. Regarding this problem, it is emphasized in the Overview and in Chapter Three, but cannot be reiterated too often, that what are needed most are repeated measurements of differences in position that are:

FREQUENT, CLOSE, AND RESPONSIVE, AS WELL AS PRECISE!!

The quantitative definition of these terms varies with the problem area. But for regions where there are measurable phenomena in a setting defined by other data (as discussed above) —e.g., California or the Mediterranean— useful frequencies are about once per year; closenesses, ten kilometers; precisions, one centimeter. These values are selected as not only clearly *useful*, but within current technological and budgetary feasibility. They are far from ideal for their purposes, as well as not what appears to be eventually achievable technically. Here "useful" is properly defined as "that which will result in a perceptible advance in understanding", *not* "the ultimate solution". Advances of understanding in problems as complicated as those in Figs. 1 and 2 are slow and painful, and cannot be defined as clear "requirements" like detergents for housekeepers or weapons systems for Pentagon generals. There often appears to be a lack of appreciation by instrument developers that science is similar to technology in this slowness and

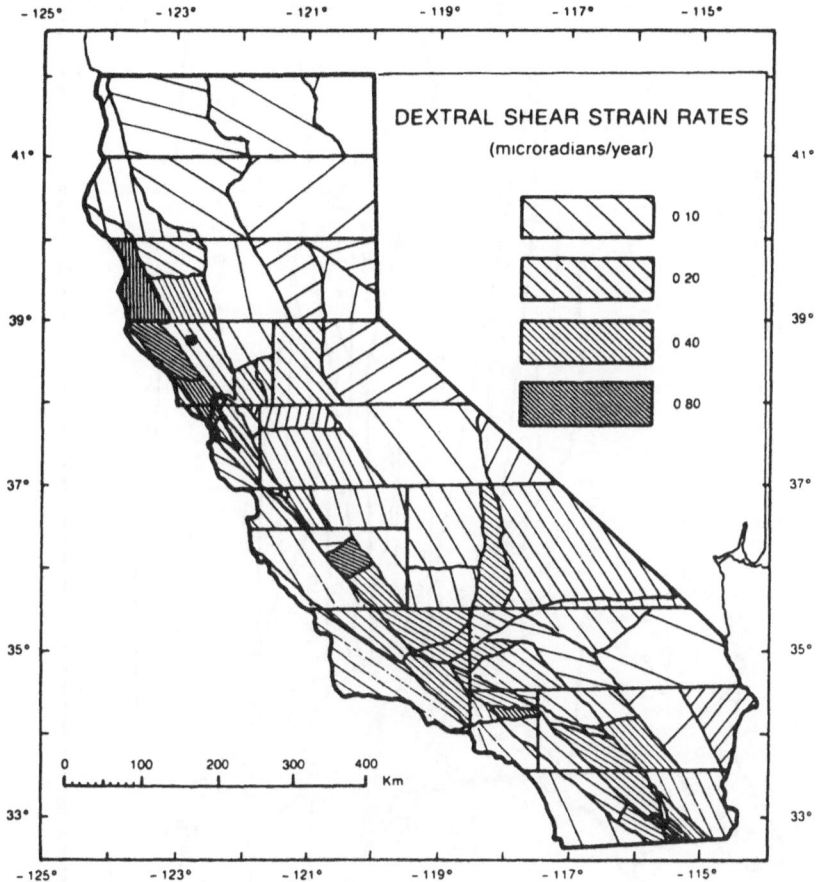

Fig. 1. Secular shear strain pattern for California as derived from historical geodetic data (Snay et al., 1986).

painfulness. But it so happens that for understanding regional tectonics and earthquakes the severe pacing element is the lack of sufficient detailed data.

Realistically, data limitations on understanding regional tectonics and earthquakes will always remain, because of limitations of technology and physical circumstance, rather than resources. The possibility of placing instrumentation at sufficient depth in all critical locations is remote: the drill holes recommended by OST (1965) —the "Press Report"— are still far from implemented. Hence resort must always be had to indirect inference requiring modeling. Some specific examples of ambiguity between adjustment in the asthenosphere or in the fault zone are given in Chapter Three. This is just a part of the ambiguity; another is the geometry of the fault system, which varies greatly from place to place. Hence for a long time to come resort must be had to models that are partly physical, partly statistical (e.g., Rundle, 1988). This statistical approach is philosophically disagreeble to some, since the physics underlying the statistics is

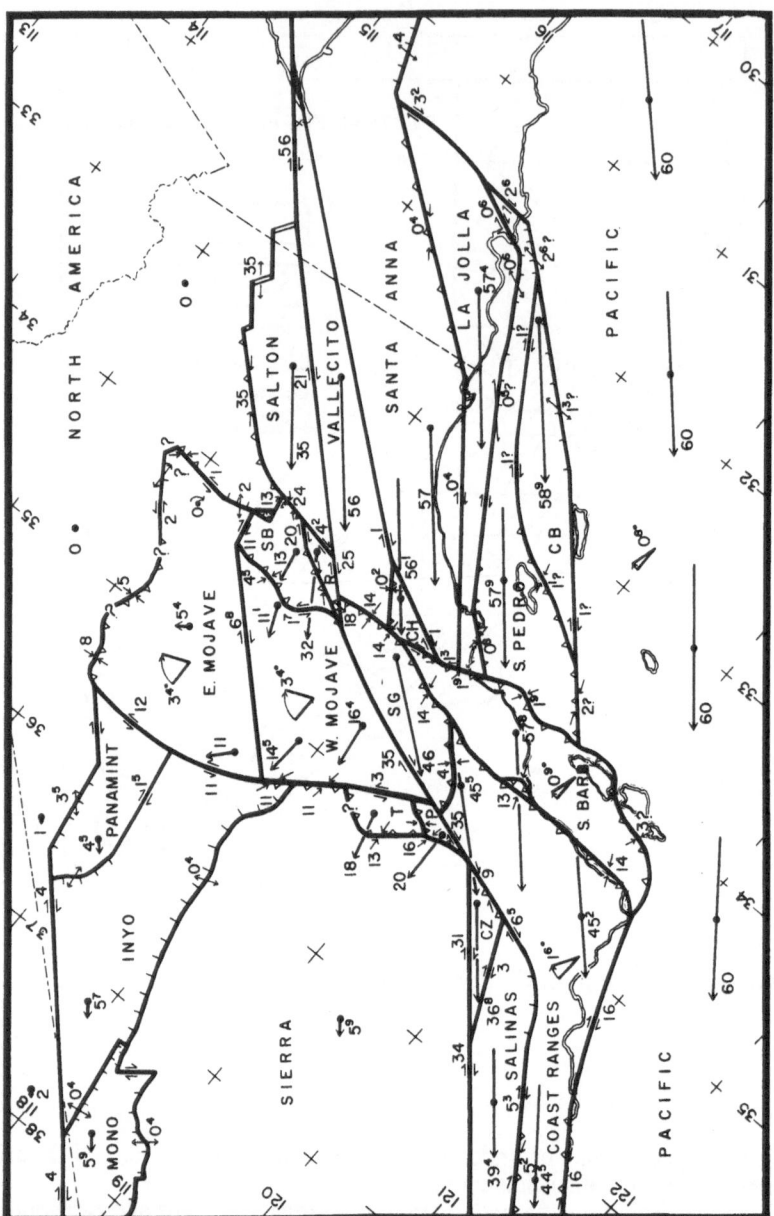

Fig. 2. Tectonic block model of southern California based on geological estimates (Bird and Rosenstock, 1984).

much coarser in its scale relative to the observations than that of quantum mechanics. But there is no alternative to statistics if predictions are to be made (as is a geophysical obligation for this societally significant problem) when the deterministic factors are incompletely known. In this situation, improvements in frequency, closeness, and precision of data can always be used.

By "RESPONSIVE" in the phrase capitalized above is meant a system (organization as well as instrumentation) capable of shifting effort rapidly in response to an earthquake, or — even more important— to premonitions of such an event from *in situ* monitoring systems: seismometers, strainmeters, etc. Aside from spatial inaccessibility, the study of regional tectonics and earthquakes suffers from a severe temporal inaccessibility; earthquakes that critically test many hypotheses rarely occur. (e.g., there has not been a really great shock in the western hemisphere since Alaska 1964). Hence if a major earthquake does occur, or if there is an ominous buildup of stress in some region, an effective geodetic monitoring system must be able to drop its ongoing program of monitoring quickly and dash to the affected scene. This responsiveness requires not only mobile instrumentation, but also shared procedures and insights among those running the *in situ* and geodetic systems. Responsiveness also includes all-weather capability: cloudy conditions prevail much of the year in some tectonically active regions.

Given the cost of instrumented three-kilometer —let alone ten-kilometer— deep drill holes wherever they are wanted, there will continue an insatiable demand for more accuracy and detail in measurements of surface strain rates and other accessible properties of deformation zones, to narrow the range over which models are needed to extrapolate from data. It will be a long road, requiring improvement in both geodynamic models and geodetic measurements. Concerning the latter, there are some aspects that may warrant reconsideration of current directions of development.

• Can satellite-borne techniques to measure the gravity field ever hope to achieve the resolution necessary to be useful constraints on tectonic models? One would expect the detail wanted in the gravity field to be about the same as that for the strain rate: i.e., a few kilometers, as discussed above. An approach to such detail shines forth in the compilation for North America by Sharpton et al. (1987). It is much more suggestive of the geologic structures and geodynamic processes than anything presented as the anticipated product of space-borne gradiometry or satellite-to-satellite range rate. Such detail is particularly desirable for tectonically active areas such as the Andes and the Alpide belt, and could be obtained by air-borne gravimetry, now that adequate altitude control is obtainable by differential GPS. The administrative difficulties of such a program would be considerable, but the scientific benefit per dollar would be significantly greater than for satellite-borne systems.

• Better accuracy in baseline measurement is obviously the prime need, to at least the millimeter level suggested in Recommendation IX of the Overview. But the laser systems evident to achieve the accuracy would not have the all-weather capability mentioned above, and hence further refinement of radiointerferometic systems seems more cost-effective.

• The NAVSTAR satellites that generate the signals for GPS is an obvious boon to geodesy. But in both signal and spacecraft characteristics it is far from optimal for precise interferometric use, and sooner or later a better successor system, one truly geodetic, should be considered. It will not be cheap.

4. OCEAN DYNAMICS

"There would be more than ocean-water broken
Before God's last *Put out the Light* was spoken."

Robert Frost
Once by the Pacific

The oceans constitute a richly varied flow continuum, its principal feature being responses to several significant external effects: insolation, wind stress, rainfall, topography (both bottom and side boundaries), and, most pervasive, rotation. The main currents— Antarctic circumpolar, Equatorial, warm western boundary (Gulf and Kushiro), and cold eastern boundary (e.g., Humboldt)— are explicable as results thereof. But, in contrast to the mantle, the ocean is quite turbulent, so that currents oscillate in location on a time scale of weeks, and generate waves and eddies 100's of kilometers in size that may persist for months. Furthermore, year-to-year variations in the external effects, together with interactions among the inertial, thermal and haline responses of the ocean (and atmosphere), lead to considerable variability on time scales appreciably longer than a year, such as the tropical oscillation "El Nino". Also superimposed are the tides and glacial and fluvial inputs. These inputs have significant effects on a 10^4 year time scale (as must plate tectonics on a 10^8 year time scale, through its influence on continental distribution). The ocean is important to the atmosphere and the land as a thermal buffer, source of water, and chemical reservoir. Better scientific understanding of the ocean is important to mankind's more effective use of it as a source of food and other resources, a means of transportation, and a dilutor of wastes.

Ocean physics interacts strongly with geodynamics in the problem areas of tides, secular change of sea level, and the definition of the steady-state flow, through the indistinguishability of ocean dynamic height from the geoid in altimetry. Indeed, as the precision and coverage of altimetry and other measurements improve, this mutual dependence is enhanced. But on the other hand there has been a growing apart intellectually since the Williamstown Report. As physical insights about very different regimes are refined, the ways of thinking that are fruitful become more different, and less is to be gained from interaction. This trend has been emphasized as the ocean science community has benefited from much more data on time-variable phenomena, so that the steady-state does not bulk as large in their thoughts. Hence the necessary interaction over projects such as Topex/Poseidon is more an engineering than a scientific matter.

5. VENUS-EARTH DIFFERENCES

"Science can neither say nor do anything about a unique occurrence."

Jacques Monod
Chance and Necessity

Venus has a special place among the planets because it is by far the most similar to the Earth, differing less than 20 percent in mass, mean density, and equilibrium black body temperature. But, despite these similarities in primary properties, it is a different as imaginable in secondary.

This difference is true for the geodynamic aspects as well. The Pioneer Venus altimeter found that the topography had only one modal level, rather than two; furthermore, that a ridge system analogous to the Earth's ocean rises is lacking, so that plate tectonics is unlikely (Kaula and Phillips, 1981). The gravity field correlates with topography at all wave-lengths, and drops off more slowly with decreasing wavelength than is true for the Earth, as though the sources were shallower. However, high admittance ratios of gravity to topography require apparent compensation depths well in excess of 100 km, and hence (given the high temperatures) support of the topography must be dynamic.

These marked differences in what should *prima facie* be a very similar body should influence our thinking about the Earth, and make us more open-minded about explanations for Earth phenomena. This need increases significantly going back in time. Archean Earth may be almost as different from Phanerozoic Earth as is Venus; in particular, plate tectonics may not always have existed.

AFTERWORD

"E passava la luna di febbraio
aperta sulla terra, ma a te forma
nella memoria, accesa al suo silenzio."

Salvatore Quasimodo
Dalla rocca di Bergamo alta

These lines from Sicily's most recent Nobel laureate were a metaphor for his failing to see reality around him during the turmoil of World War II while concentrating on his research in ancient Greek literature. They could just as well serve as a metaphor for the need for a better communication in geodynamics. Theoreticians get the impression that experimenters operate according to problem definitions that are a decade or more obsolete, while experimenters probably think that theoreticians are sporadic in their attention and underestimate technical difficulties. Better communication is particularly important now that measurement capabilities can yield valuable constraints, provided that they are applied in sufficient detail.

*"And the February moon passed
Open over the Earth, but to you a
Form in the memory, alight in its silence."

Acknowledgement. Section 4, "Ocean Dynamics", has benefited appreciably from the notes of D. E. Harrison for his presentation at the workshop, but the choice of topics and final verbiage are purely the author's responsibility.

REFERENCES

Anderson, D. L., 1987, *J. Geophys. Res., 92*, 13,968.
Bird, P. and Rosenstock, R. W., 1984, *Bull. G. S. A., 95*, 946.
Gwinn, C. R., Herring, T. A., and Shapiro, I. I., 1986, *J. Geophys. Res., 91*, 4755.
Hager, B. H., 1984, *J. Geophys. Res., 89*, 6003.
Hofmann, A. W. and White, W. M., 1982, *Earth Plan. Sci. Let., 57*, 421.
Kaula, W. M., 1979, *Proc. 9th GEOP Conf., 280*, Dept. Geod., Ohio State Univ., 345.
Kaula, W. M. and Phillips, R. G., 1981, *Geophys. Res. Let., 8*, 1187.
Lambeck, K., 1980, *The Earth's Variable Rotation: Geophysical Causes and Consequences*, Cambridge Univ. press, 449 pp.
Marsh, J. G., and 19 others, 1988, *J. Geophys. Res., 93*, 6169.
Meier, M. F., 1984, *Science, 226*, 1418.
MIT, 1970, *The Terrestrial Environment: Solid-Earth and Ocean Physics*, NASA Contractor Report CR-1579, Massachusetts Inst. of Technology, Cambridge, MA, 147 pp.

Munk, W. H. and MacDonald, G. J. F., 1960, *The Rotation of the Earth: a Geophysical Discussion*, Cambridge Univ. Press, 323 pp.

Nakada, M. and Lambeck, K., 1988, *Nature*, **333**, 36.

Newell, R. E., Kidson, J. W., Vincent, D. G. and Boer, G. J., 1974, *The General Circulation of the Tropical Atmosphere*, **2**, MIT Press, Cambridge, MA, 371 pp.

OST, 1965, *Earthquake Prediction: a Proposal for a Ten Year Program of Research*, Office of Science and Technology, Washington, DC, 134 pp.

Peltier, W. R., 1988, *Science*, **240**, 895.

Richards, M. A. and Hager, B. H., 1984, *J. Geophys. Res.*, **89**, 5987.

Rundle, J. B., 1988, *J. Geophys. Res.*, **93**, 6237 and 6255.

Sabadini, R., Yuen, D. A., and Gasperini, P., 1988, *J. Geophys. Res.*, **93**, 437.

Sharpton, V. L., Grieve, R. A. F., Thomas, M. D., and Halpenny, J. F., 1987, *Geophys. Res. Let.*, **14**, 808.

Snay, R. A., Cline, M. W., and Timmerman, E. L., 1986, *Roy. Soc. N. Z. Bull.*, **24**, 131.

Tapley, B. D., Shum, C. K., Yuan, D. N., Ries, J. C. and Schutz, B. E., 1989, *J. Geophys. Res.*, **94**, submitted.

Wyllie, P. J., 1988, *Revs. Geophys.*, **26**, 370.

Yoder, C. F., Williams, J. G., Dickey, J. O., Schutz, B. E., and Tapley, B. D., 1983, *Nature*, **303**, 757.

Yoder, C. F., and Ivins, E. R., 1985, *EOS Trans. AGU*, **66**, 245.

Chapter 2

SHORT-TERM DYNAMICS OF THE SOLID EARTH

1. INTRODUCTION

J. O. Dickey, T. A. Herring, R. J. O'Connell and D. E. Smylie

The earth is a mechanical system with separate parts: the solid earth, the atmosphere, the oceans, the liquid core and the solid inner core. These systems exchange energy and angular momentum among themselves, through a rich variety of geophysical processes, and studies of the motion of the Earth as a whole, as well as its parts, shed light on these processes, as well as on the properties of the Earth.

The phenomena included within 'Short Term Dynamics of the Earth' are those motions with significant components at periods less than about 20 years. These fall into several categories: (1) the rotation of the Earth about its axis of spin (or its time derivative, the length of day or LOD); (2) polar motion, i. e. the motion of the Earth with respect to the spin axis of the Earth; (3) nutations and precession, the motion of the Earth with respect to inertial space; (4) Earth tides; (5) temporal variation in the Earth's mass distribution and its geopotential. This division is somewhat arbitrary and is not unique. The topics described here and in the following sections of this chapter are by their very nature interdisciplinary, drawing from and contributing to the fields of geodynamics, seismology, meteorology, oceanography, astronomy and celestial mechanics. As such, we cross-reference the reader to several complementary sections in this text including: Chapter 3- Interactions with Other Disciplines and Programs, especially the sections entitled Geodynamics, Earth Interior, Atmosphere and Climate; and Chapter 2-Long-Term Dynamics of the Earth, the section entitled Temporal Variations in Gravity (a topic which is of common interest to both the Short-Term and Long-Term Dynamics Panels). We refer the reader to the individual sections for references.

Changes in Earth orientation - a term which includes the Earth rotation as a three-dimensional vector (axial spin of the Earth, polar motion, precession and nutation) - are caused by the deformation of the solid Earth and by the exchanges of angular momentum between the solid and fluid parts of the Earth, as well as by exchanges of angular momentum with extraterrestrial objects. Changes in Earth rotation and polar motion can be regarded as the response of a linear differential system to a three-dimensional excitation vector. Earth rotation, when analyzed in combination with other parameters such as atmospheric angular momentum (AAM), the Southern Oscillation Index, and torque estimates of core-mantle coupling, permits a better understanding of the geophysical processes involved. Intercomparisons between results from the various space techniques indicate that Earth rotation is routinely determined at the 0.1 millisecond level (approximately 5 cm at the equator), with higher accuracy being achieved in some cases. The interaction between the atmosphere and solid Earth is most apparent in Earth rotation. From intercomparisons of AAM and length of day, we find that Earth rotation variations over time scales of a year or less are dominated by atmospheric effects, with a dominant seasonal cycle and significant variability on the intraseasonal (30 to 60 day) time scale. The correlation between length of day and AAM is so well established that forecasts of atmospheric angular momentum, using the large numerical models from the major forecast centers, are now being investigated for Earth rotation prediction purposes. From the geophysical point of view, this now permits the effect of the atmosphere to be stripped away, revealing the angular momentum transfer effected by the other mechanisms. As a consequence, variations on interseasonal time scales have been related to the El Niño / Southern Oscillation phenomena.

Further removing the interannual signature in the LOD, the longer-scale "decade" fluctuations remain in the Earth rotation time series. Torques between the core and mantle are accepted by geophysicists to be the most probable cause for these variations. Because we have no direct access to the core, our observations of these and related effects of the fluid and solid inner cores on Earth's mechanics are of prime importance for insights into the Earth's interior and the interactions between its various subsystems. These interactions are effected through inertial, electromagnetic, topographic and gravitational coupling. Two candidate mechanisms under investigation for these decade variations are: (1) core-mantle torques which are probably largely due to dynamic pressure forces associated with core motions acting on topographic undulations of the core-mantle boundary and the equatorial bulge, and (2) torques of electromagnetic origin arising through Lorentz forces, which might have a significant contribution if the electrical conductivity of the lower mantle is high enough. Estimates of topographic torques can be made from a combination of core motion models from geomagnetism, core-mantle boundary maps from seismic tomography, and long-wavelength gravity anomaly data. Geodetic torque estimates provide a means of checking these results and imposing constraints on the models used. Results are consistent with undulations of the core-mantle boundary of ~ 500 m. This size is in accord with recent calculations based on corrections to the Standard 1980 Nutation Model using Very Long Baseline Interferometry data as is discussed below.

The Earth's fluid and solid inner cores add an infinite suite of degrees of freedom to the mechanical behavior of our planet. One of the simplest and most elegant descriptions of the effect of the fluid core on Earth's rotational dynamics has roots in the classical literature but was not well-understood in its geophysical context until the 1970's. This is the Poincaré uniform-vorticity flow model which departs only slightly from a rigid rotation. With allowance for the elasticity of the mantle and crust, in this description, a rotational mode with nearly diurnal period emerges (the departure of the period from one sidereal day being set by the well-known elasticity of the mantle and crust and by the flattening of the core-mantle boundary), in addition to the fourteen-month Chandler motion. The nearly-diurnal wobble is often referred to as the free core nutation (FCN).

While attempts to detect the FCN directly have not been successful, it has recently been observed indirectly through the anomalous nutational and tidal response it produces (confirming earlier, much more tenuous, anomalous tidal gravimetric observations). The nutation corrections obtained from the VLBI technique, and the tidal observation, using the superconducting gravimeter, have been interpreted in terms of a departure of the flattening of the core-mantle boundary from the hydrostatic value. Decade variations in Earth rotation also indicate that the core-mantle boundary deviates from its hydrostatic equilibrium shape. Interest is now focused on further theoretical developments which would allow computation of other modes arising from the outer fluid and inner solid cores, and their observation by VLBI and gravimetric methods.

Turning to the polar motion, this phenomenon consists mainly of oscillations at periods of one year (the annual cycle) and about 433 days (the Chandler wobble), with amplitudes of about 100 and 200 milliarcseconds (mas), respectively, together with a long-term drift of a few milliarcseconds a year. In addition, analysis of geodetic data reveals rapid polar motions, with a peak-to-peak variation of approximately 2 to 20 mas, fluctuating on times scales between two weeks and several months. Comparisons with meteorological excitation estimates show that these motions are at least partially driven by surface air pressure changes as modified by the response of sea level to atmospheric loading. The yearly oscillation is driven mainly by the atmosphere, while the Chandler wobble is a free oscillation of the Earth. The source of the Chandler wobble is uncertain; the two major candidates are the atmosphere and seismic deformation. Another source of variation in rotation is the changing distribution of water on the surface, in particular ground water and ice. The redistribution of water, with associated changes in sea level, will cause a drift of the mean rotation pole that may be diagnostic of significant rearrangements of ice and sea water. The longer-term "decade" changes are caused by the coupling between the Earth's liquid metallic core and the solid mantle. The secular trend is thought to be caused in large part by mass redistribution.

Success in modeling axial atmospheric angular momentum has given some confidence that equatorial components can be similarly removed from the polar motion path. However, current comparisons of polar motion excitation deduced from geodetic observations and computed from atmospheric excitations indicate that the role of the atmosphere is not as clearly defined for polar motion as it is for LOD. The pole path is now determined by space techniques at the one to two millisecond level of angular displacement. If the removal of the atmospheric contribution was accomplished, this would allow examination of possible equatorial angular momentum exchanges on the interannual and decade changes with the core, and the determination of changes in the off-diagonal components of the inertia tensor of the shell (mantle and crust) associated with seismic, aseismic and other tectonic events. However, the removal of the atmosphere will require a detailed analysis of the coupling of the atmosphere to the solid Earth and of the ocean response to atmospheric variations. The effects of the atmosphere are modulated by the oceanic response to barometric pressure changes. In general, this is an equilibrium response, but there are indications of dynamic interactions with long period oceanic waves; these affect the accuracy of the calculation of the atmospheric effect, as well as shed light on the dynamics of the oceans. The period of the Chandler wobble is influenced by the presence of the oceans as well. A better theoretical calculation of this would allow an improved determination of both the damping of the wobble and its excitation to be extracted from the data; identification of the damping in the solid Earth or within the core would have important implications for mantle rheology and core-mantle interactions.

Changes in the solid Earth directly influence Earth rotation and polar motion. Mass displacements associated with earthquakes and slower tectonic events may directly excite the Chandler wobble and changes in the LOD at the currently observable level; the current accuracy of measurement of pole position coupled with corrections for atmospheric effects should allow the next major earthquake (magnitude ~8) to be seen. More exciting, however, is the possibility of observing the effects of slower, aseismic displacements; those associated with earthquakes may even be precursors, or, if not, may produce displacements and strains unaccounted for at present. Post-seismic anelastic relaxation should also contribute to polar motion; if observed, this would illuminate the anelastic response of the crust and lithosphere, and the nature of strain propagation following earthquakes.

The oceans also affect the tidal response of the Earth; the most obvious example is the rich spectrum of oceanic tides. Identifying the source of energy dissipation associated with tidal friction that causes a secular increase in LOD is important for understanding the anelasticity of the Earth and the evolution of the Earth-Moon system. This requires better global models of oceanic tides than presently available. The loading of the solid Earth by the oceanic tides causes displacements that affect all geodetic measurements; their exact nature depends on the heterogeneous elastic properties and structure of the solid Earth over the range of tidal frequencies.

Longer term mass redistribution due to post-glacial rebound and current ice melting is also apparent in present measurements of polar motion; it can also be seen in temporal changes in the components of the gravity field. Improvements in modeling and measuring post-glacial rebound will lead to a separation of this effect from current mass redistributions, allowing better assessment of the likelihood of sea level changes due to melting of ice sheets and glaciers. It also provides a constraint on the rheology of the solid Earth, which is of importance to understanding convection, plate tectonics and the longer term evolution of the Earth.

In summary, highly accurate geodetic measurements and coupled observations of related geophysical phenomena are key to a better knowledge of the short-term dynamics of the Earth; on the other hand, we note that an accurate understanding of short term motions (both rotational and deformation) is crucial for geodetic measurements, especially in light of the one millimeter accuracy goal (see Section 7 and the Overview).

2. PRECESSION AND NUTATION

T. A. Herring

The external torques applied to the Earth by other bodies in the solar system (mainly the sun and the moon) cause a change in the direction of the Earth's rotation axis, its figure axis, its angular momentum axis, and an axis attached to the mean surface of the Earth (often referred as a "body axis"). These motions are loosely referred to as precession and nutation with precession being the secular part of the motion, and nutation the periodic part. The response of the Earth to these externally applied torques is approximately that of a rigid body. However, there are observable differences between the nutations of a rigid Earth and a deformable Earth with fluid-outer core, solid-inner core, oceans, and atmosphere. These differences will be the subject of this section. In particular, we will discuss the properties of the Earth which have already been revealed by the studies of the Earth's nutations, the questions about the response of the Earth to external torques which still remain unresolved, and the properties of the Earth which we may learn about in the future.

Before we continue, we will define precession and nutation, and the axes to which they refer. As discussed in the reference frames section, techniques such as VLBI observe the motions of the points on the surface of the Earth with respect to "fixed" extragalactic radio sources. The rotational components of these motions with nearly diurnal frequencies are the combination of the precession and the nutation of the Earth's body axis. When viewed from inertial space these nearly diurnal motions of the body axis appear as long period motions which are linked directly to the long-period orbital motions of the sun and the moon. Luni-solar precession which is the motion of the Earth's body axis about an axis normal to the orbital plane of the Earth (the elliptic) has a period of about 26,000 years. Nutations are the superposition of a large number of periodic terms on this precessional motion. The largest nutation has a period of 18.6 years and an amplitude of 9.2 arcseconds.

Currently, nutation series are computed assuming that the individual periodic components of the nutations can be linearly superimposed, and thus that the complete motion of the body axis with respect to inertial space can be obtained from the sum of all of the periodic nutations, precession, and the long-period motions due to polar motion. The individual nutations with different periods form a nutation series. The IAU 1980 nutation series, based on the works of Kinoshita (1977) for the rigid Earth series and of Wahr (1981) for the "real Earth" modification of the rigid Earth results, forms the basic series to which observations of the nutations are compared. Modern space geodetic techniques have already disclosed errors in this series which are related to the incompleteness of the geophysical models on which the theories are based. Most notable of these corrections is to the retrograde annual nutation (Herring et al., 1985; Eubanks et al., 1985; Herring et al., 1986) and is of ≈2 mas amplitude. This correction has been interpreted as being due to the flattening of the core-mantle boundary (CMB) deviating from its hydrostatic equilibrium value by about 5% (Gwinn et al., 1986). A deviation of this magnitude to the flattening of the CMB is also evident in the topography of the CMB obtained from seismic tomography results (Hager et al., 1985; Morelli and Dziewonski, 1987) and in observations of the gravimetric factor (Neuberg et al., 1987). Topography of this size on the core-mantle is also consistent with the decade length variations in the length of day (see Section 1.4).

As we probe deeper into the nutations, what will we discover? We address this question in several items which deal with both theoretical and observed corrections to the IAU 1980 nutation series. Since the adoption of the IAU 1980 nutation series, two sets of corrections based on theoretical considerations have been suggested. Wahr and Sasao (1981) have proposed corrections due to the effects of ocean tides that are as large as 1.1 mas for the 18.6 year principal nutation, and 0.6 mas for the semi-annual nutation. Wahr and Bergen (1986) have proposed corrections for the effects of anelasticity of the mantle. In addition, VLBI observations of the nutations have disclosed corrections to the retrograde annual nutation, prograde semi-annual, prograde 13.7 day, and the long period nutations (the combination of

Table 1. Corrections to the Circular Amplitudes of the IAU 1980 Nutation Series

Forcing freq. (cpsd)	Period (days)	Obs. VLBI In phase (mas)	Obs. VLBI Out of phase (mas)	Ocean tides In phase (mas)	Ocean tides Out of phase (mas)	Anelasticity In phase (mas)	Anelasticity Out of phase (mas)
-1.1092	-9.13	0.05	-0.02	-	-	-	-
-1.0730	-13.66	-0.03	0.05	-	-	0.00	0.00
-1.0362	-27.55	0.03	-0.02	-	-	-	-
-1.0313	-31.81	0.07	0.01	-	-	-	-
-1.0082	-121.75	0.00	-0.00	-	-	-	-
-1.0055	-182.62	-0.17	-0.06	-	-	0.04	0.01
-1.0027	-365.25	-2.04	0.29	0.16	-0.17	0.18	0.07
-1.0023	FCN	0.25	-0.13	-	-	-	-
-1.0001	-6798.37*	-	-	-1.05	1.06	-0.35	-0.14
-0.9999	6798.37*	-	-	0.13	-0.13	0.05	0.02
-0.9977	433.50	-0.07	-0.13	-	-	-	-
-0.9973	365.25	0.04	0.13	-	-	-0.01	-.00
-0.9945	182.62	0.43	-0.50	0.61	-0.62	0.28	0.11
-0.9918	121.75	-0.08	0.01	-	-	-	-
-0.9687	31.81	-0.07	-0.01	-	-	-	-
-0.9638	27.55	-0.02	0.01	-	-	-	-
-0.9270	13.66	-0.34	0.06	0.02	-0.02	0.07	0.03
-0.8908	9.13	-0.11	0.05	-	-		

The forcing frequency is given in an Earth fixed frame; the period is given in inertial space. Ocean tides are for the Schwiderski model; Anelasticity for the QMU model with $\alpha=0.15$. The standard deviations of the VLBI corrections before accounting for noise in the rigid Earth nutation series is 0.04 mas, with noise from the rigid Earth series it is 0.10 mas. VLBI results based on 588 experiments, spanning 7.5 years (July, 1980 to February, 1988).

* Estimates of the 18.6 year nutation amplitudes are available from the analysis of lunar laser ranging (LLR) data (Dickey, private communication, 1988). For the retrograde 18.6 year nutation, these estimates are -4.7 mas in phase and 2.1 mas out-of-phase, and for the prograde 18.6 year nutation these estimates are 1.0 mas in phase and -3.4 mas out-of-phase. The standard deviations of these estimates is about 3 mas and is too large for us to currently differentiate between the various predictions of the corrections to the 18.6 year nutation. This same analysis yields corrections to the precession constant of between -0.16 and -0.24 $\pm 0.10"$/century.

the 18.6 year, 9.3 year, and the precession constant). These latter long period corrections can not yet be separated accurately because of the short (7.5 year) span of Mark III, dual-frequency band VLBI data. Lunar Laser Ranging data, now spanning a nineteen year period with range accuracies varying from 20 cm in the early seventies to 2-3 cm currently, are useful to the determination of these long period terms (King et al., 1988; Newhall et al., 1988, and see Table 1). A combination of LLR and VLBI is beneficial.

From Table 1, we see that overall the agreement between the IAU 1980 nutation series and the VLBI results is very good considering that the IAU series is truncated to 0.1 mas. Of the 32 corrections to the IAU series that were estimated, 11 corrections exceed 0.1 mas, and only 7 corrections exceed 0.15 mas. For all but one of these 7 corrections, there are corrections to the IAU series expected based on the extension of the geophysical models used to derive this series. However, the theoretical corrections to the IAU series do not agree with the observed corrections. The resolution of these differences will yield information about the anelasticity of the mantle, the dissipative coupling of the fluid-core to the mantle, and the effects of the oceans and the atmosphere on the nutations. The one observed correction to the IAU series for which no geophysical model has been proposed which would predict a large correction is the 13.66 day prograde term. For this term the VLBI derived correction is -0.34 mas. Such a correction could arise if the changes in the principal moments of inertia of the mantle due to either tilting the mantle's rotation axis or an external potential, are not correct in the current nutation models. However, it is not clear that this relationship could be incorrect given the direct effect of such a parameter on the k_2 Love number. However, the effect may enter through an effect of the ocean's response. One other correction which is anomalous by its absence is the freely excited core nutation. Until recently it was expected that the amplitude of this free mode, which is analogous to the Chandler Wobble but with a free period of about one day, would be 2 mas or greater. The current bound of its amplitude placed by VLBI data is <0.4 mas. The question naturally arises as to why this mode is not excited to the expected level? From the solution to the forced nutation problem, we have tightly constrained estimates of the eigenfrequency of this mode and a lower bound on its damping time, and thus we must conclude that either the theory of the excitation of this mode is currently not correct, or that there is little power in the diurnal band which could excite this mode.

Even now there are questions about the structure of the Earth which need to be answered to obtain agreement between the observed and theoretical nutations. The constraints which will be placed on the Earth models should become even stronger as more VLBI and LLR data are obtained. In the future, accurate estimates of the precession constant (uncertainty <0.001"/century, about two orders of magnitude smaller than current uncertainties) and the 18.6 year nutations (uncertainty <0.1 mas) should be obtained when there is a sufficient duration of VLBI and LLR data to allow these long period terms to be separately estimated. Based on observational evidence and the expected effects of changes in the flattening of the core-mantle boundary, ocean tides, and the anelasticity of the mantle, we know that corrections of the order of 1-2 mas for the 18.6 year nutation and 0.1-0.3"/century for the precession constant are needed to these long period terms.

In addition to VLBI data, other ancillary data will be needed in the future to fully exploit the advantages of using nutations to study the dynamic response of the Earth to externally applied torques. Already it is clear that the effects of ocean tides are not negligible, and in the future we will need to know their contribution to better than 0.01 mas (about 2% of the effect on the semiannual nutation). Also, the models for the dynamic effects of currents in the ocean will be needed. The long-term stability of the VLBI reference frame also needs to be established. While currently it is known that components in the brightness distributions of quasars can exhibit "proper motions" of greater than 1 mas/year, it seems that these motions will not pose a major problem because the superluminal components decrease in brightness as they move away from the "core" of the quasars. In one radio source which exhibits superluminal motion, the core has been shown to have a proper motion of <20 μarcsec/year (Bartel et al., 1986).

The analysis of the nutations will yield information about the anelasticity of the mantle, dissipative coupling of the fluid core to the mantle, coupling of the solid-inner core to the fluid-outer core and the mantle, the principal moments of inertia of the Earth, and the excitation of the Earth with periods near one day (particularly those from the atmospheric and oceanic loading). In general, the presence of the fluid-outer core, and possibly the solid-inner core, have major effects on the rotation of the Earth, and thus studies of the Earth's rotation may provide sensitive probes of the internal dynamics and structure of these parts of the Earth. It is also clear that as the accuracy of observations of the rotation of Earth improve, the distinction between nutations and short-period polar motion will become less clear. In the future, we will need to treat Earth rotation as a process having a continuous spectrum with power at all frequencies. Superimposed on this continuous spectrum will be the spectral lines at frequencies near -1 cpd associated with the forced nutations. In particular, the contribution of the continuum spectrum will need to be removed from nutation results before these latter results can be fully exploited for studying the response of the Earth. Some of the information about Earth structure discussed above is already attainable from the current data but the nutation theories currently lag behind the observations in quality. It does not seem unreasonable at this time to expect that nutation series coefficients with uncertainties of about 10 μarcsec will be determined within the next decade by VLBI. It is yet to be seen the complexity of the models of the Earth that will be needed to explain such data.

REFERENCES

Bartel, N., Herring, T. A., Ratner, M. I., Shapiro, I. I. and Corey, B. E., 1986, *Nature*, **319**, 733-737.

Gwinn, C. R., Herring, T. A., and Shapiro, I. I., 1986, *J. Geophys. Res.*, **91**, 4755-4765.

Eubanks, T. M., Steppe, J. A., and Sovers, O. J., 1985, *Proc. Inter. Conf. Earth Rotation and Terrestrial Reference Frame*, ed. by I. Mueller, Ohio State Univ., 326-340.

Hager, B. H., Clayton, R. W., Richards, M. A., Comer, R. P., and Dziewonski, A. M., 1985, *Nature*, **91**, 541-545.

Herring, T. A., Gwinn, C. R. and Shapiro, I. I., 1985, *Proc. Inter. Conf. Earth Rotation and Terrestrial Reference Frame*, ed. by I. Mueller, 307-325.

Herring, T. A., Gwinn, C. R., and Shapiro, I. I., 1986, *J, Geophys. Res.*, **91**, 4745-4754.

King, R. W., Murray, M. H., Chandler, J. F. and Whipple, A. L., 1988, *EOS*, **69**, 331.

Kinoshita, H., 1977, *Celestial Mechanics*, **15**, 277-326.

Morelli, A., and Dziewonski, A. M., 1987, *Nature*, **325**, 678-683.

Neuberg, J., Hinderer, J. and Zurn, W., 1987, *Geophys. J. R. Astr. Soc.*, **91**, 853-868.

Newhall, X X, Williams, J. G. and Dickey, J. O., 1988, *Proc. Union Symp. No. 128*, 159-164

Wahr, J. M., 1981, *Geophys. J. R. Astron. Soc.*, **64**, 705-727.

Wahr, J. M., and Sasao, T., 1981, *Geophys. J. R. Astron. Soc.*, **64**, 747-765.

Wahr, J. M. and Bergen, Z., 1986, *Geophys. J. R. Astron. Soc.*, **87**, 633-668.

3. POLAR MOTION

S. R. Dickman

3.1 WHERE ARE WE?

In the years since the Williamstown report, we might say that problems concerning the major polar motion features — the Chandler wobble (amplitude ~200 mas, period about 433 days), the annual wobble (amplitude ~100 mas), and the secular trend (3 mas/yr toward Hudson Bay) — still exist, but have been narrowed down considerably.

Theoretically, for example, the period of the Chandler wobble is better understood and more predictable now. The contributions of static (e.g. Dahlen, 1976) and dynamic (e.g. Dickman, 1988a) oceans to the Chandler period have by now been accurately computed; the effects of core decoupling, both on the period and on oceanic contributions to the period, are now fully acknowledged (e.g. Smith and Dahlen, 1981); and, the range of possible contributions by mantle anelasticity can be easily estimated, using idealized representations of anelastic relaxation mechanisms (e.g. Anderson and Minster, 1979).

Because of the small size (~5 days) of anelastic effects, it is important that the observed Chandler period be correspondingly well-determined (see below); however, even with the quality of the new observing techniques — accuracies approaching 1 - 2 mas for both SLR, at 3-day intervals, and VLBI, at 5-day intervals — their duration is not yet long enough. Analysis of prior data sets emphasizes also that it is critical for the wobble excitation spectrum to be known, because of its ability to distort the location of the wobble peak (thus, its period) and its width (thus, the wobble Q).

The wobble frequency of an oceanless elastic earth with a fluid core, σ_e, represents theoretically the most important part of the wobble period; yet, its calculation is somewhat problematic: independent calculations of σ_e by Smith and Dahlen (1981) and Yoder and Ivins (1987) are sufficiently different (a 5 day difference in period) that they imply — when the predicted total period is compared with the observed — moderate anelasticity (Q $\sim\sigma^\alpha$ with α ~0.1) and insignificant anelasticity ($\alpha \rightarrow -\infty$), respectively.

It is now established that deep-ocean dissipation of wobble energy is negligible. The problem of Chandler wobble damping is thus narrowed down to mantle anelasticity versus shallow-sea dissipation. If the predicted Chandler period turns out to require mantle anelasticity, then appreciable wobble damping will be a by-product, and the subject may be considered closed. On the other hand, highly anomalous pole tide observations in the North and Baltic seas suggest that shallow-sea dissipation should predominate; however, there is a possibility of the data being a meteorological artifact (e.g., O'Connor 1987), and the observations remain to be substantiated theoretically.

Numerous efforts have concentrated on discovering the primary excitation source(s) of the Chandler wobble. At present, meteorological and seismic processes are considered the two most likely candidates. Correlation studies involving earthquakes have been equivocal; some individual events have been identified with polar motion path or excitation "glitches", but not contemporaneously nor with theoretically small enough amplitudes. Seismicity time series also correlate equivocally with long-term Chandler wobble amplitude variations. Weak correlations have been found between polar motion and some ENSO events; correlations involving global atmospheric processes are suggestive.

Excitation computations have shown, firstly, that individual earthquake events are not sufficiently energetic nor frequent enough to maintain the wobble and, secondly, that cumulative seismic energy has probably not been able to sustain the major portion of the wobble. No work has yet elucidated the level, individual or cumulative, of pre-seismic or aseismic excitation. Computations of excitation by the atmosphere/hydrosphere, or portions of it, are encouraging but have not yet been able to account for the bulk of wobble energy; the importance of groundwater (including also lakewater and ice storage) has often been indicated (e.g., Merriam, 1982; Chao, 1988; Kuehne and Wilson, 1988) and remains potentially decisive, but has proven difficult to estimate. Thus, despite all efforts to date, it is not yet clear whether cumulative seismotectonic processes or the totality of atmospheric/hydrospheric processes dominates the excitation of the Chandler wobble, or whether a combination of both is sufficient, or even whether a third, so far unidentified, phenomenon is the primary excitation source.

In the case of the annual wobble, there is still a general consensus that atmospheric pressure variations provide the major excitation (see e.g. Lambeck, 1980). Attention has focused recently on second-order excitation processes — winds, ocean currents, atmospheric self-gravitation and loading, etc — which has led to an improved agreement between data and theory. However, the match is imperfect, implying the need for additional (or more accurate) excitations; as with the Chandler wobble, the role of groundwater + lakewater + ice remains to be settled.

The secular polar trend is now recognized as a true polar wander (Soler and Mueller, 1978; Dickman, 1977), not just a consequence of the drift of the observatories. Within the past decade, its reality has been confirmed by analysis of data obtained using the new techniques (Markowitz, 1982; Gross and Chao, 1985). The most widely accepted cause of the trend is mass redistributions accompanying and following the termination of Pleistocene glaciation 20,000 years ago (Nakiboglu and Lambeck, 1980; Sabadini and Peltier, 1981); that connection has been employed to constrain mantle viscosity structure (e.g. Sabadini et al., 1982; Yuen et al., 1982). Other mechanisms of mass redistribution whose effect on the polar trend has been estimated include cumulative earthquake activity (e.g. Gross and Chao, 1985; Souriau and Cazenave, 1985), a global rise in sea level (Barnett, 1982), trends in groundwater storage (Chao, 1988) and recent glacial activity (Yoder and Ivins, 1985; Gasperini et al., 1986). At present it is unclear how much of the observed secular trend is a result of these alternative mechanisms.

In the years since Williamstown, smaller polar motion features have also been detected. Despite the surprisingly strong signals of the long-period "Markowitz wobble" (amplitude ~30 mas, period 31 yr) in both the retrograde and prograde (ILS) polar motion spectra, its existence and even physical meaning is still uncertain (Okamoto and Kikuchi, 1983). The theory accounting for such a wobble is similarly uncertain; wobble of the inner core was briefly considered (Busse, 1970) but is no longer tenable (Kakuta et al., 1975; Dickman, 1976), while the possibility of coupled wobble of the ocean/earth system (Dickman, 1983) is unresolved (Wahr, 1985; Dickman, 1985).

The semi-annual wobble — much smaller in amplitude (~10 mas) than the annual despite its comparable excitation (because it is farther from the Chandler resonance), and in principle much less controversial than the Markowitz wobble — has recently been measured using the new observational techniques (Eubanks et al., 1985). Extensive investigation and interpretation of the results have not yet been made, however.

One of the most exciting developments since Williamstown has been the detection of small-amplitude (2 - 20 mas), rapid polar motion, with periods in the range 10 — 50 days (e.g. Robertson et al., 1985; Dickey and Eubanks, 1987). Correlation with meteorological excitation estimates (Eubanks et al., 1988) implies that such motion is at least partially driven by surface air pressure variations.

3.2 WHERE ARE WE GOING, AND HOW?

In the next decade, we can expect the Chandler period to be fairly well resolved in space-geodetic data; the wobble Q will be less well determined. To complete our understanding of the Chandler period, the calculation of the oceanless earth wobble frequency (σ_e) must be verified; it may also be important to re-calculate σ_e accounting for lateral variations in mantle properties. Any refinements in core shape and core-mantle coupling may affect σ_e significantly as well. The total predicted Chandler period will then require one final adjustment, accounting for mantle anelasticity, to bring it into agreement with the observed period. Combined with studies of tidal signals in Earth rotation (UT1), using independent predictions of the ocean tide components, our picture of mantle anelasticity at sub-seismic frequencies will be detailed enough to assess the validity of the power law representation of the frequency dependence of Q.

Once the predicted and observed Chandler periods are reconciled, the single primary sink of wobble energy will have been pinpointed. If it turns out to be shallow seas, tidal theorists will be challenged to develop appropriate, and inevitably complicated, shallow-sea tide models (cf. Dickman and Preisig, 1986). After the amount of wobble attenuation has been quantified, it may be possible to place bounds, comparable of those by Gwinn et al. (1986), on core viscosity; this would eventually lead to constraints on the light-alloy composition of outer core fluid (Poirier, 1988).

Despite the difficulty, as stated earlier, of obtaining directly satisfactory estimates of groundwater excitation, it may be possible to deduce the Chandler wobble excitation by taking advantage of the high accuracy of space-geodetic polar motion data. Using a more rigorous determination of atmospheric angular momentum — based on improved global atmospheric circulation models and a more exact oceanic response to atmospheric pressure (see Merriam, 1982; Dickman, 1988b) — the atmospheric component can be subtracted out. Using extensive space-geodetic measurements of crustal deformation [RECOMMENDATION: over REGIONAL areas, on a time scale of 10 days (see related recommendation below)] to detect possible pre-seismic and/or aseismic activity, attempts can be made to correlate the residual wobble excitation with seismic, aseismic, and meteorological events (the latter including El Niño).   Successful correlation would identify the remaining excitation source; failure to discover significant correlations might leave groundwater (+related) as the remaining major source.

Now that the semi-annual wobble has been reliably detected, it may be possible to improve our understanding of the various contributions to both the annual and semi-annual wobble excitation by calculating and comparing their excitation functions.

Because the various alternative excitation mechanisms of the secular polar trend are likely to have very different gravitational signatures, it may be possible to accurately constrain their contributions to the observed polar trend by monitoring changes in the geopotential field over time [RECOMMENDATION: harmonic coefficients through degree and order 40 (since ~500 km resolution is needed to discriminate source regions, e.g. lakes versus glaciers) on an annual, or even semi-annual, time scale]. The result would be a better determination of the deglaciation-induced secular trend and, with a more accurate drift rate, a better estimate of mantle viscosity and viscosity structure. Similar benefits would be derived from using a secular trend which has been corrected for site plate motions (RECOMMENDATION: IERS and other related activities should make polar motion data available with careful attention given to the plate motion models used). Space-geodetic plate models, being more accurate than long-term seismotectonic plate models, should be preferred for observation corrections; these corrections might amount to a ~30% improvement in the drift rate (e.g. Dickman, 1977). Finally, it is recommended that future gravity models treat C_2^1, S_2^1 as unknown coefficients to be solved for; such determinations of the figure axis directly (e.g. Lerch et al., 1985) would provide complementary information on the drift rate, as well as resolving any separation between the rotational barycenter and the figure axis.

Further investigation of the rapid polar motion should (Salstein et al., 1988) take account of the dynamic oceanic response (Dickman, 1988b) and the crustal response (Van Dam and Wahr, 1987) to any atmospheric pressure forcing considered. It should also be worth considering pre-seismic/aseismic excitation (see first recommendation) since they may well share the same time scale (e.g. Smylie and Mansinha, 1971; Gross and Chao, 1985; Preisig, 1988). If a connection with seismotectonic activity is established, it may well lead to a viable means of earthquake prediction, and a way to test elastic dislocation theories.

Covariance studies (Freedman and Dickey, 1987) have indicated that a full GPS system coupled with a global network can provide high-quality sub-daily polar motion at the centimeter level.   Such high-frequency wobble data would represent a whole new frontier, possibly answering questions about earthquake processes and weather systems — and their abilities to affect rotational dynamics on the shortest rotational time scale.

REFERENCES

Anderson, D. L. and Minster, J. B., 1979, *Gephys. J. Roy. Astr. Soc.*, **58**, 431-440.
Barnett, T. P., 1982, preprint (SIO ref. 82-10) [see also Barnett, T. P., 1983, *Climat. Change*, **5**, 15-38].
Busse, F. H., 1970, *Earthquake Displacement Fields and the Rotation of the Earth*, ed. Mansinha, Smylie and Beck, 88-98.
Chao, B. F., 1988, *EOS*, **69**, 326.
Dahlen, F. A., 1976, *Geophys. J. R. Astr. Soc.*, **46**, 363-406.
Dickey, J. O., and Eubanks, T. M., 1987, *Proc. of Inter. Symp.: Figure and Dynamics of the Earth, Moon and Planets, special issue of the Monograph Series of the Research Institute of Geodesy, Topography and Cartography*, (Prague, Czechoslovakia), ed. P. Holota, 907-930.
Dickman, S. R., 1976, *EOS*, **57**, 896.
Dickman, S. R., 1977, *Geophys. J. Roy. Astr. Soc.*, **51**, 229-244.
Dickman, S. R., 1983, *J. Geophys. Res.*, **88**, 6373-6394.
Dickman, S. R., 1985, *J. Geophys. Res.*, **90**, 11553-11556.
Dickman, S. R., 1988a, *Geophys. J. Roy. Astr. Soc.*, **94**, 519-543.
Dickman, S. R., 1988b, *J. Geophys. Res.*, in press.
Dickman, S. R., and Preisig, J. R., 1986, *Geophys. J. R. Astr. Soc.*, **87**, 295-304.
Eubanks, T. M., Steppe, J. A., and Dickey, J. O., 1985, *Proc. Int. Conf. Earth Rotation and the Terrestrial Reference Frame*, publ. Dept. of Geodetic Sci. and Surveying, Ohio State Univ.
Eubanks, T. M., Steppe, J. A., Dickey, J. O., Rosen, R. D., and Salstein, D. A., 1988, *Nature*, **334**, 115-119, 1988.
Freedman, A. P., and Dickey, J. O., 1987, *EOS*, **68**, 1245.
Gasperini, P., Sabadini, R., and Yuen, D. A., 1986, *Geophys. Res. Lett.*, **13**, 533-536.
Gross, R. S., and Chao, B. F., 1985, *J. Geophys. Res.*, **90**, 9369-9380.
Gwinn, C. R., Herring, T. A., and Shapiro, I. I., 1986, *J. Geophys. Res.*, **91**, 4755-4765.
Kakuta, C., Okamoto, I., and Sasao, T., 1975, *Publ. Astron. Soc. Japan*, **27**, 357-365.
Kuehne, J. W., and Wilson, C. R., 1988, *EOS*, **69**, 328.
Lambeck, K., 1980, Cambridge Univ. Press.
Lerch, F. J., Klosko, S. M., Patel, G. B., and Wagner, C. A., 1985, *J. Geophys. Res.*, **90**, 9301-9311.
Markowitz, W., 1982, *Rept. to IAU Comm. 19 and 31*, Patras.
Merriam, J. B., 1982, *Geophys. J. Roy. Astr. Soc.*, **70**, 41-56.
Nakiboglu, S. M., and Lambeck, K., 1980, *Geophys. J. R. Astr. Soc.*, **62**, 49-58.
O'Connor, W. P., 1987, NASA technical report.
Okamoto, I., and Kikuchi, N., 1983, *Publ. Int. Lat. Obs. Mizusawa*, **16** (2), 35-40.
Poirier, J. P., 1988, *Geophys. J. R. Astr. Soc.*, **92**, 99-105.
Preisig, J. R., 1988, *EOS*, **69**, 326.
Robertson, D. S., and Carter, W. E., 1985, *Proc. Int. Conf. Earth Rotation and the Terrestrial Reference Frame*, publ. Dept. of Geodetic Sci. and Surveying, Ohio State Univ., 296-306.
Sabadini, R., and Peltier, W. R., 1981, *Geophys. J. Roy. Astr. Soc.*, **66**, 553-578.
Sabadini, R., Yuen, D. A. and Boschi, E.V., 1982, *J. Geophys. Res.*, **87**, 2885-2903.
Salstein, D. A., Rosen, R. D., and Wood, T. M., 1988, *EOS*, **69**, 327.
Smith, M. L., and Dahlen, F. A., 1981, *Geophys. J. R. Astr. Soc.*, **64**, 223-281.
Smylie, D. E., and Mansinha, L., 1971, *Geophys. J. R. Astr. Soc.*, **23**, 329-354.

Soler, T., and Mueller, I. I., 1978, *Bull. Geodes.*, **52**, 39-57.
Souriau, A., and Cazenave, A., 1985, *Earth Planet. Sci. Lett.*, **75**, 410-416.
Van Dam, T. M., and Wahr, J. M., 1987, *J. Geophys. Res.*, **92**, 1281-1286.
Wahr, J. M., 1985, *J. Geophys. Res.,* **90**, 11557.
Yoder, C. F., and Ivins, E. R., 1985, *EOS*, **66**, 245.
Yoder, C. F., and Ivins, E. R., 1987, *Proc. IAU Symp. #129.*
Yuen, D. A., Sabadini, R., and Boschi, E. V., 1982, *J. Geophys. Res.*, **87**, 10745-10762.

4. AXIAL RATE OF SPIN OF THE EARTH

J. O. Dickey

4.1 INTRODUCTION

The rotation of the solid Earth, as monitored from observatories fixed on the Earth's crust, is not constant. The measurements reveal minute but complicated changes of up to several parts in 10^8 in the speed of the Earth's rotation, corresponding to several milliseconds in the length of the day (LOD). Earth studies have embarked on a new era with the advent of highly accurate space geodetic techniques and the availability of complementary geophysical data sets. Techniques utilized include laser ranging to the Moon and artificial satellites (SLR and LLR), and very long baseline interferometry (VLBI). Intercomparisons indicate that Earth rotation is routinely determined at the 0.1 millisecond level (approximately 5 cm at the equator), with higher accuracy being achieved in some cases. Hence, geophysically interesting variations are detectable at these levels. The analysis and understanding of these phenomena draws upon and contributes to the fields of meteorology, oceanography, astronomy, celestial mechanics, seismology, tectonics, and geodynamics.

Changes in Earth orientation are caused by the deformation of the solid Earth and by the exchanges of angular momentum between the solid and fluid parts of the Earth, as well as by exchanges of angular momentum with extraterrestrial objects. Changes in Earth rotation and polar motion can be regarded as the response of a linear differential system to a three-dimensional excitation vector. Earth rotation, when studied in combination with other parameters such as atmospheric angular momentum (AAM) and the Southern Oscillation Index, allow new and unique insights into geophysical processes. From intercomparisons of AAM and length of day (Fig. 1), we find that Earth rotation variations over time scales of a year or less are dominated by atmospheric effects, with a dominant seasonal cycle and significant variability on the intraseasonal (30 to 60 day) time scale. The correlation between length of day and AAM is so well established that forecasts of atmospheric angular momentum, using the large numerical models from the major forecast centers, are now being investigated, with the view to using them for the purpose of meeting objective predictions of short-term changes in the Earth's rate of rotation. Variations on interseasonal time scales have been related to the El Niño / Southern Oscillation phenomena. Turning to the longer-scale "decade" fluctuations in the LOD, torques between the core and mantle are accepted by geophysicists to be the most probable cause for these variations. Two candidate mechanisms are under investigation: (1) core-mantle torques which are probably largely due to dynamic pressure forces associated with core motions acting on topographic undulations of the core-mantle boundary and the equatorial bulge, and (2) the electromagnetic origin, which might make a significant contribution if the electrical conductivity is high enough. Estimates of topographic torques can be made from a combination of core motion models from geomagnetism, core-mantle boundary maps from seismic tomography, and long-wavelength gravity anomaly data. Geodetic torque estimates provide a means of checking these results and imposing constraints in the models used. Trends found on even longer time scales, the "secular" changes, are due to tidal dissipation torques, which produce a steady increase in the LOD at a rate estimated from ancient eclipse records to lie between 1 and 2 milliseconds per century. Contributions to LOD changes on the same time scale are also produced by internal sources, such as changes in the moment of inertia of the solid Earth resulting from the melting of ice after the last major ice age. The reader is referred to several review articles (Rochester, 1984; Dickey and Eubanks, 1986; Hide, 1986; and Wahr, 1986) and the references within for further details.

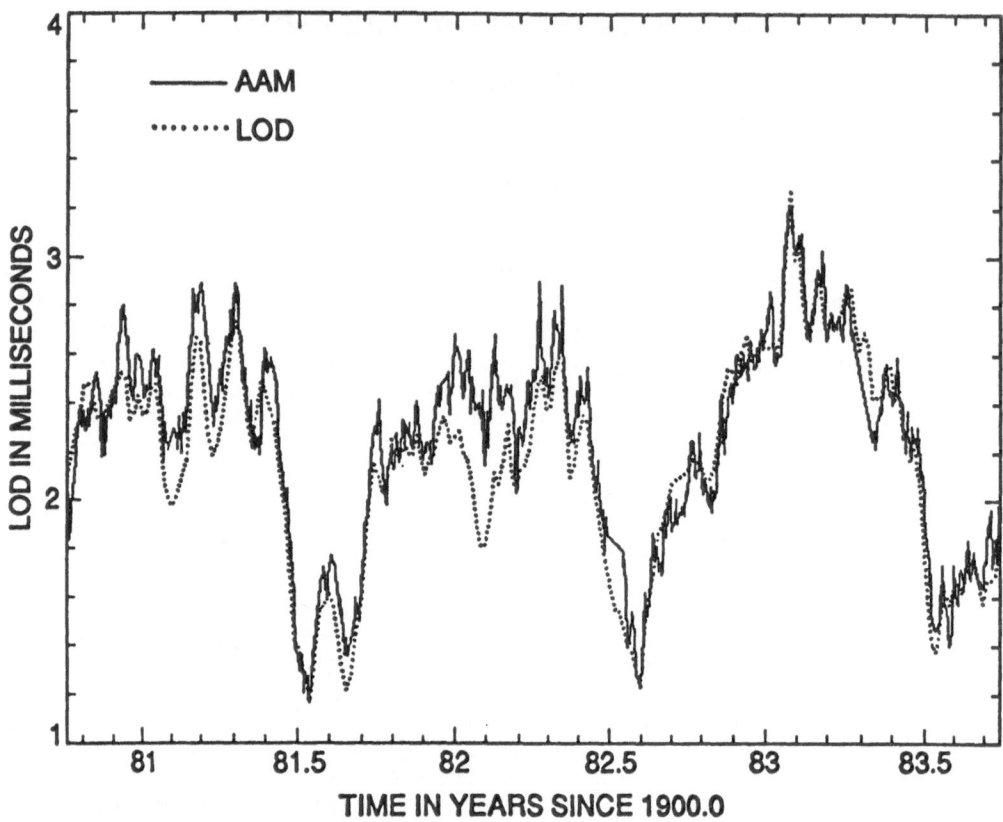

Fig. 1. AAM changes as estimated by the National Meteorological Center (solid line) together with Kalman-smoothed estimates of the LOD (dotted line) [after Dickey et al., 1986].

4.2 SHORT-PERIOD CHANGES

Comparison of astronomical measurements with axial atmospheric angular momentum first indicated the significance of the atmospheric contribution to Earth rotation. Various studies (e.g., Rudloff, 1973; Lambeck and Cazenave, 1977, and Lambeck and Hopgood, 1981) have related LOD variations to changes in atmospheric angular momentum on time scales ranging from months to a few years using mean monthly or longer period atmospheric data. Hide *et al.* (1980), using the zonal component of angular momentum evaluated at 12-h intervals from the First GARP Global Experiment (FGGE) and LOD data, demonstrated the good agreement between these series, indicating that on these short time scales that the angular momentum transfer between Earth and the atmosphere could fully account for the observed variation. The correlation of atmospheric data with modern length of day results has been reported by several authors (e.g., Langley et al., 1981; Carter et al., 1984; Eubanks et al., 1985a; and Dickey et al., 1986); recent atmospheric and LOD results are shown in Fig. 1. The high correlation between the two data sets is evident. Changes in both sets contain a large seasonal cycle, dominated by the annual term. Superimposed on this is the irregular "50-day" oscillation which varies in period roughly from 40 to 60 days. A number of studies (Eubanks et al., 1983 and 1985a; and Morgan et al., 1985) indicate that significant, but relatively small, imbalances in

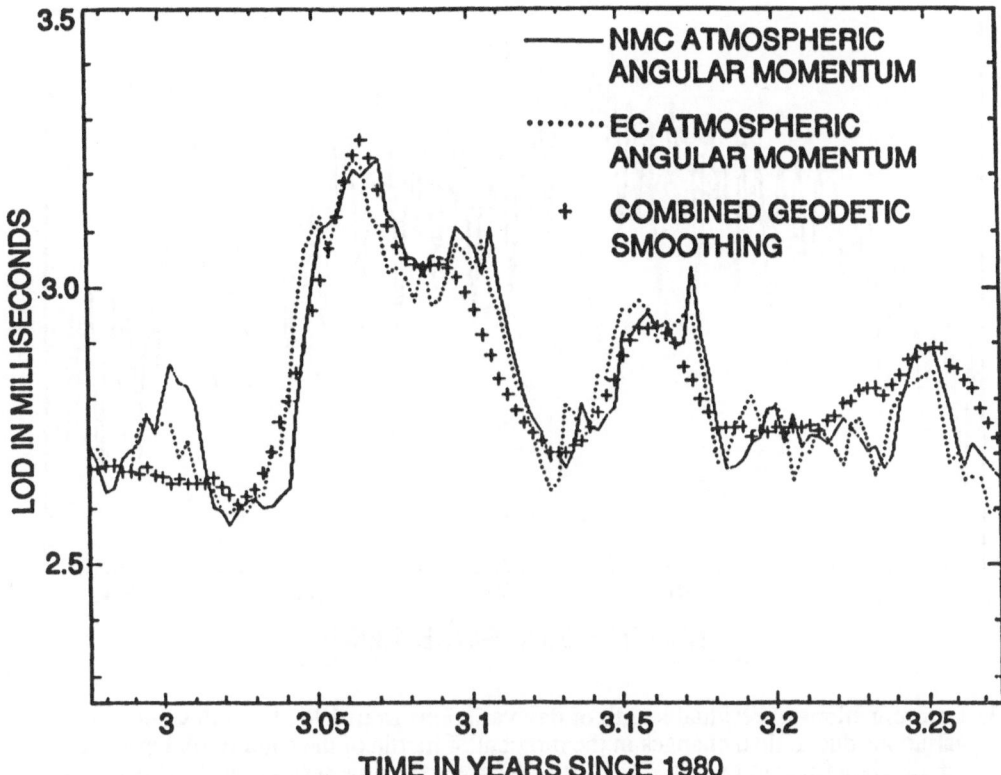

Fig. 2. European Centre for Medium Range Forecasts - EC (dotted line) and National
Meteorological Center - NMC (full line) together with the combined Kalman smoothed
LOD estimates from the space geodetic technique (+) (after Dickey and Eubanks,
1987).

the angular momentum exist at the annual and semiannual periods, with the semiannual
discrepancy being larger. Most of the annual imbalance can be attributed to the neglect of the
contribution from stratospheric winds (Rosen and Salstein, 1985), but changes in the oceanic
circulation may contribute to the unexplained seasonal changes in the length of day (Eubanks et
al., 1983 and 1985a; Brosche and Sundermann, 1985).

Oscillations in zonal winds and other meteorological quantities with periods of from 40
to 60 days were first discovered by Madden and Julian (1971 and 1972) in wind data from the
equatorial Pacific. Since that time, higher-frequency oscillations have also been seen (Miller,
1974), and the 40 to 60 day oscillation has been observed on a global scale (Krishnamurti and
Subrahmanyam, 1982; Yasunnari, 1981; Anderson and Rosen, 1983). The corresponding
Earth rotation changes were first detected by Feissel and Gambis (1980) in four independent
UT1 data types; Langley et al. (1981) later reported this effect in both the AAM and lunar laser
ranging length of day. Feissel and Nitschelm (1985) showed the periodic variability of this
oscillation. The peak amplitude seems to be about 0.25 msec of length of day, though the
oscillation exhibits changes in both period and amplitude on a year-to-year basis. Morgan et al.
(1985) showed that any nonmeteorological contribution to the 40-60 day oscillation is not
significantly larger than the uncertainties in the observations (about 0.06 ms). Contributions

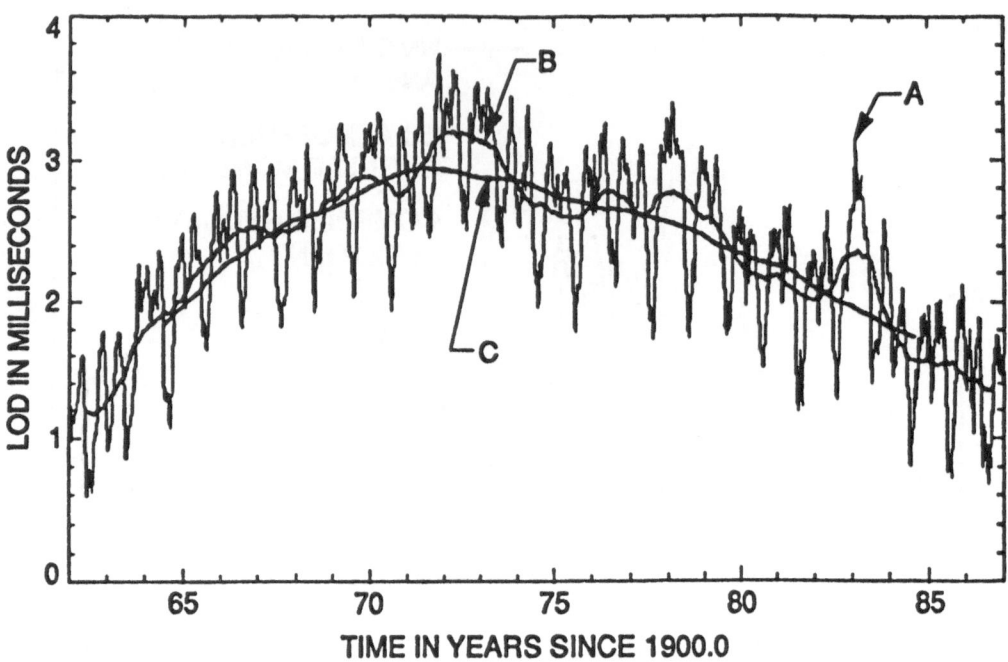

Fig. 3. Determinations of residual length of day variations from 1962 to 1986 when periodic
variations due to tidal changes in the moment of inertia of the solid Earth have been
removed (defined to be LOD*). Curve (A) gives the time series of a "Kalman
Smoothing" of data obtained by means of the techniques of optical astrometry and
space geodesy (Very Long Baseline Interferometry and Lunar Laser Ranging). They
indicate variations on time scales upwards of several days. Curve (B) is the results of
eliminating the shortest period contributions to curve (A) by taking a "rectangular" 365-
day running mean, leaving interannual and decade contributions. Curve (C) is the
results of eliminating interannual variations from curve (B) by taking a 5-year
"rectangular" running mean (after Dickey et al., 1988).

from the winter hemisphere are generally stronger (Eubanks et al., 1986). Currently, there are
ongoing studies utilizing the results from LOD, AAM and Atmospheric General Circulation
Models to the determine.the origin of these intraseasonal oscillations (Marcus et al., 1988).
    The period of early 1983, when abrupt changes in both AAM and LOD occurred and the
AAM reached record high value, represent a particular good time to test the detailed agreement
between the two series of independent data types, on a fine scale (Fig. 2). The comparison is
striking. There is no significant lag or lead between the two series. A potential future role for
AAM results is its use as a proxy index of length of day changes in near real time; this would be
extremely useful where accurate timely results are important (e.g. spacecraft navigation).
Investigations into the feasibility of using AAM forecasts in Earth rotation prediction has
already begun (Rosen et al., 1987 and Hide, 1988a). Discussions on the role of the AAM
"technique" in the new International Earth Rotation Service have been held and, as a result, a
sub-bureau for AAM will be established working in conjunction with the Central Bureau of the
IERS.

4.3 INTERANNUAL VARIATIONS

Interannual variations of the atmosphere presumably excite some of the interannual fluctuations in the LOD (Lambeck, 1980 and Stephanic, 1983). The Southern Oscillation involves a large-scale interannual redistribution of atmospheric mass, and is also associated with the El Niño phenomenon and the resulting widespread changes in the atmospheric and ocean circulation (Philander, 1983). The seasonally adjusted difference of the sea-level pressure at Tahiti and Darwin, Australia, which is known as the Southern Oscillation Index (SOI), reached a record low in January 1983 (Rasmusson and Wallace, 1983). There were unusually large changes in the LOD and the axial atmospheric angular momentum during January and February 1983 (Rosen *et al.*, 1984). These changes coincided with the most intense period of the 1982-83 El Niño and with unusual changes in atmospheric pressure over the eastern and western tropical Pacific. The unusual maxima in January and February 1983 are clearly visible in both the LOD and AAM data sets (Figs. 1 and 2). This peak represents an atmospheric absorption and then loss of ~ 3 x 10^{25} kg m^2 s^{-1} of eastward angular momentum over a period of less than a month. Evidence presented indicates that most of the anomalous torque is in the tropics and subtropics. Studies suggest that there may have been similar LOD changes during previous El Niños and that some of the interannual changes in the LOD are related to the Southern Oscillation (Chao, 1984 and 1988; Eubanks *et al.*, 1985b and 1986). These studies reveal correlations ranging from -0.5 to -0.7 between the interannual fluctuations in the SOI and the LOD. Perfect correlation between the Earth rotation and the SOI should not be expected since the SOI is derived from local measurements, while the AAM and LOD are global quantities. However, these initial results are very promising; historical records of LOD changes may serve as a proxy index of global wind fluctuations and may be of interest to meteorologists, oceanographers, and climatologists (Salstein and Rosen, 1986).

4.4 DECADE FLUCTUATIONS

Torques at the core-mantle interface are due to time varying fluid motions in the core; several mechanisms have been forwarded as sources for the implied stress at the core-mantle interface (see for example, Lambeck, 1980 and Hide, 1986). They include: (1) tangential viscous drag in the Ekman-Hartmann boundary layer at the top of the core; (2) the action of normal pressure forces on topographic undulations on the core-mantle boundary; and (3) electromagnetic coupling due to Lorentz forces between the conducting fluid core and partially conducting lower mantle. Clearly, the strength and the effectiveness of each of these mechanisms is key to a broad number of disciplines. The Earth rotation data presented in Fig.3 provide important constraints on models of the internal structure and dynamics of the core and lower mantle.

Viscous coupling is likely to be comparatively very weak and would only be significant under extreme assumptions about the viscosity of the core. Hence, most modern research is concerned with topographic and electromagnetic coupling. The electromagnetic coupling may make a significant contribution if the electrical conductivity is high enough (Paulus and Stix, 1986 and Hide, 1986). Matters of concern include assumptions made concerning the strength of the toroidal part of the geomagnetic field in the outer reaches of the core and the distribution of electrical conductivity in the lower mantle. According to the concept of topographic coupling [introduced by Hide (1969)] the magnitude of the stresses implied by the amplitude and time scale of the decade variations in the LOD might be accounted for if there were undulations on the core-mantle boundary with a height of less than one kilometer. These bumps could be supported by the viscous stresses associated with deep convention in the mantle. A method for obtaining estimates of the topographic core-mantle coupling has recently been proposed by Hide (1986), which combines core motion models and core mantle boundary maps to produce estimates of these torques. Topographic maps of the core-mantle interface are based on the data of seismic tomography and the low degree harmonics of the geo-gravitational field together with various assumptions concerning the rheology of the upper and lower mantle. Also required in the topographic coupling calculation are pressure fields at the core-mantle interface. Here, geomagnetic secular variation data are used to calculate the horizontal motion just below the core-mantle interface through the use of the frozen flux theorem of Alfven and additional assumptions used to close the system of equations. The horizontal pressure gradients are computed from these velocity fields assuming geostrophic balance exists between the Coriolis acceleration and the horizontal pressure gradient in the outer reaches of the core.

The preliminary findings of the work (Hide, 1988b) are encouraging, as a consistent picture of the core-mantle boundary is emerging based on the results of Earth rotation and research into seismic tomography and geomagnetism. Earth rotation results indicate a torque of ~ 0.2 Hadley units in the axial component during the 1980 period (Dickey *et al.*, 1988). When torque calculations are computed with the combined results of seismic tomography and geomagnetism for the 1970-80 decade, axial torque estimates are an order of magnitude too large with "bumps" on the core-mantle interface at the 3 km level. When a D" layer is included in the analysis (a boundary layer in the bottom 100-300 km of the mantle, which is thought to have low viscosity), the magnitude of the torque is consistent with those of Fig. 3 and the corresponding undulations are ~ 500 m (Hager, private communication). The size of these bumps is in accord with recent calculations based on corrections to the Standard IAU 1980 Nutation Model using Very Long Baseline Interferometry data (Eubanks *et al.*, 1985c; and Herring *et al.*, 1986). The corrections are dominated by a large retrograde annual term of ~ 2 mas. These observations can be interpreted as implying a core-mantle boundary that is more elliptical than one for an Earth in static rotational equilibrium; the total required deviation from the figure of equilibrium is 500 m (Gwinn *et al.*, 1986). Collectively, these results indicate undulations at the core-mantle boundary of about 500 m in size and the existence of a D" layer in the mantle.

4.5 PROSPECTS FOR THE FUTURE

The advent of the interdisciplinary view of Earth System Science and Global Change is key to Earth rotation studies. The continual improvement in the accuracy and density of data from the new techniques will permit the study of the Earth's exchange of angular momentum over an even shorter time scale. We urge the attainment of 1 mm accuracy systems as a goal in the techniques of laser ranging and VLBI. Besides the techniques of VLBI, SLR and LLR, covariance studies indicate that the full constellation of GPS satellites coupled with a global network of stations will provide highly accurate sub-daily determination of Earth rotation (Freedman and Dickey, 1987). Further, we recommend that variations of the rotation vector of the solid Earth be determined with the highest possible accuracy [at least 0.1 milliarcsecond (equivalent to 0.3 cm on the surface of the Earth)] and temporal frequency (at least several cycles per day), and that complementary geophysical, oceanographic, and atmospheric data be taken to facilitate the interpretation of geophysical processes.

These combined data types would allow for a better understanding of the total Earth system. For example, atmospheric data as analyzed and assimilated into General Circulation Models will lead to an improvement in knowledge of the dynamics and physics of the atmosphere. Measurements of interest include outgoing longwave radiation, and improved determination of atmospheric pressure, and stratosphere and tropospheric winds. Further, these analyses, coupled to highly accurate geodetic measurements of the Earth rotation vector, would permit new and unique insights into the coupling mechanisms between the solid Earth and the atmosphere. The EOS concept of an interdisciplinary mission is ideal for such investigations. A coordinated intensive geodetic and interdisciplinary measurement campaign should be held in conjunction with WOCE, TOPEX, and other relevant international programs; the resultant data series would allow new insights into the coupled solid-Earth /atmosphere /ocean system. In addition, the benefits of high-frequency measurements will extend to seismology (source phenomena and possible earthquake prediction). On the interannual timescales, continued studies of global phenomena, such as the El Niño/Southern Oscillation, will lead to further advances. Continued interaction and joint studies with geomagnetic results and seismic tomographic findings will increase our knowledge of the longer-term "decade" variation and the Earth's interior. The dividends of the interannual and decade studies will extend to the oceanographic community as well with implications for tidal friction, tides and oceanic circulation and mass distribution anomalies during El Niño.

REFERENCES

Anderson, J. A. and Rosen, R. D., 1983, *J. Atmos. Sci.*, **40**, 1584-1591.

Brosche, P., and Sundermann, J., 1985, *Dt. Hydroqv Z.*, **38**, 1-6.

Carter, W. E., Robertson, D. S., Pettey, J. E., Tapley, B. D., Schutz, B. E., Eanes, R. J. and Miao Lufeng, 1984, *Science*, **224**, 957-961.

Chao, B. F., 1984, *Geophys. Res. Lett.*, **11**, 541-544.

Chao, B. F., 1988, *J. Geophys. Res.*, **93**, B7, 7707-7715.

Dickey, J. O. and Eubanks, T. M., 1986, *Space Geodesy and Geodynamics*, Academic press, editors, A. J. Anderson and A. Cazenave, 221-269.

Dickey, J. O., Eubanks, T. M., and Steppe, J. A., 1986, *NATO Advanced Research Workshop, (co-sponsored by the Council for Europe), Earth Rotation: Solved and Unsolved Problems, NATO Advanced Institute Series C: Mathematical and Physical Sciences,* ed. A. Cazenave, D. Reidel, Boston, **187**, 137-162.

Dickey, J. O., and Eubanks, T. M., 1987, *Proc.of the International Symposium: Figure and Dynamics of the Earth, Moon and Planets, special issue of the Monograph Series of the Research Institute of Geodesy, Topography and Cartography,* (Prague, Czechoslovakia), ed. P. Holota, 907-930.

Dickey, J. O., Eubanks, T. M., and Hide, R., 1988, *Geophysical Monograph Series of the American Geophysical Union, Proceedings of the International Union of Geodesy and Geophysics (IUGG), Interdisciplinary Symposium, Variations in the Earth's Rotation, IUGG XIX General Assembly* (Vancouver, August 1987), in press.

Eubanks, T. M., Dickey, J. O. and Steppe, J. A., 1983, *Proc. Int. Assoc. of Geodesy Symp.*, International Union of Geodesy and Geophysics XVIIIth General Assembly (Hamburg, Germany, Aug. 15-27, 2983),Ohio State University, Columbus, **2**, 122-143.

Eubanks, T. M., Steppe, J. A., Dickey, J. O. and Callahan, P. S., 1985a, *J. Geophys. Res.*, **90**, 5385-5404.

Eubanks, T. M., Dickey, J. O., and Steppe, J. A., 1985b, *Tropical Ocean- Atmosphere Newsletter*, **29**, 21-23.

Eubanks, T. M., Steppe, J. A., and Sovers, O. J., 1985c, *Proc. Inter. Conf. Earth Rotation and Terrestrial Reference Frame*, Ohio State University, Columbus, **1**, 326-340.

Eubanks, T. M., Steppe, J. A., and Dickey, J. O., 1986, *NATO Advanced Research Workshop (co-sponsored by the Council of Europe), Earth Rotation: Solved and Unsolved Problems, NATO Advanced Institute Series C: Mathematical and Physical Sciences,* ed. A. Cazenave, D. Reidel, Boston, **187**, 163-186.

Feissel, M. and Gambis, D., 1980, *Comptes Rendus Hebdomadaires es Sceances de l'Academie des Sciences, Series B*, **291**, 271-273.

Feissel, M., and Nitschelm, C., 1985, *Ann. Geophys.* **3**, 181-186.

Freedman, A. P., Dickey, J. O., 1987, *EOS*, **68**, 1245.

Gwinn, C. R., Herring, T. A., and Shapiro, I. I., 1986, *J. Geophys. Res.*, **91**, 4755-4765.

Herring, T. A., Gwinn, C. R., and Shapiro, I. I., 1986, *J. Geophys. Res.*, **91**, 4745-4754.

Hide, R., 1969, *Nature*, **222**, 1055-1056.

Hide, R., Birch, N. T., Morrison, L. V., Shea, D. J. and White, A. A, 1980, *Nature*, **286**, 114-117.

Hide, R., 1986, *Quart. J. Roy. Astron. Soc.*, **27**, 3-20.

Hide, R., 1988a, *Geophysical Monograph Series of the American Geophysical Union, Proceedings of the International Union of Geodesy and Geophysics (IUGG), Interdisciplinary Symposium, Variations in the Earth's Rotation, IUGG XIX General Assembly* (Vancouver, August 1987), in press.

Hide, R., 1988b, *Phil. Trans. Roy. Soc. A*, in press.

Krishnamurti, T. N. and Subrahmanyan, D., 1982, *J. Atmos. Sci.*, **39**, 2088-2095.

Lambeck, K. and Cazenave, A., 1977, *Phil. Trans. R. Soc. Lond. A*, **284**, 495-506.

Lambeck, K., 1980, *The Earth's Variable Rotation*, London: Cambridge University Press.

Lambeck, K. and Hopgood, P., 1981, *Geophys. J. R. Astr. Soc.*, **71**, 67-84.

Langley, R. B., King, R. W., Shapiro, I. I., Rosen, R. D. and Salstein, D. A., 1981, *Nature*, **294**, 730-733.

Madden, R. A. and Julian, P. R., 1971, *J. Atmos. Sci.*, **28**, 702-708.

Madden, R. A. and Julian, P. R., 1972, *J. Atmos. Sci.*, **29**, 1109-1123.

Marcus, S. L., Dickey, J. O., Eubanks, T. M., Ghil, M., 1988, *EOS*, **69**, 327.

Miller, A. J., 1974, *J. Atmos. Sci.*, **31**, 720-726.

Morgan, P. J., R. W. King, and I. I. Shapiro, 1985, *J. Geophys. Res.*, **90**, 12645-12652.

Paulus, J. and Stix, M., 1986, *Earth Rotation: Solved and Unsolved Problems*, A. Cazenave, ed., D. Reidel Publishing Co., Dordrecht, 259-267.

Philander, S. G. H., 1983, *Nature*, **302**, 295-301.

Rasmusson, E. U. and Wallace, J. M., 1983, *Science*, **222**, 1195-1202.

Rochester, M. G., 1984, *Phil. Trans. R. Soc. Lond. A*, **313**, 95-105.

Rosen , R. D. and Salstein, D. A., 1983, *J. Geophys. Res.*, **88**. no C9, 5451-5470.

Rosen, R. D., Salstein, D. A., Eubanks, T. M., Dickey, J. O. and Steppe, J. A., 1984, *Science*, **225**, 411-414.

Rosen, R. D. and Salstein, D. A., 1985, *J. Geophys. Res.*, **90**, D5, 8033-8041.

Rosen, R. D., Salstein, D. A., Nekrkorn, T., McCalla, M. R. P., Miller, A. J., Dickey, J. O., Eubanks, T. M., and Steppe, J. A., 1987, *Monthly Weather Rev.*, **115**, 9, 2170-2175.

Rudloff, W., 1973, *annln Meteorol. neue folge*, **6**, 221-223.

Salstein, D. A. and Rosen, R. D., *J. Clim. and Appl. Meteorol.*, **25**, 1870-1877.

Stephanic, M., 1983, *J. Geophys. Res.*, **87**, 428-432.

Wahr, J. M., 1986, *Space Geodesy and Geodynamics*, A. J. Anderson and A. Casenave, eds., Academic Press, London, 281-313.

Yasunari, T., 1981, *J. Meteor. Soc. Japan*, **59**, 336-354.

5. EARTH TIDES

J. Zschau

5.1 INTRODUCTION

Earth tides are the response of the solid Earth and its liquid outer core to the combined luni-solar gravitational attraction. About 20 years ago this response was assumed to be static, and independent of what happens in the ocean and in the atmosphere. Tidal displacements and gravity changes were, therefore, described by three parameters only, the three Love numbers h, l, k for the radial and horizontal displacements, and for the potential variations, respectively. Great progress has been made since then in understanding that the Earth responds to the tidal forces as a coupled dynamic system which exhibits important interactions between the different subsystems, the solid Earth, the liquid core, the atmosphere, and the ocean. There is no way of measuring the tidal response of a decoupled solid Earth alone. Thus, if one is interested in what happens in this part of the system one has to have good models for the other parts of the system in order to correct their influences from the measurements. The separation between the different contributions is the major goal for future Earth tide research. It is the absolutely necessary precondition which has to be reached if one wants to look deeper into the problem of tidal dynamics within the liquid core, and if one wants to obtain a better understanding of how the tides are influenced by the Earth's rotation as well as by elliptical stratification, anelasticity and heterogeneity in the mantle. These are some of the problems which are just starting to be considered. Theoretical estimations show that the tidal perturbations introduced by these phenomena should be at the level of 1%. In order to extract them from measurements, instrumental techniques as well as theoretical models must therefore have an accuracy at the level of tenths of a percent. This is roughly an order of magnitude better than present space techniques can achieve. However, improvements can be expected from all three modern space techniques which are presently used for measuring the tidal response: Satellite Laser Ranging (SLR), Lunar Laser Ranging (LLR), and Very Long Baseline Interferometry (VLBI). They either detect the tidal perturbations in the orbits of close Earth satellites (SLR), the tidal induced variations of the length of the day (LLR, VLBI), or the tidal displacement below the station (VLBI, SLR).

5.2 SPACE TECHNIQUES VERSUS TERRESTRIAL MEASUREMENTS

All the problems addressed above can in principle be studied with both space techniques and terrestrial instruments such as gravimeters. There are, however, two major problems with terrestrial measurements:

(1) The long-term stability of the instruments as well as of their installations is not sufficient to get good information on long period tides above the semidiurnal and diurnal tidal band. Thus, studying frequency dependence of the Love numbers, as for instance introduced by mantle anelasticity, becomes extremely difficult. The only exception from this is probably the superconducting gravimeter, which has already provided important results on long period gravity variations (see for instance Richter, 1985).

(2) Terrestrial tidal instruments such as gravimeters and tiltmeters cannot measure the Love numbers separately. They provide only information on Love number combinations, for instance, on the gravimetric factor and on the diminishing factor. In these quantities, the Love numbers, h and k, unfortunately enter with a different sign, and as a consequence several interesting influences, such as those of mantle inelasticity and mantle heterogeneity cancel to a certain extent. This is not true for modern space techniques. They are capable of obtaining the

Love numbers separately, and they also provide information on long period tides. In this respect, space techniques have clear advantages compared to terrestrial methods.

5.3 OCEAN TIDES

A limiting factor for studying the above addressed problems with space techniques and/or with terrestrial methods has always been the ocean tides. Their indirect effect on Earth tide measurements had long been recognized, although a rigorous computation of the ocean load influences did not exist before the beginning of the seventies. Then it was shown that ocean loading effects could reasonably amount to 10%, 20%, and 90%, respectively, of the body tide in gravity, strain and tilt (Farrell, 1972). The total oceanic influence on close Earth satellite orbits amounts roughly to 10% of the body tide effect. A similar figure holds for the tidal variations of the length of the day, and the ocean loading effect is also extant in the tidal displacements of the solid surface.

In order to accurately correct Earth tide measurements for the indirect effect of ocean tides, detailed ocean tide models are needed. Rapid progress was made with computer-generated models, from which Parke and Hendershott's (1979), and Schwiderski's (1980) models seem to fulfill the requirements best. Today, Schwiderski's model is more often used than any other global ocean tide model. It is given in great detail (1° by 1°) and provides amplitudes and phases not only for the main semidiurnal and diurnal constituents but also for the long period tides.

Although the Schwiderski ocean tide maps have proved to be able to predict gravity load tides to an accuracy of better than 1 microgal (Ducarme and Melchior, 1978), they are by far not good enough to extract the solid Earth tide signal from tidal measurements with the required accuracy of a few tenths of a percent or better. A new perspective in this connection is the determination of ocean tide height distribution from satellite altimetry. First results on a global solution for the principal tidal wave M_2 have already been obtained from SEASAT altimetry (Mazzega, 1986). They are very promising for satellite altimetry as the future source of detailed ocean tide knowledge. Considering the fact that accurate ocean tide corrections in Earth tide measurements are the key to understanding the process of tidal deformation in the solid Earth and liquid core, and having in mind that viewing the ocean surface from space is the best way of getting global information on tides in the open ocean, it is strongly recommended that high precision satellite techniques including dynamic satellite solutions as well as satellite altimetry get as much support as possible.

5.4 TIDAL FRICTION

Astronomical as well as paleontological evidence suggests that the Earth's rotation rate is steadily slowing down, resulting in a lengthening of the day by about 1-2 ms within 100 years. This is commonly attributed to internal friction within and especially between the Earth's solid and fluid layers during tidal deformation causing the tidal bulge to be retarded by about 10 minutes with respect to the primarily acting tidal potential. The nonequilibrium bulge is held by the gravitational forces of the Moon and acts as a brake block on the Earth which is rotating underneath. The Earth's rotation rate is thus slowing down, and conservation of angular momentum in the Earth-Moon system ensures that at the same time the Earth-Moon distance as well as the Moon's mean angular position is changing. There is a similar effect due to the tidal forces of the Sun which, however, amounts only to 20% of the Moon's influence.

In principle, the experimental verification of this phenomenon can be obtained from observing the secular changes in all three parameters: the length of the day (L.O.D.), the

Moon's mean longitude, and its geocentric distance. In practice, however, the expected linear trend in the L.O.D. of 1-2 ms per century is masked by irregular decade fluctuations of about 4-5 ms over time scales of from 20 to 30 yr and atmospherically driven perturbations over shorter time scales. Even 400 years of optical astrometric data are not enough to separate a linear trend from the irregular decade fluctuation (Dickey and Eubanks, 1986). Newton (1985) points out that the Earth's rotation may not be uniform even on time scales of thousands of years.

Thus, it remains to determine the secular changes in the Moon's mean longitude and in its orbital distance. The second can be accomplished by lunar laser ranging. The first can be accomplished either

(1) by the analysis of astronomical observations of the Moon's motion
    with respect to the Earth,
(2) by the analysis of lunar laser ranging data, or
(3) by the analysis of the motion of close Earth satellites giving the tidal
    parameters directly applicable to the lunar problem.

Rapid progress has been made in areas (2) and (3). A very accurate result was obtained for the Moon's acceleration in longitude ($\dot{n}$) from a simultaneous solution of tracking observations to a high number of satellites (Christodoulidis et al., 1988): $\dot{n} = -25.27 \pm 0.61$ arcsec/(100 yr)^2. It could even be separated into the contributions coming from the semidiurnal, the diurnal, and the long period tides. The value agrees within a few percent with the one obtained from lunar laser ranging: $\dot{n} = -24.9 \pm 1.0$ arcsec/(100 yr)^2 (Newhall et al., 1988).

One of the consequences of tidal friction is the dissipation of tidal energy, i.e. the conversion of tidal energy into heat or into some other form of non-tidal energy. It adds up to the enormous energy flux of about 4×10^{19} erg/s, which is more than 10% of the total heat flux at the Earth's surface, about 10 to 100 times the total energy release due to earthquakes, and about equal to the present energy consumption of mankind. Ocean tide modelling suggests that most of this energy is consumed in the oceans and only a small amount, most probably less than 5%, in the solid Earth. The above values of $\dot{n}$ include both , dissipation in the ocean and in the solid Earth. A separation between ocean tide and Earth tide dissipation has not been possible so far, because the solid Earth contribution to $\dot{n}$ is probably smaller than the uncertainties in the determination of $\dot{n}$ (Zschau, 1986). More crucial, however, are the shortcomings in the present ocean tide models. They introduce uncertainties in the value of the solid Earth dissipation rate which may be several times the solid Earth dissipation rate itself. Significant improvement in the lunar laser ranging, in the satellite tracking, and especially in the ocean tide knowledge is required before the small amount of solid Earth dissipation can be separated from the total amount.

5.5 CORE RESONANCE AND ELLIPTICAL STRATIFICATION

A complete theory of how the Earth's rotation and the ellipticity of its internal layers affect the Earth tide response became available only a couple of years ago. It was shown by Wahr (1981), that these influences cause tidal perturbations at the level of 1% which enter in all three components of the Earth tide response: deformation (the body tide), changes in the Earth's angular orientation in space (precession and forced nutations), and changes in the Earth's rotation rate. The most obvious effect of rotation on Earth tides are resonance perturbations in the Love numbers at diurnal frequencies due to the nearly diurnal free wobble of the liquid core. They have been confirmed observationally by strain measurements (Levine, 1978), gravity measurements (Warburton and Goodkind, 1978), and recently by Very Long Baseline Interferometry (Eubanks et al., 1986; Herring et al., 1986). Moreover, it was found that the experimentally determined resonance frequency is considerably different from the theoretically predicted one. In trying to find an explanation for this shift in the eigenfrequency, the influence of the ocean tides and dispersion due to mantle anelasticity can be excluded, because both would give a shift in the wrong direction. However, the result agrees well with a 6% increase in core flattening, which corresponds to a deviation from the hydrostatic figure of the core-mantle boundary of about 500 m (Neuberg et al., 1987). No explanation so far has been given for the observed quality factor of the liquid core resonance. The recently determined value of below 3000 (Neuberg et al., 1987) is much too low to be caused by mantle anelasticity alone. Other mechanisms such as core-mantle boundary effects and ocean dissipation may be important. The value is, however, not determined well enough to allow a solution of the problem.

Even more difficult than to determine the characteristics of the liquid core resonance is the verification of the latitude dependence in the Love numbers introduced by an elliptical stratification of the layers throughout the Earth. As it is expected to be at the order of one percent, its verification will require measuring the Earth tides with an accuracy of a few tenths of a percent or better. At the moment the standard deviations of the station dependent tidal parameters from VLBI are typically 1% at good stations. For further improvement, a good coverage of different latitudes, and the inclusion of a higher number of stations are necessary. Additionally, it should be considered how satellite laser ranging can contribute to solving both the problems of Love number latitude dependence and liquid core resonance quality factor.

5.6 MANTLE RHEOLOGY

It is known from the damping of seismic waves as well as from a variety of other phenomena that the Earth's mantle exhibits deviations from ideal elasticity. This in principle has two consequences for the tidal response of the Earth: (1) a small phase shift in the tide with respect to the forcing function, and (2) a frequency dependence of the tidal parameters due to dispersion of the Love numbers. Model calculations (Zschau and Wang, 1986) suggest that the latter leads to slightly higher Love numbers (~1-2%) at the semidiurnal periods than those which one obtains from seismological models. The Love numbers increase with increasing period, and they are expected to be higher by about 10% at the 18.6-year tide, this value, however, being entirely dependent on the rheological model chosen for the mantle.

In order to check the models of mantle rheology against observations, one needs highly accurate measurements of the tidal response in a broad frequency band including long period tides. Terrestrial techniques, except perhaps the superconducting gravimetry, are not appropriate for this purpose because of stability problems with instruments and installations. Early investigations of the tidal perturbations of satellite orbits have mostly provided information on the principal M_2-tide. It was not before the middle of the eighties that a broad spectrum of tidal waves, including the most important semidiurnal and diurnal ones as well as the zonal tides M_f, M_m, Ssa, could be extracted from Starlette laser ranging to about 1% in the effective Love number k (Williamson and Marsh, 1985). The effective Love number comprises the influences of the body tide and of the ocean tide. Observations of LAGEOS node rate residuals (Yoder et al., 1983; Rubincam, 1984) allowed the extension of the observation spectrum to very long tidal periods (S_{Sa}, S_a, and 18.6 years). A simultaneous interpretation of tracking data for a high number of different satellites has further improved the results for most of these tidal waves (Christodoulidis et al., 1988).

Similar information on the effective Love number k for long period tides, although not yet that accurate, has also been obtained from observing the tidal induced variations in the Earth's rotation with LLR (Dickey et al., 1984). First attempts have been made to get the same kind of information from LOD observations by VLBI (Schuh, 1986). Thus, good tidal data from a broad spectrum of periods are available. On the other hand, because the expected order of magnitude of the anelastic dispersion is a few percent, the present accuracy of the methods involved is not sufficient. Again, a few tenths of a percent accuracy in measuring the Earth tide response is required, and like before there is an urgent need for precise ocean tide height data in order to correct for the oceanic effect. Also, the atmospheric effect and possible core-mantle processes will have to be considered at certain frequencies.

5.7 MANTLE HETEROGENEITY

Seismic tomography has clearly shown that the Earth's mantle is laterally inhomogeneous to a few percent in the seismic velocities. As a consequence, the tidal parameters should be dependent on latitude as well as on longitude. The exact magnitude of this influence is not known. But it was suggested by Wang and Zschau (1986) that - provided the lateral heterogeneities are partly due to temperature differences - these parameters might show dispersion, and the lateral heterogeneities could be considerably stronger at tidal periods than at seismic periods.

First model calculations based on the results of seismic tomography predict tidal perturbations on the level of one percent for the radial displacements, and five percent for the horizontal displacements, respectively. Comparing this with the station dependent Love numbers as observed with VLBI (Carter et al., 1985) reveals the interesting fact that the observed differences between the radial Love numbers h at different stations are indeed at the level of one percent, and thus could already reflect lateral heterogeneities in the mantle.

However, an improvement of the accuracy of these determinations by about one order of magnitude to a tenth or at least a few tenths of a percent is necessary to arrive at a conclusive judgment. A good global coverage with VLBI stations in different tectonic environments would be a further precondition. This applies to laser ranging stations as well, because they have the same capability of providing information on station dependent Love numbers.

REFERENCES

Carter, W. E., Robertson, D. S. and MacKay, J. R., 1985, *J. Geophys. Res.*, **90**, B6, 4577-4587.

Christodoulidis, D. C., Smith, D. E., Williamson, R. G. and Klosko, S. M., 1988, *J. Geophys. Res.*, **93**, B6, 6216-6236.

Dickey, J. O., Eubanks, T. M., 1986, *Space Geodesy and Geodynamics*, Academic Press, eds., A. J. Anderson and A. Cazenave, 221-267.

Dickey, J. O., Williams, J. G., Newhall, X X and Yoder, C. F., 1984, *Proc. IAG Symp., IUGG XVII the General Assembly*, Hamburg FRG, August 15-27, 1983, **2**, 509-521.

Ducarme, B. and Melchior, P., 1978, *Phys. Earth Planet. Inter.*, **16**, 257-276.

Eubanks, T. M., Steppe, J. A. and Sovers, O. J., 1986, *Proc. Int. Conf. on Earth Rotation and the Terrestrial Reference Frame,* publ. Dept. of Geodetic Sci. and Surveying, Ohio State Univ., ed. Mueller, I.I., 326-340.

Farrell, W. E., 1972, Rev. Geophys. Space Phys., 10, 761-797.

Herring, T. A., Gwinn, C. R. & Shapiro, J. J., 1986, *J. Geophys. Res.*, **91**, 4745-4754.

Levine, J., 1978, *Geophys. J.R. astr. Soc.*, **54**, 27-41.

Mazzega, P., 1986, Proc. 10th Int. Symp. Earth Tides, Madrid, 609.

Neuberg, J., Hinderer, J. and Zurn, W., 1987, *Geophys. J.R. astr. Soc.*, **91**, 853-868.

Newhall, X X, Williams, J. G. and Dickey, J. O., 1988, *Proc. Union Symp. No. 128*, 159-164

Newton, R. R., 1985, *Geophys. J.R. Astr. Soc.*, **80**, 313-328.

Parke, M. E. and Hendershott, M., 1979, *Marine Geod.*, **3**, 379-408.

Richter, B., 1985, *Proc. 10th Int. Symp. Earth Tides*, 131-140.

Rubincam, D. P., 1984, *J. Geophys. Res.*, **89**, 1077-1087.

Schuh, H., 1986, *Terra Cognita*, **6**, 3, 466.
Schwiderski, E. W., 1980, *Rev. Geophys. Space Phys.*, **18**, 243-268.
Wahr, J. M., 1981, *Geophys. J.R. astr. Soc.*, **64**, 677-703.
Wang, R. and Zschau, J., 1986, *Terra Cognita*, **6**, 3, 465.
Warburton, R. and Goodkind, J., 1978, *Geophys, J.R. astr. Soc.*, **52**, 117-136.
Williamson, R. G. and Marsh, J. G., 1985, *J. Geophys. Res.*, **90**, B 11, 9346-9352.
Yoder, C. F., Williams, J. G., Dickey, J. O., Schutz, B. E., Eanes, R. J. and Tapley, B. D.,
    1983, *Nature*, **303**, 757-762.
Zschau, J., 1986, Space Geodesy and Geodynamics (eds., A. J. Anderson and A. Cazenave),
    Academic Press, 315-344.
Zschau, J. and Wang, R., 1986, *Proc. 10th Int. Symp. Earth Tides*, Madrid, 2379-382.

6. TIME VARIATIONS IN THE GRAVITY FIELD

Roberto Sabadini

Future LAGEOS-type satellites will allow improved measurements of the temporal variations of the gravity field of the Earth. Precise detection of time-dependent perturbations in geopotential components is needed to improve our knowledge of the dynamic processes occurring in the Earth's interior.

In the last several years, the analyses of temporal changes in the gravity field through the observation of the orbital motions of LAGEOS-I (Yoder et al., 1983), have permitted and stimulated calculations that have produced crucial information regarding the viscosity structure of the mantle (Yuen et al., 1982; Peltier, 1985; Sabadini et al., 1985). In these investigations, the source of the forcing responsible for perturbations has been attributed solely to melting from Pleistocenic deglaciation. Coupling of steady-state and transient creep clearly emphasized, in these previous studies, the potential impact of short-term dynamic processes on the interpretation of long-term viscosity structure. Preliminary findings show that the interplay between short-term rheologies and short time scale forcing sources, such as the redistribution of surface water due to ongoing glacial activities, must be responsible of observable signatures in the Earth geopotential (Gasperini et al., 1986).

Recently it has been recognized by Yoder and Ivins (1985) that part of the present-day variations in the Earth's rotation rate, or equivalently J_2 may be produced by the ongoing retreat of temperate latitude glaciers (Meier, 1984). This point was quantified by Yuen et al. (1987), who found that present-day glacial activities and the current variability of the Antarctic ice volume can cause variations in the long-wavelength components of the gravity field as a consequence of transient viscoelastic responses in the mantle. The assumption of steady-state rheology may not be correct for time scales of $O(10^2)$ yr, as in recent glacial retreats. The issue of transient rheology has been revived by the discrepancy in the lower mantle viscosity estimates inferred from post-glacial rebound signatures (Peltier and Andrews, 1976; Yuen et al., 1986) and long-wavelength geoid anomalies (Hager et al., 1985). Sabadini, Yuen and Gasperini (1985) called the attention to the problem of the contamination of the steady-state viscosity from transient rheology. It has been recognized that this contamination can have an impact on the convection style within the planet. These recent developments have indeed affected our thinking concerning the interaction between short- and long-term mantle rheology and our means of extracting mantle viscosity from rebound signatures. Results from other fields, such as laboratory work on creep experiments, are required to put bounds on the parameter values that enter the constitutive equations of the geophysical models.

On the basis of recent developments, we expect that, in the future, the analysis of satellite data, together with the monitoring of current cryospheric activities by modern space methods, will allow us to address important questions in mantle rheology.

Earthquakes represent another class of terrestrial sources that are potentially capable of perturbing the Earth's geopotential. We think that the issue of the effects of seismic sources on the Earth rotation and gravity field has not been sufficiently addressed in the past. Basically, the models that have been used to study the consequences of seismic excitations are purely elastic and do not account for the amplification in the perturbations in the moment of inertia due to post-seismic relaxation (Chao and Gross, 1987). Preliminary findings derived from viscoelastic Earth models (Boschi et al., 1985), clearly show that post-seismic motions can enhance the perturbations in the geopotential components with respect to simpler elastic modelling. On the short time scale, the effects of non-linear rheologies can be relevant for the perturbations in geopotential components due to earth quakes (Melosh and Raefsky, 1983). Nonlinearity tends to speed-up the process of post-seismic relaxation after the occurrence of the dislocative event. This effect can enhance the amount of perturbation in the gravity field and Earth rotation.

It is thus recommended that future satellite missions be designed to account for the possibility of measuring the effects of seismic sources on the gravity field. Due to the strong dependence of the gravity signals from Earth's rheology, it must be expected that the detection of geopotential perturbations induced by seismic sources can help the geophysicists to answer crucial questions on the physical properties of the mantle.

Both the redistribution of surface water associated with cryospheric activity and seismic sources are expected to produce a strong azimuthal dependence in the gravity signals. This is due to the non-axisymmetric contributions from recent glacial retreat for cryospheric forcing (Meier, 1984). As far as the seismic contribution is concerned, we expect that the latitude and longitude distribution of the earthquakes, coupled with the characteristics of the source, will be responsible for the perturbations of geopotential components of different degree and order. By making use of resonances in satellite orbits, one may potentially measure the perturbations in tesseral harmonics. The detectability of perturbations in geopotential due to the redistribution of material within the Earth induced by surface load and seismic sources, may also enable to improve our knowledge on the laterally varying structure of the planet. It is obvious that an azimuthal dependence in the response, is also induced by lateral variations in the rheological properties of the planet. Preliminary findings suggest that lateral heterogeneities in mantle viscosity strongly affect the response of the Earth to surface loads (Sabadini et al., 1986; Gasperini and Sabadini, 1988). We thus expect that observational data derived from satellites with orbital inclination different from LAGEOS-I can eventually be employed to put bounds on the amount of lateral variations in mantle rheology.

REFERENCES

Boschi, E., Sabadini, R. and Yuen, D.A., 1985, *J. Geophys. Res.*, **90**, 3559-3568.

Chao, B.F. and Gross, R.S., 1987, *Geophys. J. Roy. Astr. Soc.*, **91**, 569-596.

Gasperini, P. and Sabadini, R., 1988, *Geophys. Journal*, (submitted).

Gasperini, P., Sabadini, R. and Yuen, D.A., 1986, *Geophys. Res. Lett.*, **13**, 533-536.

Hager, B.H., Clayton, R.W., Richards, M.A., Comer, R.P. and Dziewonski, A.M., 1985, *Nature*, **313**, 541-545.

Meier, F. J., 1984, *Science*, **226**, 1418-1421.

Melosh, H.J. and Raefsky, A., 1983, *J. Geophys. Res.*, **88**, 515-526.

Peltier, W.R., 1985, *Nature*, **318**, 614-617.

Peltier, W.R. and Andrews, J.T., 1976, *Geophys. J.R. astr. Soc.*, **46**, 605-646.

Sabadini, R., Yuen, D.A. and Gasperini, P., 1985, *Geophys. Res. Lett.*, **12**, 361-364.

Sabadini, R., Yuen, D.A. and Portney, M., 1986, *Geophys. Res. Lett.*, **13**, 337-340.

Yoder, C.F. and Ivins, E.R., 1985, *EOS*, **66**, No. 18, 245.

Yoder, C.F., Williams, J.G., Dickey, J.O., Schultz, B.E., Eanes, R.J., and Tapley, B.D., 1983, *Nature*, **303**, 747-762.

Yuen, D.A., Sabadini, R., and Boschi, E., 1982, *J. Geophys. Res.*, **87**, 10745-10762.

Yuen, D.A., Gasperini, P., Sabadini, R., and Boschi, E., 1987, *Geophys. Res. Lett.*, **14**, 812-815, 1987.

Yuen, D.A., Sabadini, R., Gasperini, P. and Boschi, E., 1986, *J. Geophys. Res.*, **91**, 11420-11438.

7. CONCLUSIONS AND RECOMMENDATIONS

Earth studies have embarked on a new era with the advent of highly accurate space geodetic techniques and the availability of complementary geophysical and meteorological data sets. Techniques utilized include laser ranging to the Moon and artificial satellites (SLR and LLR), and very long baseline interferometry (VLBI). Studies indicate that Earth rotation and polar motion are routinely determined at the one to two milliarcsecond level, with higher accuracy being achieved in some cases. Periodic corrections to the standard nutation model have been determined to 0.1 milliarcsecond for many terms. These unprecedented accuracies have allowed the detection and analysis of geophysically interesting variations in all areas of the short-term dynamics of the Earth. Several examples include: the detection and measurement of changes in the J_2 of the Earth, indirect analysis of undulations / "bumps" on the core-mantle boundary, new landmark measurements of tidal braking, detection of short-term variations in polar motion associated with atmospheric pressure variations, and analysis of intraseasonal oscillations which are seen in both the atmospheric and length-of-day measurements. The analysis and understanding of these phenomena draws upon and contributes to the fields of geodynamics, meteorology, oceanography, astronomy, celestial mechanics, seismology, and tectonics.

The future is even more promising with the anticipated technological advances envisaged for space geodesy and developments that are being planned in related areas. Looking ahead, the panel has formulated several resolutions to guide future planning. These recommendations address the major areas of short term dynamics of the Earth and are related to topics addressed in other sections (Resolution I is included in the Overview). For example, the goal of measurement of the rotation vector of the solid Earth with the highest possible accuracy at least 0.1 milliarcsecond (see Resolution I) are of major impact to the technology effort and couples well with the goal of one millimeter measurements (see Overview).

RESOLUTION I

Recommendation. The variations of the three-dimensional rotation vector of the solid Earth should be determined with the highest possible accuracy (at least 0.1 milliarcsecond) and temporal frequency (at least several cycles per day), and complementary geophysical, oceanographic, and atmospheric data be taken to facilitate the interpretation of geophysical processes.

Rationale. Changes in the Earth rotation vector are a consequence of changes of the inertia tensor of the solid Earth and its response to both periodic and irregular applied torques. The recommended measurements will exploit the periodic torques applied to the Earth by external planetary bodies, the resonances in Earth rotation due to the fluid-outer and solid-inner cores, and additional aperiodic variations to study the dynamic response of the solid Earth to external force and forces applied by its fluid parts (atmosphere, oceans, and fluid core). For example, changes in atmospheric angular momentum are expected to affect Earth rotation by >1 mas, even at periods near one day; earthquake excitations are also likely to be >1 mas. The coupling, transfer, and storage of angular momentum involving the fluid and solid parts of the Earth would be determined with the aid of complementary geophysical, atmospheric and oceanographic data. The response of the solid Earth depends on its internal rheological properties which are not amenable to direct study. These measurements would also support the maintenance of a terrestrial coordinate system in which millimeter point positions could be expressed.

RESOLUTION II

Recognizing that the excitation of the Chandler wobble has remained a problem of global importance in geophysics and geodesy; and

whereas,   meteorological processes, which can largely be accurately modelled, have been shown to contribute a significant fraction of the required excitation, but may never be completely understood because of the difficulty in obtaining comprehensive groundwater storage data; and

whereas,   co-seismic excitation is theoretically expected to contribute only a minor portion of the required excitation; and

whereas,   observation and theory both suggest that aseismic (pre- and post-seismic) activity, including aseismic tectonic slip and postseismic relaxation on time scales as short as ~10 days, may affect polar motion at a level far exceeding the theoretical prediction for seismic events; and

* that there is evidence suggesting that significantly large aseismic activity may accompany earthquakes of moderate magnitude; and

* that seismotectonic and hydrospheric activities are expected to perturb a variety of harmonic components in the geopotential, as a consequence of the azimuthal dependence in the forcing and in mantle properties;

The panel recommends that

1.   crustal deformation associated with seismotectonic activity be monitored by space-geodetic techniques,

over regional distances (100 km or greater),

on time scales of 10 days,

to an accuracy of 1 cm,

so that such activity can be detected, implied changes in the inertia tensor can be modelled, and the amount of wobble excitation can be computed; and, at the same time,

2.   polar motion be determined to an accuracy of at least 0.1 mas, so that the residual (non-atmospheric) polar motion can be accurately correlated with both aseismic activity and earthquakes exceeding magnitude 7, and compared with the theoretical seismotectonic excitation; and, at the same time,

3.   the feasibility of using satellite-derived geopotential coefficients, extended (e.g. using satellites in different inclinations) so that time-variations in components of degree and order possibly as high as 30 can be determined, so that all the various excitation sources can be monitored and short-term rheology and lateral variations in the mantle can be constrained.

RESOLUTION III

Recognizing that mantle anelasticity is important in the understanding of the connections between dynamic processes in the mantle at different time scales, and

* that significant lateral heterogeneities exist throughout the mantle and within the core, and

* that considerable progress has been made in theoretically modelling their influence on Earth's tides, revealing perturbations of the order of one per cent, and

Taking into account that the Earth responds to the tidal forces as a coupled dynamic system which exhibits important interactions between the different sub-systems, the solid Earth, the liquid core, the ocean and the atmosphere,

The panel recommends
* that full use should be made of modern space techniques to determine the potential Love number k as well as station dependent Love numbers h and 1 to an accuracy of a few tenths of a percent, and
* that all efforts towards improving our present knowledge on global ocean tides and atmospheric forcing should get as much support as possible

RESOLUTION IV

Recommendation. The major effort in the next twenty years should be to achieve global positioning at the one millimeter level or better with spatial resolution to order and degree thirty and from subseismic to postglacial rebound time scales. These positioning measurements must be complemented by concurrent monitoring of temporal variations of the external gravity field to the same resolution in time.

Rationale. This data type can contribute to four main categories:
1.   Periodic variations due to the solid Earth and ocean tides.
2.   Seasonal variations due to mass displacements in the hydrosphere and atmosphere.
3.   Quasi-secular changes driven by postglacial rebound and present-day glacial melting and other long-term processes.
4.   Nearly step-wise changes caused by crustal displacements associated with major earthquakes.
It is expected that this will elucidate both the radial and lateral elastic as well as anelastic structure of the Earth's interior.

RESOLUTION V

Recognizing the interdisciplinary nature of Earth rotation and polar motion measurements and the advantage of joint studies utilizing other geophysical parameters, the panel **recommends** that, in addition to improving the accuracy of the existing technologies (VLBI, LLR and SLR), alternative synergistic and innovative approaches be pursued for Earth rotation and polar motion measurements and related geophysical measurements be coordinated. Of particular interest is the emerging global network of GPS stations and superconducting gravimeters which could, if colocated, tie the space-based temporal variations to the geopotential with ground observation of gravity. Methods to be considered would also include gyroscopic techniques.

RESOLUTION VI

Recognizing a common requirement for all geodetic and geodynamic investigations is the necessity of well defined conventional inertial and terrestrial reference frames to which all relevant observations can be referenced and in which theories or models for the dynamic behavior of the Earth can be formulated, the panel *recommends* that :

1.   The determination of the positions of radio sources by means of radio interferometry should be continued and accelerated so as to achieve the best possible all sky coverage and overall stability and accuracy of the conventional inertial reference frame.

2.   Colocations of SLR, LLR, VLBI and GPS instruments should be continued, accelerated and extended so as to achieve the best possible connections between the conventional terrestrial reference frames inherent in the above systems and to allow intercomparisons between these techniques.

Chapter 3

LONG TERM DYNAMICS OF THE SOLID EARTH

D.C. Agnew, K. Burke, A. Cazenave, T. Dixon,
B.H. Hager, J.R. Heirtzler, T.H. Jordan,
J.B. Minster, M.K. McNutt, L.H. Royden,
D.T. Sandwell, and D.L. Turcotte

1. INTRODUCTION

Space based geodetic observations can contribute to the understanding of the long-term dynamics of the solid earth in a variety of ways. Measurements of surface displacements have shown that plate motions deduced from magnetic anomalies on time scales of millions of years are also applicable on time scales of years. Continued observations of surface displacements can extend our knowledge of plate motions but, more significantly, can contribute important data on the distribution of deformation within and across zones of intense tectonic activity. A direct goal is to understand the cyclic accumulation and release of stress on great faults.

Space based determinations of the ocean geoid have made many contributions to our understanding of geodynamic processes in the last 15 years. Continued improvement in the available resolution of altimetry coverage is an important goal. Unfortunately, the gravity and geoid over the continents is far less well determined than over the oceans. If future gradiometry missions could provide a 1-5 mgal accuracy with a 50-100 km resolution this would contribute significantly to our understanding of both mantle and crustal processes. However, in order to interpret gravity it is necessary to know the topography to a corresponding resolution. Global determinations of topography with a 1 m accuracy and 100 m resolution are the stated goal.

2. IMPLICATIONS OF PRECISE POSITIONING

Arguably, one of the most exciting development in crustal kinematics over the last two decades has been the birth of space-geodetic positioning techniques capable of achieving accuracies of a cm or better. The principal motivation for using these techniques for precise point positioning is to take advantage of the significant increase in technological capabilities that they represent in order to address directly important tectonic problems that cannot be tackled economically at present by ground-based geodetic techniques and classical field geology. The main techniques which have matured over the past decade or so include the *Very Long Baseline Interferometry* (VLBI) and *Satellite Laser Ranging* (SLR).

More recently, the less burdensome —from the point of view of field operations— *Global Positioning System* (GPS) has gained considerable popularity for the study of regional deformation problems. Over the technological and scientific horizon, we may count as future candidates the *Geodynamic Laser Ranging System* (GLRS) which is being considered as a facility instrument on the EOS platform, as well as alternatives developed in Europe, such as the French DORIS and the German PRARE systems. Convenience considerations aside, the main advantage of these new techniques, from the geologist's point of view, is that they should permit (in principle) *frequent* resurveys of *dense* networks. This, together with improved control of vertical displacements, represents a completely new, enhanced capability, which will allow geologists to address problems that have so far eluded them.

It may seem difficult to believe that the ability to measure the relative positions of two points on the earth separated by 100 or even 10,000 km has a measurable socio-economic impact. The applications to earthquake prediction and volcanic surveillance are, of course, often invoked, but somehow seem too remote to justify an immediate development effort. Nevertheless, in the past decade, space geodesy has begun to provide useful constraints on the solution of difficult geological problems of immediate importance, such as the distribution of crustal deformation both east and west of the San Andreas Fault in central California (e.g. Jordan and Minster, 1988b). This issue is an interesting one from a scientific point of view, but it also has great practical importance, in view of the high population density and numerous critical facilities found along the Pacific coast of the U.S. In the next two decades, we must extend our experience to other tectonically active areas which have significant human as well as scientific importance.

2.1.   CRUSTAL MOTIONS AND DEFORMATIONS

Deformation of the earth's lithosphere covers a broad spectrum of temporal and spatial scales, from seconds to aeons and from mineral grains to planetary dimensions. We rely below on the discussion of Jordan and Minster (1988a).

Table 1 categorizes a subset of lithospheric motions that cause geologically and geophysically significant deformations. It is convenient to discriminate secular motions persisting on geological time scales of thousands to millions of years from transients associated with, for example, seismic and volcanic events. Practical research is more concerned with the transients, because they tend to disturb human activities. Secular motions also warrant vigorous study, however, since they provide the kinematical framework for describing transients and understanding their driving mechanisms.

The most significant long-term deformations are those related to plate tectonics. Although local tectonic movements near plate boundaries display large vertical components and time-dependent behavior, the net motions between the stable interiors of large blocks are forced by viscous damping and gravity to be nearly steady and horizontal. The characteristic tangential velocity of the plate system is about 50 mm/yr, which gives rise to displacements easily measured by *geodetic* methods. Horizontal secular motions have been observed both by ground-based networks (e.g., Savage, 1983) and by space-geodetic systems (e.g., Christodoulidis et al., 1985; Herring et al., 1986). Though their application to geodesy is relatively new, space-based techniques have already revolutionized the science of terrestrial distance measurement. They are contributing new information about active tectonics, particularly on the planetary scales previously inaccessible to ground-based surveys (Figure 1).

Table 1. Types of Motions at the Earth's Surface.

	SECULAR	TRANSIENT
HORIZONTAL	Plate motions Boundary-zone tectonics Intraplate deformation	(Pre,co,post)-seismic Fault creep Stress redistribution
VERTICAL	Tectonic motions Thermal subsidence Diapirism Crustal loading Post-glacial rebound Cratonic epeirogeny	(Pre,co,post)-seismic Magma inflation Tidal loading

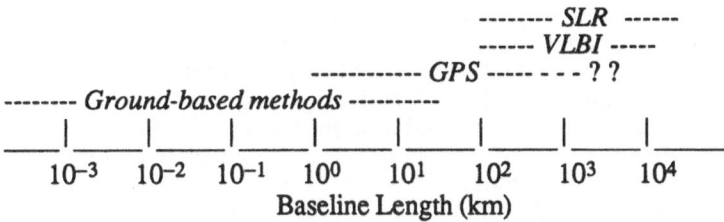

Fig. 1   Spatial scales sampled by various geodetic methods.

Rigid–plate motions. In the ocean basins, most of the deformation related to horizontal secular motion occurs in well-defined, narrow zones that are the boundaries of a dozen or so large lithospheric plates. The current plate velocities are constrained by three basic types of data collected along these submerged boundaries: (1) spreading rates on mid-ocean ridges from magnetic anomalies, and directions of relative motion from (2) transform-fault azimuths and (3) earthquake slip vectors. The first self-consistent global models were synthesized soon after the formulation of plate tectonics (Le Pichon, 1968), and significant refinements were made throughout the next decade (Chase, 1972; Minster et al., 1974). Third-generation models were published by 1978 (Chase, 1978; Minster and Jordan, 1978) and are still in use. Work has been recently completed at Northwestern University on an improved fourth-generation plate-motion model named NUVEL-1 (DeMets et al., 1987), which remedies most of the problems identified with earlier models.

Since the reference frame is arbitrary, the angular velocity vectors describing the instantaneous relative motions among M rigid plates are specified by $3(M - 1)$ independent components, which are derived by least-squares inversion of a carefully selected, globally distributed data set. As shown in Table 2, the increasing sizes of the data sets upon which successive generations of models have been based reflect continued vigorous research activity in geology and geophysics since the advent of plate tectonics.

Table 2. Data Sets Used in Successive Generations of Plate Motion Models

	Magnetic rates	Transform faults	Slip vectors	Total data
2nd generation, e.g. RM1[1], 1974	68	62	106	236
3rd generation, e.g. RM2[2], 1978	110	78	142	330
4th generation, NUVEL-1[3], 1988	277	121	724	1122

[1] Minster at al., (1974). [2] Minster and Jordan, (1978). [3] Gordon et al., (1988).

Table 3 lists the instantaneous rotation vectors ("Euler vectors") that describe the relative motions of the plates in NUVEL-1. It is reproduced from the presentation by R. Gordon and his co-workers at the October 17-21, 1988 NASA Crustal Dynamics Principal Investigators Meeting, held in Munich, Germany (Gordon et al., 1988). The convention adopted in this table is that the first plate moves counterclockwise relative to the second plate. The nomenclature is as follows: af - Africa, an - Antarctica, ar - Arabia, au - Australia, ca - Caribbean, co - Cocos, eu - Eurasia, in - India, na - North America, nz - Nazca, pa - Pacific, sa - South America. Table 3 also lists the one-sigma error ellipses (marginal distributions) attached to the Euler vector estimates. The two-dimensional marginal distribution of the pole position is specified in each case by the angular lengths of the principal axes and the azimuth ζ_{max} of the major axis, and the one-dimensional marginal distribution of the angular rotation rate is specified by its standard deviation σ_ω.

The magnetic anomalies employed in the various rigid-plate motion models mentioned above average the rates over the last 2-3 million years, about the shortest time span for which good spreading rates can be obtained on a global basis. Although this interval is hardly "instantaneous" from a geodetic point of view, it is geologically brief, and the small plate displacements that take place during it are well described by infinitesimal (as opposed to finite) rotations. It will probably be some time before the global plate-tectonic models can be significantly improved by space geodesy. Because the geological data sets are large and the inverse problem is strongly overdetermined, the formal uncertainties in the angular velocity components are already quite small, and correspond to formal uncertainties of 1 or 2 mm/yr in the predicted rates of relative motions. More importantly, the fourth generation NUVEL-1 model listed in Table 3 is *consistent*, at the 1-2 mm/yr level, with the hypothesis that major plates behave *rigidly* over a million-year time scale. Moreover, there is growing evidence that the rates-of-change of geodetic baselines spanning plate boundaries are consistent with the geological estimates (e.g. Herring et al., 1986), provided that the endpoints are located within stable plate interiors. (This is comforting, both as a check on the techniques and as corroboration of the geophysical expectation that the instantaneous velocities between points in stable plate interiors are dominated by secular plate motions.) This means that, for those plates whose motions are well constrained by geological observations, direct geodetic measurements will not add significant constraints to the estimate of secular velocities. In any case, given the level of internal consistency of the geological models, to contribute to the improvement of existing models of present-day motion among the major plates, the tangential components of relative velocities on interplate baselines with endpoints located within stable plate interiors must be resolved to an accuracy on the order of 1 mm/yr.

It is therefore clear that geologically-based global plate motion models provide in fact a *kinematic reference frame* in which to analyze short-term geodetic observations, as well as the *kinematic boundary conditions* which must be satisfied by models of plate boundary deformation zones. In other words, the most important and interesting geodetic signals with characteristic time scales ranging from 1 hour to 100 years are detected as *departures* from predictions of million-year average rates based on geological rigid-plate models.

However, there exist examples for which the reliability of currently available global models are difficult to assess, and some improvement could be made from space-geodetic data. For example, Southeast Asia is assumed to be part of the large Eurasian plate, but the active tectonics of China imply it should be moving as a separate entity. Because it is completely surrounded by complex zones of deformation, its motion relative to Eurasia will be difficult to quantify without space-geodetic observations. Similarly, we note that geological rates are lacking across convergent boundaries, such as trenches. This is one reason why the motion of the Philippine Plate relative to its neighbors is not well known. Again, space-geodesy should provide useful constraints. Closer to home is the case of the Pacific-North America plate pair, whose relative rate of motion can be directly measured only on a tiny ridge segment in the mouth of the Gulf of California; The recent modeling work of DeMets et al. (1987) (reflected in NUVEL-1) yields a relative plate velocity along that boundary that is 15-20% slower than earlier estimates. Since this rate is critical to models of deformation in the western United States, a geodetic check on the Pacific-North America angular velocity will be very valuable.

With these exceptions and a few others, however, the global networks of VLBI and SLR stations are just too sparse to control plate motions as tightly as the geological observations. The impact of geodetic observations on the development of these plate-motion standards will be relatively minor, at least for the next few years. Of course, this does not say that interplate observations made by space-geodetic methods will not reveal new and interesting phenomena associated with other categories of motion listed in Table 1; particular attention should be focused on time-dependent signals, including the possibility that plate speeds and directions have changed significantly during the 2-My averaging period of the geological data. Our point is to emphasize that *the exciting issues for space geodesy lie beyond the now-classical descriptions of major plate motions.*

Table 3. EULER Vectors for the Rigid-Plate Model NUVEL-1 (from Gordon et al., 1988)

Plate Pair	Latitude °N	Longitude °E	ω (deg/My)	Error Ellipse			
				σ_{max}	σ_{min}	ζ_{max}	σ_ω (deg/My)
Pacific Region							
na-pa	48.7	-78.2	0.78	1.3	1.2	-61	0.01
co-pa	36.8	-108.6	2.09	1.0	0.6	-33	0.05
co-na	27.9	-120.7	1.42	1.8	0.7	-67	0.05
co-nz	4.8	-124.3	0.95	2.9	1.5	-88	0.05
nz-pa	55.6	-90.1	1.42	1.8	0.9	-1	0.02
nz-an	40.5	-95.9	0.54	4.5	1.9	-9	0.02
nz-sa	56.0	-94.0	0.76	3.6	1.5	-10	0.02
an-pa	64.3	-84.0	0.91	1.2	1.0	81	0.01
pa-au	-60.1	-178.3	1.12	1.0	0.9	-58	0.02
eu-pa	61.1	-85.8	0.90	1.3	1.1	90	0.02
co-ca	24.1	-119.4	1.37	2.5	1.2	-60	0.06
nz-ca	56.2	-104.6	0.58	6.5	2.2	-31	0.04
Atlantic Region							
eu-na	62.4	135.8	0.22	4.1	1.3	-11	0.01
af-na	78.8	38.3	0.25	3.7	1.0	77	0.01
af-eu	21.0	-20.6	0.13	6.0	0.7	-4	0.02
na-sa	16.3	-58.1	0.15	5.9	3.7	-9	0.01
af-sa	62.5	-39.4	0.32	2.6	0.8	-11	0.01
an-sa	86.4	-40.6	0.28	3.0	1.2	-24	0.01
na-ca	-74.3	-26.1	0.11	25.5	2.6	-52	0.03
ca-sa	50.0	-65.3	0.19	15.1	4.3	-2	0.03
Indian Ocean and African Regions							
au-an	13.2	38.2	0.68	1.3	1.0	-63	0.00
af-an	5.6	-39.2	0.13	4.4	1.3	-42	0.01
au-af	12.4	49.8	0.66	1.2	0.9	-39	0.01
au-in	-5.5	77.1	0.31	7.4	3.1	-47	0.07
in-af	23.6	28.5	0.43	8.8	1.5	-74	0.06
ar-af	24.1	24.0	0.42	4.9	1.3	-65	0.05
in-eu	24.4	17.7	0.53	8.8	1.8	-79	0.06
ar-eu	24.6	13.7	0.52	5.2	1.7	-72	0.05
au-eu	15.1	40.5	0.72	2.1	1.1	-45	0.01
in-ar	3.0	91.5	0.03	26.1	2.4	-58	0.04

Departures from rigid–plate motions.   The various types of departures from the predictions of the rigid plate models fall into two major categories: 1) large scale and regional scale non-rigid behavior (plate deformation and plate boundary zones of deformation), and 2) nonsteady motions, including in particular post-seismic strains and aseismic deformations.

Plate deformation. There are conspicuous instances where the ideal rigid–plate model fails to describe adequately the complexities of present-day tectonic interactions, especially within the continents and along their margins. Examples can be found in regions undergoing compression due to plate collision, such as the Alpine Belt, and more particularly Tibet, or the northern margin of the South America plate, as well as regions dominated by extensional tectonics, such as the African Rift Zone and the western US. Such regions are often characterized by spectacular landscapes and have long attracted the attention of geologists.

Perhaps the most outstanding example of large-scale continental deformation is Tibet. Collision of the Indian subcontinent with the Asian continent about 50 million years ago has been followed by roughly 2000-4000 km of convergence across Tibet and the Himalayas, resulting in the most impressive region of young and active deformation on earth (Molnar and Tapponnier, 1975, 1978). Although it is generally accepted that deformation within Tibet has resulted in movement of crustal fragments, with dimensions of tens to hundreds of kilometers, in directions oblique or even orthogonal to the overall direction of convergence between India and Asia, the details of present rates and directions of motion are still sketchy at best [e.g. Lyon-Caen and Molnar, 1985; Molnar et al., 1987]. The use of space geodetic techniques to unravel the complex kinematical picture of this region, even at the reconnaissance level of detail, should certainly be recognized as an important target for the next two decades.

Another young example of active tectonics which space geodesy will help understand better is the Mediterranean region. There we have "back arc" type basins adjacent to zones of coeval subduction and convergence, such as the Aegean-Hellenic system, the Tyrrhenian-Appennine/Calabrian system, and the Pannonian-Carpathian system (see for example Malinverno and Ryan, 1986; McKenzie, 1972, 1977; Mercier, 1977; Royden et al., 1983a,b; Scandone, 1979). These continental systems present an excellent opportunity to study the interaction of active extensional and convergent processes because, unlike most oceanic systems, a reasonably large amount of the region is exposed above sea-level, so that geodetic and geologic field studies are practical. Again the importance of reaching a more precise understanding of the current tectonic evolution of the area is enhanced by the high population density. Accordingly, a substantial long-term multi-national effort, the Wegener-Medlas Project, was undertaken in 1984 to refine the kinematic picture of the Mediterranean region, and has begun to yield SLR data capable of placing useful constraints on the geological models (e.g. Wilson, 1987). The continuation of this project on a regional scale, and the densification of the network in critical areas (using, for example, GPS campaigns) will remain an important component of tectonic studies of the region in the coming years.

A particularly interesting third example is the western United States, where the interaction between the North American plate and the northwestward moving Pacfic plate is spread out over a broad zone of deformation, and where the available geodetic data sets are substantially larger. Although California's San Andreas Fault can be identified as a major locus of movement on the Pacific-North America plate boundary, the likelihood that significant crustal deformation is occuring both east and west of the San Andreas has long been recognized by geologists. Geological and geodetic observations of the present rate of slip along the fault in central California (about 34 mm/yr) is significantly lower than the rate predicted from successive generations of rigid-plate motion models, including NUVEL-1 (about 50 mm/yr). This so-called "San Andreas discrepancy" has been analyzed by Minster and Jordan (1984, 1987), using both geological and geodetic data. Their conclusion was that even the short 4-year record of VLBI observations available to them for the relevant baselines (see Fig. 2) was sufficient to place useful constraints on the integrated deformation east of the San Andreas across the Basin and Range, and, by vector addition, on the integrated deformation west of the San Andreas across the California margin (see also Weldon and Humphreys, 1985).

Jordan and Minster (1988a) reviewed the use of space geodetic observations to solve geological problems, with a focus on the various types of *secular horizontal* motions listed in Table 1. They concluded that many geological and geophysical problems related to the such motions are indeed currently being addressed by space–geodetic experiments, provided that critical measurements are made at accuracies not feasible by conventional techniques. In particular, they state that measuring the velocities between crustal blocks to ± 5 mm/yr can place geologically useful constraints on the integrated deformation rates across continental plate-boundary zones, such as the western U.S., the Mediterranean and Tibet.

However, it must be emphasized that baseline measurements in geologically complicated zones of deformation are useful only to the extent that the relationship of the endpoints to geologically significant crustal blocks is understood. Some antennas have a long history of participation in VLBI experiments, so that their motions in the VLBI reference frame are becoming well known; but they lie within complex zones of faulting, and their motions in kinematical frames fixed to local geology are not at all known. For example, the baseline rates for the Owens Valley Radio Observatory (OVRO) in California relative to the Westford, Massachusetts, and Ft. Davis, Texas, antennas have been measured to a precision of about 2 mm/yr (Fig. 2). In their analysis of Basin and Range extension, Minster and Jordan (1987) have assumed OVRO moves with the Sierra Nevada-Great Valley block, but unfortunately, it is separated from the Sierra Nevada by a major system of faults, one of which broke in the great 1872 Owens Valley earthquake. Until the position of OVRO is regularly resurveyed in a local geodetic network which includes stations planted firmly on the Sierra block, the geological implications of the VLBI data will remain in doubt. Consequently, we can recommend that the establishment of frequently (or even continuously) surveyed local geodetic networks of sufficient density, around major geodetic sites in active areas should receive high priority.

The discussion has been focused so far on rather localized deformation, that is, on the occurrence of deformations zones with horizontal scales significantly less than overall plate dimensions. Whether the major tectonic plates, which are found to behave rigidly to an excellent approximation over million-year time scales, are actually undergoing nonlocal steady or episodic deformation on shorter time scales, is another important scientific question which bears directly on our understanding of the mechanical behavior of the lithosphere on a large scale, as well as our models of the force systems that drive plate tectonics. In order to resolve this issue, we will require geodetic coverage on a *global* scale, using techniques capable of delivering mm/yr accuracy for baselines 10,000 km long. Only space geodesy can meet such requirements, and, in the absence of any kind of "ground truth", we must rely on intercomparisons between independent techniques to evaluate the actual performance of the systems.

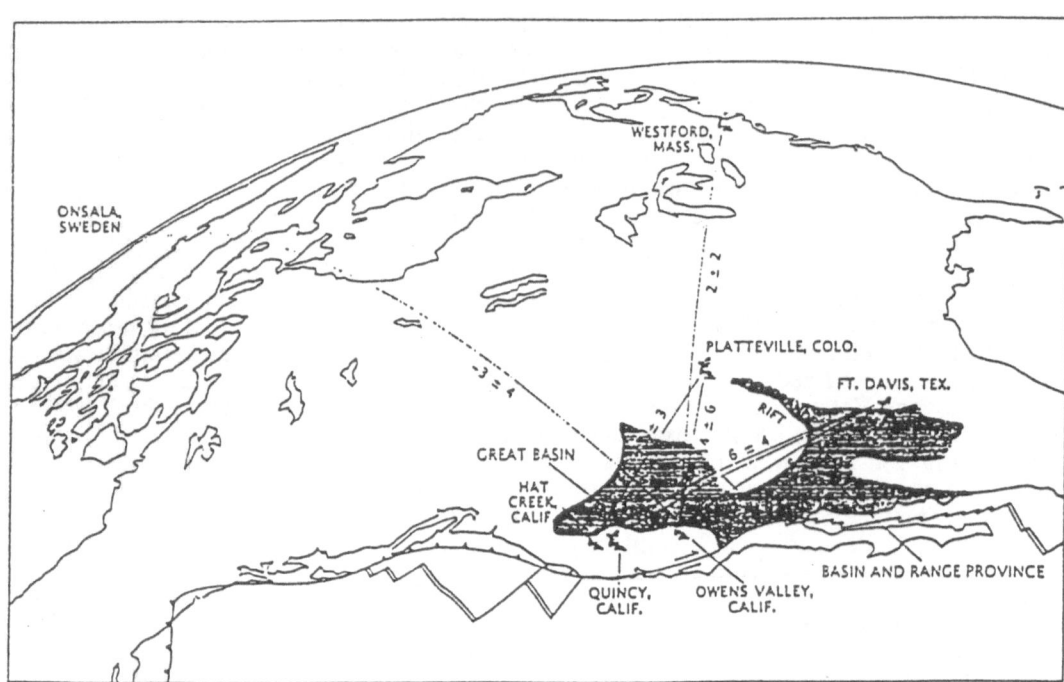

Fig. 2.   The San Andreas fault in central California is one element of the western U.S. plate
           boundary zone. In addition, we must account for contributions from crustal
           deformation both east and west of the fault. The integrated deformation west of the
           fault, which consists of NW-SE extension across the Great Basin, can be measured
           directly from the rates-of-change of VLBI baselines monitored by NASA's Crustal
           Dynamics Project. Combining this estimate with the geologically and geodetically
           observed rate of slip on the San Andreas, and the rigid-plate estimate of total motion
           between the Pacific and North America plates, we can derive geologically useful
           constraints on the integrated rate of deformation across the California margin, west of
           the fault. (From Jordan and Minster, 1988b. Reproduced with permission and by
           courtesy of *Scientific American*.)

Nonsteady motions. The problems of time dependence of the motions and nonrigid behavior of
the plates, two issues which lie beyond the now-classical descriptions of major plate motions,
are exciting applications of current and future space geodetic systems. The fundamental
underlying scientific problem is the *transmission of strain (or stress)* in the lithosphere. This
question is intimately related to the problem of coupling between earthquakes, evolution of
volcanic eruption, and, ultimately to the problem of *predicting* catastrophic events.

The time scales involved range from a fraction of an hour to centuries or longer, and
span a range in which the physical phenomena are very poorly understood, primarily because
the measurements are sparse, infrequent, and mostly very recent. Thus the evidence for
episodic (as opposed to steady) motions along plate boundaries and within plate boundary
deformation zones is insufficient at the present time to map the time and spatial scales involved.

A space geodetic system capable of high sampling rate and dense spatial coverage over
large areas will make possible a nearly unprecedented exploration of how crustal deformation
varies with time. Much too little is known about this, the only data so far available comes from
geodetic measurements that are too infrequent, too insensitive, or too localized to provide
conclusive evidence. Space geodesy will provide information that is crucial to understanding
the physics of the earthquake process. The motivation for these measurements can be

understood in the context of a simple model of the mechanical properties of the crust and upper mantle (Fig. 3).

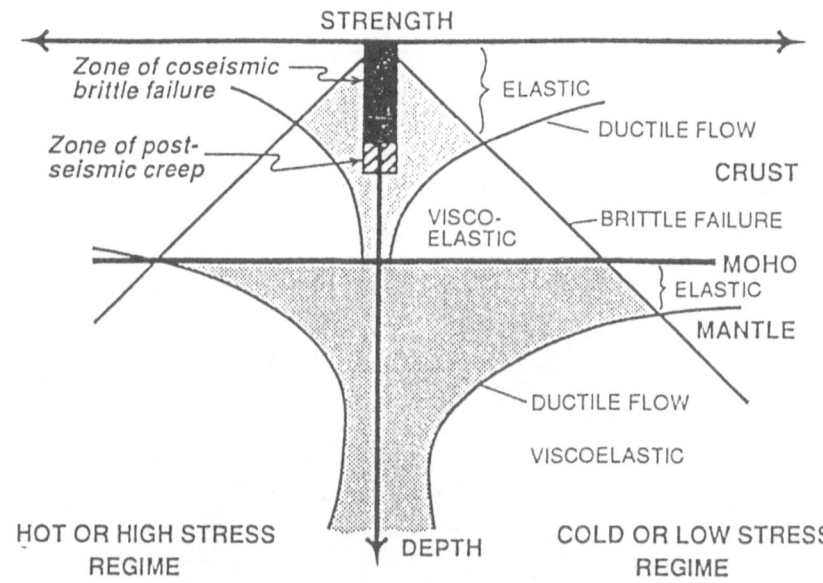

Fig. 3.   Schematic diagram of the strength and mode of deformation of the crust and upper mantle. Postseismic deformation occurs as the result of creep on the extension of the fault plane or stress relaxation in the viscoelastic regions.

Rocks respond to applied stress in a way very similar to the children's toy Silly Putty, i.e., they are viscoelastic. They respond elastically to rapid variations in stress, but undergo irreversible deformation by creeping flow under low but sustained applied stresses. If the stresses are high enough, rocks fail brittlely; the result of this sudden failure is an earthquake. Low temperature and low confining pressures favor brittle failure, while high temperatures promote creep by decreasing the effective viscosity of rocks. The rheological behavior also depends upon rock type, crustal rocks being more prone to creep than mantle rocks at the same ambient conditions. Pore fluids and volatiles also have effects. Although the details are poorly understood, an important goal of geodetic monitoring is to *obtain the observations needed to understand the details* ; the general variation in effective strength of the crust and upper mantle is illustrated schematically in Fig. 3.

Large earthquakes generally nucleate near one of the maxima in the strength versus depth curve and propagate rapidly through the adjacent brittle region. However, the deformation associated with an earthquake is not limited to this seismic rupture. Aftershocks typically relieve stress on slip-deficient areas of the fault plane and extend the rupture to adjoining regions. Aseismic creep on the fault plane, perhaps also extending into the ductile regime, may increase the total amount of displacement associated with an earthquake. Viscoelastic adjustments of the Earth's crust and mantle cause redistribution of strain after a major earthquake. Stress is relieved by flow in the ductile regions, allowing elastic strain to accumulate in the stronger, more elastic regions.

A general feeling for the interaction of the viscous and elastic properties of rocks in the crust and mantle may be obtained by considering the simple model originally developed by Elsasser (1969) and extended by others (e.g. Melosh, 1976, 1977, 1983; Cohen, 1984; Rundle and Jackson, 1977; Rundle, 1988a,b). This model consists of an elastic layer of thickness h_e

overlying a viscous layer of thickness h_v and a rigid base. An initial sudden displacement of a fault diffuses outward with a diffusivity $\kappa = h_e h_v / \tau$, where τ is the Maxwell time of the system. Repeated jerky offsets on plate boundary faults result in very smooth motion some distance away, corresponding to the steady velocities of plate interiors.

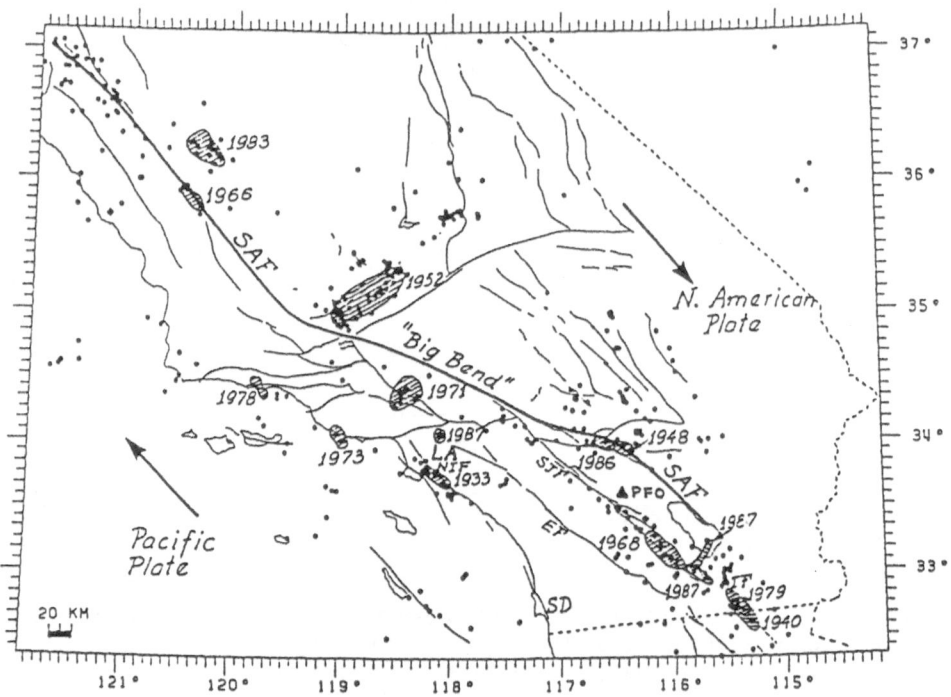

Fig. 4. Major fault lines and earthquake epicenters in southern California. All events of magnitude > 4.5 in the past 55 years are shown, with rupture zones of the largest events shaded. Fault abbreviations are SAF, San Andreas; SJF, San Jacinto; EF, Elsinore; NIF, Newport-Inglewood; GF, Garlock; and IF, Imperial. PFO is the Piñon Flat Observatory.

For regional scale problems, such as southern California (Fig. 4), the elastic layer can be taken to be the upper crust, the viscous layer the lower crust, and the rigid substratum is the upper mantle. The displacement from an earthquake on a given fault will diffuse through the crust, causing regional strain and perhaps influencing other faults, on a timescale that depends upon the effective viscosity of the lower crust (e.g. Turcotte et al., 1984). Crustal rheology is poorly constrained, but reasonable estimates indicate that strains may propagate 30 km in 50 years. As can be seen from Fig. 4, there have been many earthquakes in southern California in the past 50 years with significant source dimensions. Based on Elsasser's model, these events would be expected to have viscoelastic strain migration associated with them. Observing this strain migration could constrain crustal rheology.

Over a scale of tens of kilometers around the fault, Thatcher (1983) showed geodetic evidence for time–dependent strain rates after great earthquakes on the San Andreas. He showed that his rather sparse data could be fit both by a model in which a viscoelastic asthenosphere underlays an elastic lithosphere, and by a purely elastic model with exponentially-decaying afterslip on the fault. Thus, the change in strain rates could be due to the constitutive properties either of the fault zone or of the asthenosphere, but we do not know which. It is crucial to separate these effects in order to better understand the physics of earthquakes and transmission of stress through the crust. The distribution of postseismic strain in time and space would provide important constraints on the physics of faulting and on the material properties of the fault zone and surrounding Earth. Redistribution of stress and strain to adjacent faults would provide important information related to earthquake prediction.

We discuss below possible time-varying strains that might be observed following earthquakes in the context of specific recent experience in southern California. We then examine another possible source of time dependent strain not associated with individual earthquakes.

Post-seismic strains. A variety of observations have made it clear that the extremely rapid strain of a fault rupture is followed by strain rates that are high compared to the long-term average measured for the fault zone, but the data are lacking to determine the physics responsible. To begin closest to the fault, ruptures that break to the surface commonly show substantial slip in the hours and days after the actual earthquake. Does this reflect equal amounts of slip at greater depth, or is the deeper part of the fault stable, this afterslip merely being caused by the gradual propagation of deep slip through the near-surface layers?

A particularly clear example of this kind of ambiguity has recently been provided in southern California by the Superstition Hills earthquake sequence of November 1987. The first large event was the Elmore Ranch earthquake (M_s~ 6.2), followed 12 hours later by the main Superstition Hills event (M_S ~6.6), and of course many aftershocks. Seismicity patterns and surface rupture showed two conjugate faults, the Elmore Ranch earthquake being on a fault conjugate to the San Jacinto fault zone and the mainshock on a fault (Superstition Hills) parallel to the zone. Fig. 4 shows the location of the Piñon Flat Observatory (PFO) relative to these conjugate events (labeled "1987").

We are fortunate to have available data from the long-base strain instruments at Piñon Flat Observatory (e.g. Wyatt et al., 1982); since these have been properly "anchored" to depth, they give results (out to periods of a year) that are better than any existing geodetic measurement. They thus provide a window, if only at one place, for what we might expect from future space-geodetic systems, and promise to be a source of "ground truth" for these measurements. Strains from these two events were recorded by two of the laser strainmeters at PFO; they are dominated by the coseismic offsets (e.g. Fig. 5). These offsets and those recorded by other instruments at PFO are in reasonable agreement with results for a dislocation in a half-space, if seismically-derived parameters are used to define the dislocation. The frequent sampling of these data allows us to look at preseismic and postseismic deformations in some detail, with the best data coming from the fully-anchored NW laser strainmeter. During the interval between the two earthquakes the largest signal on this record consists of microseisms plus some small residual drift in the instrument. We can certainly rule out any anomalous strains during this time above the level of about 3% of the eventual coseismic offset, and for the final 1000 seconds above about the 0.5% level.

It is hard to believe that the Superstition Hills earthquake was not triggered by the Elmore Ranch earthquake; but it is clear that no simple model of elastic strain and brittle failure can explain the 12-hour delay between events: if the Superstition Hills fault were close to failure it should have ruptured during the dynamic strains generated by the Elmore Ranch earthquake, or at least soon after it, when the elastic stress changes had been fully imposed. We can think of several models, all speculative, to account for this.

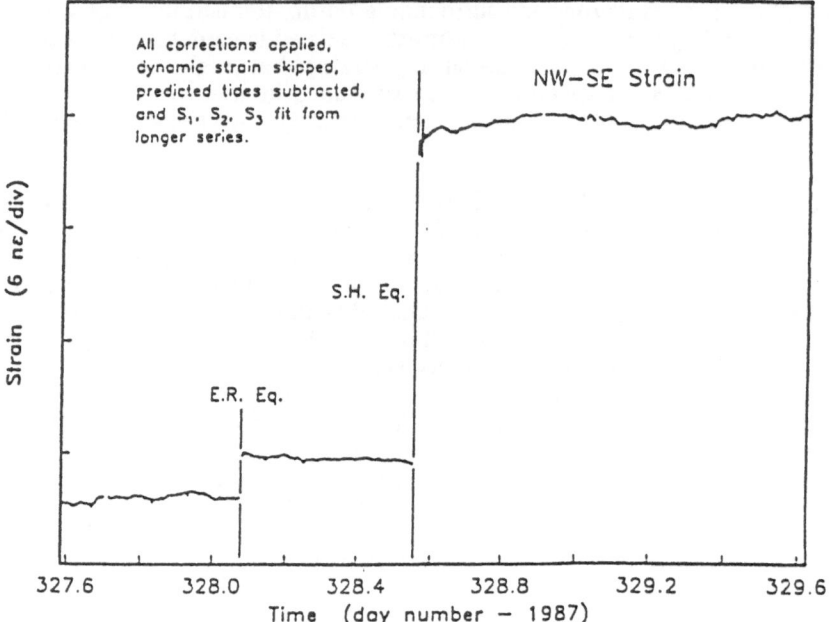

PFO Residual Strain — Superstition Hills Sequence

Fig. 5.   Superstition Hills events recorded at The Piñon Flat Observatory (see Fig. 4)

1. All the surrounding material is elastic, and the Superstition Hills fault failed by brittle rupture at a critical stress level. The 12-hour delay was caused by afterslip on the Elmore Ranch fault; only after this had gone far enough was the applied stress sufficient for failure. This model would appear to be ruled out by the PFO data, which show no obvious afterslip (this would appear in proportion to the coseismic strain); any stress changes from this cause could be at most 10% of the stress changes at the time of the first earthquake. It could be that a small additional stress was enough, but this seems unduly *ad hoc*.

2. All parts of the system could have responded elastically, but the fault actually failed by some kind of stress corrosion. The model results of Tse and Rice (1986) show something like this; for a realistic friction law, earthquakes in their fault model begin with slow slip over a very small depth range, rapidly accelerating to seismic slip, which would probably be consistent with the strainmeter data. However, this assumes a steady increase of applied stress; whether it still would result in inter-earthquake slip undetectable by the strainmeters would require additional modeling.

3. The Superstition Hills fault could have failed brittlely (as in the first hypothesis) with the delay between earthquakes being due to stress diffusion (Elsasser 1969; Melosh 1976, 1977, 1983) outward from around the Elmore Ranch fault. The physical model proposed recently by Rundle (1988a,b) may be invoked to account for the effects of interactions between faults (e.g. Rundle and Kanamori, 1987). We may assume that the first earthquake occurred in the brittle upper crust (Fig. 4), underlain by a ductile lower crustal "asthenosphere." Just after the earthquake, the stresses in both regions are described by elasticity (for otherwise the coseismic strain step at PFO would not match the halfspace solution), but the ductile zone then begins to flow, interacting with the overlying elastic layer to cause a diffusion of stress outward from the fault, and hence increasing the stress on the Superstition Hills fault. For the usual estimates of crustal viscosity, we would expect little change in stress over 12 hours. If, however, the rheology is time dependent or obeys power-law flow, the effective viscosity close to the fault (where the largest stress changes occur) could be quite low, allowing rapid diffusion of stress, which would tend to slow down as stress levels smooth out. Again, more detailed

modeling will be needed to see if such stress diffusion could occur without causing strains in the overlying lithosphere large enough to be detected by the distant strainmeters at PFO.

Space Geodesy could have helped distinguish between these models if measurements had been collected in the time interval between events. Multiple surveys of an array of monuments around these faults would have provided the best evidence possible on changes of strain between them; had none been seen, we would have good evidence for some kind of delayed failure on the fault itself, rather than a delay from stress propagation. *It is crucial that planning for space geodetic systems allow a fast enough response time to make critical observations like these possible.*

Aseismic Deformations. Aseismic deformation, in the form of strain episodes not associated with seismic events, have been hypothesized. One of the classic examples is the now infamous Palmdale Bulge. Repeat leveling of the region near Palmdale on the "Big Bend" segment of the San Andreas fault (see Fig. 5) showed an apparent uplift of up to 30 cm, followed by subsidence (Castle et al., 1976). This feature has been interpreted to result from episodic slip on a horizontal slip zone in the lower crust (Thatcher, 1979; Rundle and Thatcher, 1980). It was later discovered that at least part of the inferred bulge was the result of atmospheric refraction errors (Holdahl, 1982), with an additional smaller error due to miscalibration of survey rods (Jackson et al., 1983). The revised amplitude of the uplift is close enough to measurement error to be ambiguous (Stein, 1987), leading some geophysicists to dismiss the entire phenomenon. The controversy still rages, pointing out the problems inherent in interpreting infrequent measurements where the tectonic signal expected is close to the noise level.

There is less controversial evidence from conventional laser trilateration surveys across the San Andreas fault near Palmdale for the existence of a strain event in 1979 (Savage and Gu, 1985). Their paper suggests a five year long period of strain accumulation perpendicular to the San Andreas which was recovered abruptly in a $\sim 10^{-6}$ strain event in 1979. Unfortunately, given the measurement error and the possible effects of aliasing on such sparse ($\sim$yearly) occupations, these results by themselves are not definitive. They are tantalizing, however, since completely independent observations suggest that these time dependent strains may be real. Gravity (Jachens et al., 1983), water level in wells (Merifield et al., 1983), and seismicity (Sauber et al., 1983) in this region all showed unusual changes in 1979. There is a good correlation between variations in these quantities, suggesting that the changes are real, despite the relatively large uncertainties of the individual measurements. There is also intriguing structure in the various (sparse) time series, suggesting that there is time variation on time scales of a month or less that is missed by the rather infrequent measurements, i.e. that aliasing is a serious problem.

Jachens et al. (1983) compared variations in gravity, in dilatational strain, and in elevation (determined by repeat leveling) over several other areas of southern California. Their results show a remarkable tracking of variations in these quantities with relative amplitudes consistent with what is expected based on simple elastic compression, suggesting that they are real, with periods of from months to over two years. Unfortunately, the sampling is sparse enough in time that once again the possible effects of aliasing are a problem. Savage et al. (1987) also found evidence for a strain fluctuation in 1984, but comparisons of geodetic and strainmeter data strongly suggest systematic error in the former. In contrast, the data from PFO have never shown a strain change as large as discussed above, and indeed the best records from fully-anchored instruments show that strain fluctuations are very small. We thus have several possibilities:

1. Strain fluctuations of the sizes seen in 1979 (perhaps?) are fairly common in some regions (although the EDM data for southern California show only one event of this size), but have not been seen at PFO. It may be significant, for example, that Palmdale lies in the Big Bend region, where mantle convergence and downwelling may be occurring (e.g., Bird and Rosenstock, 1984; Humphreys et al., 1984), while PFO is in a simpler strike-slip regime. Or perhaps PFO is atypically quiet, somehow decoupled from its surroundings.

2. Large strain fluctuations occur, but are rare in time and localized over distances shorter than 100 km. Their absence at PFO thus is no more than a consequence of small–sample statistics.

3. The very small changes in strain seen at PFO are typical of most parts of southern California: strain accumulation is mostly steady. The 1979 episode is merely due to larger than expected instrumental error.

There is no way to settle this question without more frequent measurements over more baselines at higher precision, and this is what the next generation of space geodetic systems will provide. Only with such systems will we be in a position to characterize and understand the *spatial distribution* and the *time dependence* of deformation within tectonic regions, from which contraints on the physics of the deformation process can be inferred.

2.2   TYPES OF GEODETIC SIGNALS AND MEASUREMENT TECHNIQUES

The arguments presented so far lead us to conclude that the most significant departures from rigid plate motions occur in zones of 100 to 1000 km width, within which differential motions are accommodated by a combination of seismic slip and aseismic deformation. However, because direct measurements of the kinematic evolution of geological systems using space geodetic techniques can only cover time intervals of a few years, we are faced with the need to extrapolate these measurements to geological time scales in order to interpret them. In so doing, it is important to emphasize the space-time organization of geological systems and the processes that control their evolution. In fact, we seek a generic understanding of complex systems and their evolution, with special attention paid to transient behavior. If we are to trust the validity of extrapolations to geological time scales, a *process-oriented* approach appears to be superior to an *observation-driven* approach. Only then can we take full advantage of other data sources. These include, for example:

1. terrestrial geodesy, for which data sets spanning the last century are available (although not everywhere!) with dense spatial coverage (e.g. Snay et al., 1986).

2. classical geological techniques, which provide constraints pertinent to 10^3-10^6 year time scales and more, and spatial scales ranging from local outcrops to continental dimension,

3. geophysical data and models, which help us understand the underlying physics, and formulate testable hypotheses.

In order to examine quantitatively the potential of present and future space geodetic systems to contribute to the solution of geological problems, we consider on Fig. 6 a simple map of various general types of geodetic signals, in terms of the spatial dimensions and temporal scales of the underlying physical phenomena. The times scales of interest range from seconds in the case of brittle seismic fracture, to 10^6 years for plate tectonics. Similarly, spatial scales to be considered range from a fraction of the crustal thickness to the dimensions of continents. We note that geological techniques are primarily useful to constrain phenomena mapped in the upper right corner of the graph, whereas the lower left corner is the realm of seismology. The vast domain occupying most of the figure is left for geodesists to study.

To compare this signal map with the capabilities of various geodetic tools, we use a very simple parametric model of the precision precision limit σ of any given instrumental technique. Specifically, we assume that it is given by $\sigma^2 = \alpha^2 + (\beta\lambda)^2$, where α is a lower limit independent of the spatial scale (or "wavelength") λ and β is a coefficient of proportionality. Given this model, we examine the likelihood that a single event will be detected *and correctly characterized as to spatial and temporal scales* by selected techniques. The events considered here correspond to geodetic signals, expressed in terms of the displacement of a monument by a fixed fraction γ of the spatial scale λ, over the time scale τ. The model for σ allows us to construct a detection map in the (λ, τ) plane. Estimates of α and β yield the precision of a single observation, and, for simplicity, we take the noise to be Gaussian. In addition, we also truncate the detection maps according to the following considerations:

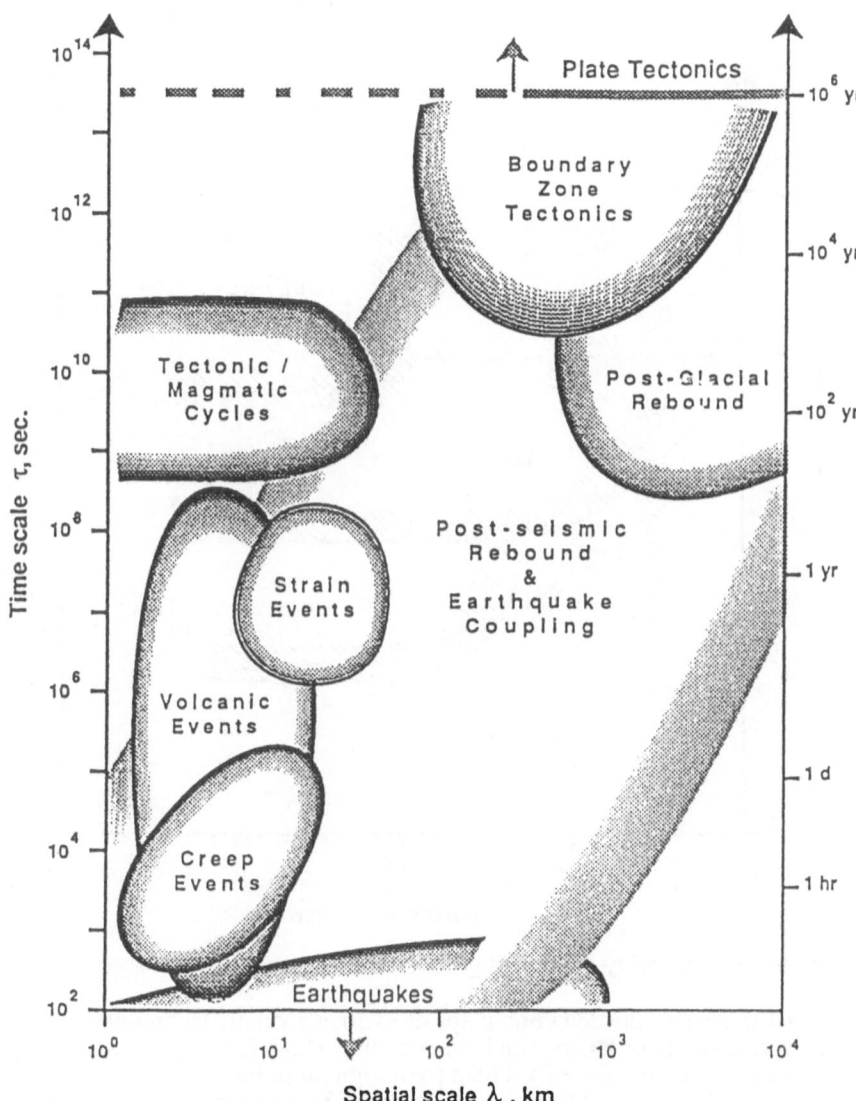

Fig. 6.   Map of geodetic signals in terms of spatial and temporal scales.

    1. For any given technique, we assign a largest value of λ, λ_{max}, beyond which the technique is inapplicable or inaccurate, and a smallest value λ_{min}, below which *routine* deployment is probably not practical, because of logistical or cost considerations. In practice, we should truncate the map for values of λ smaller than about twice the site spacing, corresponding to the spatial "Nyquist" wave number that could be resolved by the observations.

    2. Similarly, the frequency of measurements allowed for each technique is not taken to be the highest frequency achievable, but a realistic re–occupation rate for a routinely re–surveyed network. Some notion of logistics and costs is therefore built into the detection map. The detection map is truncated for time scales smaller than that corresponding to the Nyquist frequency of the measurement time series, that is, for time scales smaller than twice the time interval between measurements.

    3. Finally, although we allow some degree of noise reduction if repeated measurements are carried out, we limit the improvement to a maximum factor of 3, in view of the empirical observation that further statistical improvements are usually negated by uncontrolled systematic errors.

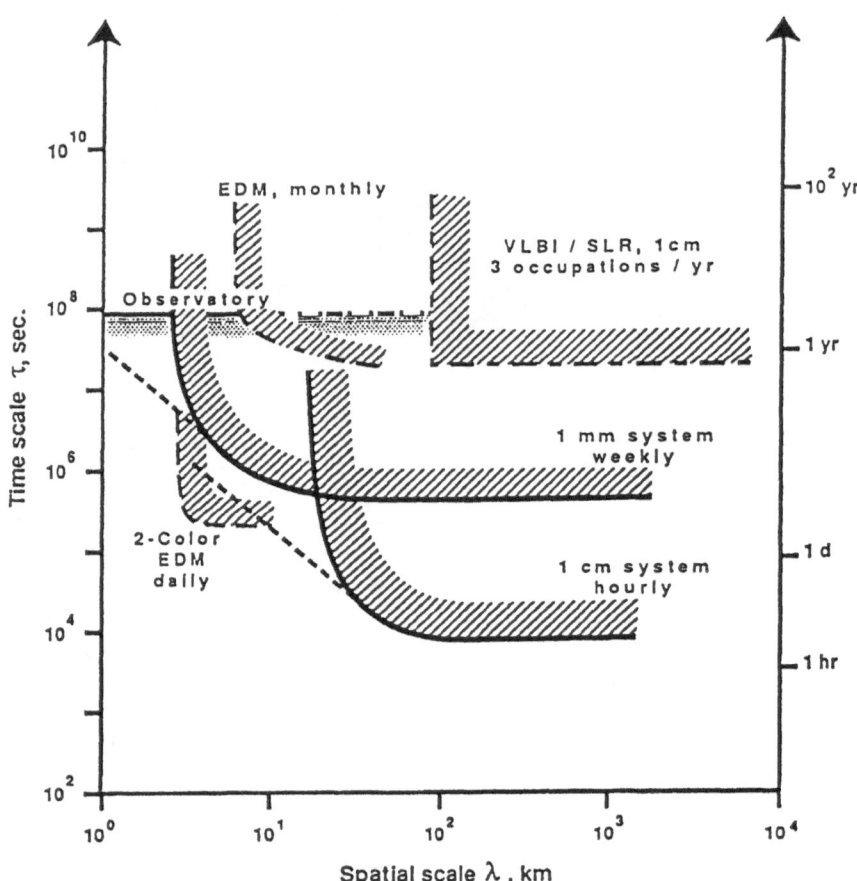

Fig. 7.   Detection capability of various geodetic techniques at the 10^{-7} strain level.

Figs. 7 and 8 show the detection maps of various geodetic techniques for $\gamma = 10^{-7}$, and $\gamma = 10^{-8}$, respectively. These figures include detection maps for several existing systems, as well as two hypothetical ones, examined here for design purposes:

1. VLBI and SLR (for which we have assumed 3 measurements/year, with $\alpha=1$ cm),
2. 1-color EDM (monthly measurements, $\alpha=3$ mm and $\beta=2 \times 10^{-7}$),
3. dedicated 2-color EDM (daily measurements, $\alpha=0.2$ mm and $\beta=10^{-7}$),
4. "observatory" measurements, i.e. continuously-recording strainmeters and tiltmeters,
5. a high precision, moderate frequency hypothetical system (weekly measurements, $\alpha=1$ mm and $\beta=10^{-8}$), and
6. a moderate precision, quasi-continuous system (hourly measurements, $\alpha=10$ mm and $\beta=10^{-8}$).

As γ decreases from 10^{-7} to 10^{-8}, most detection curves appear to migrate toward the upper right, with the conspicuous exception of the observatory instruments. These instruments have red noise spectra, which is another way of saying that they drift by larger amounts over longer times; they are thus too unstable to detect small strains with long characteristic time scales. There is in principle no restriction in spatial scale, in the sense that an instrument measuring strain over a short distance will also respond to a strain change with much larger characteristic spatial scale. However, there is no way to recognize that a strain event is wide–spread with a single observatory of this type, so we truncate the curve for λ on the order of the thickness of the elastic portion of the crust.

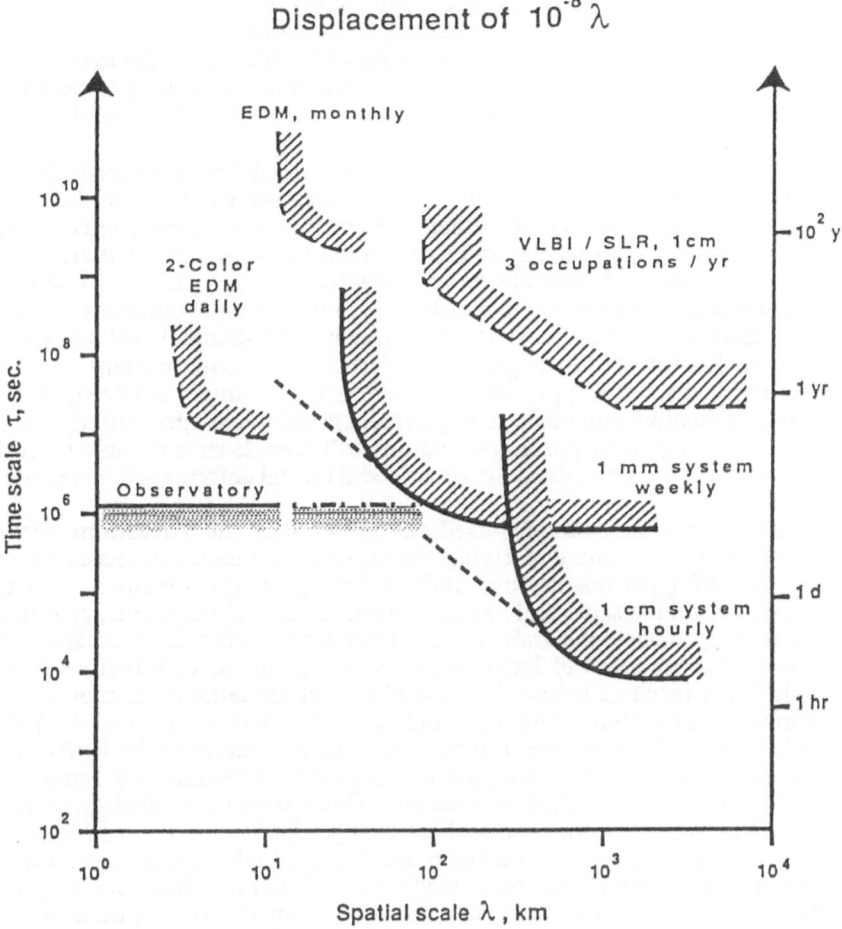

Fig. 8.   Detection capability of various geodetic techniques at the 10^{-8} strain level.

In both figures, it is clear that the hypothetical systems would fill a niche that no currently available geodetic technique is capable of filling. In particular, *if we can afford them,* either or both would permit investigations of the physics of earthquake coupling, and refinements of physical models of the spatial and temporal distribution of crustal strains that give rise to earthquakes. Whether we should prefer one over the other depends in part on the tradeoff that exists between the spatial and temporal scales we wish to investigate, but also very much on cost and ease of deployment. These diagrams illustrate the fact that space-geodetic techniques already have opened new windows for us to investigate tectonics phenomena, and that these windows will be even wider in the future.

2.3   EXTRAPOLATION TO GEOLOGIC CONTEXT

Unlike the situation in oceanic lithosphere, plate boundary activity within the continental lithosphere is commonly distributed over broad zones up to several thousand kilometers wide, which consist of complex networks of faults and folds. In these areas, it is not unusual for extensional, thrust and strike slip faulting to occur contemporaneously within relatively small neighborhoods. Rotation of crustal fragments is common. Even seemingly undeformed areas are probably affected by considerable internal deformation and are only "undeformed" in relation to the more intense deformation occurring along their boundaries. Moreover, not all or even most faults within a given plate boundary deformation zone are active at any one time. Studies in the western U.S. and in China have shown that displacement may shift from one set of faults to another on a time scale of less than one million years, and perhaps as short as 100,000 years, even when rates of displacement on the individual faults are 10 mm/yr or greater.

This degree of spatial complexity and temporal variability limits the insight that even the most accurate geodetic measurements can, by themselves, provide into the nature of deformation of the continental crust. When deployed in such an environment, regional strain nets can at best resolve present-day rates of motion between various parts of a broad plate boundary, and without doubt this has to be a significant achievement in its own right. What regional strain nets cannot provide are the answers to fundamental questions about how strain is in fact accommodated, in what ways and how quickly the spatial distribution of deformation may change with time within a geologic system, and how mechanical and dynamical connections unify disparate types of deformation into a single coherent tectonic picture. However, in combination with other geological and geophysical information, accurate geodetic measurements using carefully positioned stations will provide an extremely powerful tool with which to address specific well-posed questions about crustal deformation in a particular tectonic setting.

Revisiting tectonic examples used earlier, we may raise questions pertaining to the dynamical mantle phenomena underlying the surface kinematics observed by geodesy. For instance, the role of upper mantle flow under Tibet is virtually unknown. It is not known if mantle flow occurs on the same scale as movements of crustal fragments, or if the scale of the mantle flow is comparable to the scale of the entire plate boundary collision zone. In the former case, one would expect zones of local mantle downwelling beneath regions of major crustal shortening in Tibet (such as below the Tien-Shan) In the latter case, movements of crustal fragments must occur within a thin sheet that is largely decoupled form the behavior of the upper mantle beneath it. If mantle flow can be linked spatially to the active movements of sizable crustal fragments, then one might reasonably infer that the same has been true throughout the history of the Tibetan Plateau. This would imply that temporal changes in mantle flow occur over the same time scale as changes in the direction and rate of movements within the crust, which can be dated using traditional geological techniques. Preliminary data from northeastern Tibet show that, on a length scale of tens of kilometers, major deformation has switched from one system of faults to another and from one type of deformation to another over time intervals of less than a million years (Zhang, 1987). If similar temporal variability in fault activity occurs at even larger scales in Tibet, then the time scale over which changes in crustal motion occur may be very short indeed. Alternatively, if mantle flow can be shown to be related to crustal motion only at the scale of the plate boundary as a whole, then little can be said about temporal variability of flow in the mantle from reconstructing crustal movement histories. One might, however, infer that such large scale flow would be likely to vary only over the same time scale as the overall plate boundary activity, roughly tens of millions of years (for example England et al. 1985).

Similarly, in the case of the Mediterranean, questions aimed at understanding the interaction between active thrusting and extension, and which are amenable to study with geodetic and geologic data might include: How do rates of active convergence vary along the subduction boundaries? How does the rate and direction of active extension in "back arc" regions vary within the overriding plate? Is there a correlation between the rates and directions of extension and the rates and directions of convergence? Do major changes in rate coincide with structural features such as lateral offset or variations in trend of the convergent boundary? The answers to these and other questions about active crustal processes in these and similar systems elsewhere are essential if one seeks to evaluate the interaction betwen active crustal and mantle processes.

Comparable arguments can clearly be made in the case of the western U.S.: What the mantle flow is beneath southern California or beneath the Basin and Range are outstanding geophysical questions. The point made here is that, through geodesy and through geophysical techniques such as seismic tomography, one obtains but a "snapshot" of the present state of complex dynamical systems. The time scales that govern such systems can be short be geological standards, but are typically long compared to the lengths of geodetic time series. In order to progress in our understanding of the physics, we must turn to a joint interpretation of geodetic observations, geophysical data and models, *and* geological reconstructions of past history. In fact, the understanding of crustal dynamics depends almost exclusively upon geologic data to characterize how deformation has evolved and been re-distributed over time. Prediction of crustal dynamic activity is not possible without a specific knowledge of regional geological context. Clearly a broad range of techniques must be brought to bear on large scale tectonic problems. For instance, conventional geologic field studies are crucial for providing a detailed geological context, but such detailed studies cannot be made over large regions. Remote sensing including structural geologic interpretation, mapping of lithologies and soils, and characterization of weathering surfaces (including relative dating of fault scarps) provides the key to an informed extension of detailed study results to regional scales.

2.4 EMERGING LINES AND FUTURE DIRECTIONS

Strategy. Selection of a strategy for the continued development of space-geodetic solutions to outstanding geological problems in the next decade or two should be based on the recognition that resources and capabilities will remain finite, and in some instances will be in fact somewhat limited. Consequently, we recommend the following guidelines:

    1. *Focus on areas with major geological issues that require quantitative answers.* This entails the selection of areas and problems for which *testable* hypotheses can be formulated, together with a preferred focus on recognizable geological problems, for which other data sets of adequate quality are available. In addition, the socio-economic importance of the problem and the potential human impact of eventual solutions should be recognized and taken into consideration.

    2. *Develop affordable technology.* This would have the desirable consequence that it would permit involvement of a greater proportion of the domestic and foreign scientific community. Further, it would allow improved spatial and temporal coverage within controlled costs, and often minimize the logistical burden of field work.

    3. *Emphasize easily deployable systems.* In particular, we require a capability for *rapid* field deployment (on time scales of hours). Moreover, such systems would lead to lower personnel requirements, would allow easier access to remote areas, and would support both low resolution reconnaissance work, as well as detailed surveys of dense networks.

    4. *Emphasize instrument calibration and measurement validation.* This entails the continued operation of the more burdensome, expensive systems at a level adequate to permit validation of results obtained by "lightweight" field systems. In addition, there will be a continued, and even enhanced need to maintain very high quality global and regional fiducial and reference networks. The systematic intercomparison of independent systems (e.g. VLBI, SLR, GPS) should be retained as an important validation technique, and we strongly recommend a careful examination of monumentation issues (particularly long-term stability), if future space geodetic systems are used to monitor very large, dense networks.

Point Positioning Objectives. Scientific objectives for Point Positioning activities during the next decade include, for example:

    1. to collect and analyze suitable "baseline" data to support the design and implementation of dense space-geodetic networks, to diagnose sources of geodetic noise, and to provide independent calibration and validation data.

    2. to develop techniques for the *geological* interpretation of space-based ranging and altimetry data in plate-boundary deformation zones;

    3. to formulate and attempt to solve significant problems in crustal deformation resolvable with the limited data (in time) that will be collected within the next decade.

Specifically, we should refine parameterized kinematical models of crustal deformation, and extend the analysis to include constraints from dense, frequently surveyed space geodetic networks. At the same time, we must improve physical models of time–dependent deformation of the crust and upper mantle, and use these models in the planning of space geodetic surveys and in the interpretation of the spatio-temporal strain patterns seen by space geodetic techniques. Of fundamental importance for interpretation purposes will be the systematic comparison of space geodetic observations to other, independent geophysical and geodetic data, such as observatory measurements of strain along short baselines, seismicity patterns, and other repeated geodetic and gravity surveys, in an effort to achieve a quantitative understanding of the phenomena which control volcanic and seismic cycles. As argued earlier, the elucidation of the time scales relevant to many important aspects of crustal dynamics will require integration of geodetic measurements with interpretations of local and regional geology. Finally, we will need to develop a systematic methodology for exporting approaches proven to be successful in a given region where accessibility is not an issue (e.g. the American West), to inaccessible regions with comparable or very different tectonic regimes (e.g. continental collision zones such as Tibet, trench boundaries such as the west coast of central and south America, strike-slip boundaries such as Turkey and New Zealand, and regions of back-arc deformation, such as the Aegean). Finally, in many instances, some of the critical structures will be sumerged, so that a capability to locate precisely points on the seafloor relative to one another, and relative to a land-based geodetic network would be extremely valuable (e.g. Spiess, 1985).

2.5 CONCLUSIONS AND RECOMMENDATIONS

Over the next decade and into the next century, the priorities we assign to various aspects of remote sensing applied to the study of the solid earth must acknowledge the tradeoff that exists between the application of recent advances in the technological aspects of space geodesy to solve urgent geological and geophysical problems, and the need to press ahead and develop new and better capabilities. The time scales associated with some of the most critical aspects of the current evolution of our environment are poorly understood, but are thought to be on the order of decades. Consequently, we may not have the luxury to defer practical applications until better technology is at hand, but should in many cases undertake systematic studies on an unprecedented global scale. Important conclusions reached in this section include:

1.   To contribute to the improvement of existing models of present-day motion among the major plates, the tangential components of relative velocities on interplate baselines with endpoints located within stable plate interiors must be resolved to an accuracy of about 1 mm/yr.
2.   The most important and interesting geodetic signals averaged over 1 hour to 100 years take the form of *departures* from predictions of million-year average rates based on geological rigid-plate models.
3.   The most significant departures from rigid plate motions occur in zones of 100 to 1000 km width, within which differential motions are accommodated by a *combination* of seismic slip and aseismic deformation.
4.   Measuring the velocities between crustal blocks to ± 5 mm/yr can place geologically useful constraints on the integrated deformation rates across continental plate-boundary zones, such as the western U.S. and Tibet.
5.   The establishment of frequently (or even continuously) surveyed local geodetic networks of sufficient density, around major geodetic sites in active areas, should receive high priority.
6.   The development and systematic deployment of affordable and easily deployable space-geodetic systems with cm to mm precision and high sampling rates—that is, "occupation" frequency ranging from hourly to weekly—will permit investigation of geophysical phenomena, particularly the earthquake cycle, in a range of spatial and temporal scales never explored before.

7. Space geodetic observations yield constraints on crustal kinematics; to achieve an improved understanding of the dynamics and thus a better grasp of the underlying physical phenomena, we must rely on a broad combination of geophysical and geological observations as a way to extend the geodetic signals to longer time scales and to extrapolate surface information to crustal and mantle depths.

Based on the discussions held at the Erice workshop, and in view of the conclusions listed above, the panel on the "Long-Term Dynamics of the Solid Earth" formulated the following recommendations concerning the continued development and application of precise point positioning techniques:

Over the next 20 years, major efforts in applying precise positioning techniques should be aimed primarily at:

a) *Continued large-scale <u>reconnaissance</u> surveys with station spacing on the order of 10^2 km, to improve our understanding of the kinematic evolution of extensive, largely unexplored zones of continental deformation.*

b) *Sustained, repeated measurements of dense networks at centimeter-level accuracy, to determine the <u>time dependence and spatial distribution</u> of deformation within and across zones of intense tectonic activity. Measurement frequencies should range from daily to annually over a decade or more, with station spacing from 3 to 30 km, and network dimensions from 10 to 1000 km.*

c) *Continued improvement of capabilities, to achieve:*
    -- millimeter-level accuracy in both horizontal and vertical components for detailed subaerial studies, system calibration, and ultimately, <u>low-cost, routine</u> deployment.
    -- centimeter-level accuracy in both horizontal and vertical components for sea-bottom systems.

3. GRAVITY AND GEOID

The Earth's geoid and gravity anomalies are sensitive to lateral variations in the internal mass distribution. Such mass inhomogeneities could be maintained by a variety of static and dynamic compensation mechanisms, involving density differences arising from changes in rock chemistry, temperature, or physical state. Thus improved resolution and accuracy in mapping the Earth's geopotential would contribute to the understanding of a number of fundamental problems concerning the structure and rheology of the lithosphere and the dynamics of the deep interior. Although the interpretation of gravity data in the absence of other physical and observational constraints is nonunique, joint inversion of gravity data within the context of a well-posed physical model offers great promise for lending new insights into Earth processes. After briefly reviewing the fundamentals of potential fields, we will highlight some outstanding problems in Earth science for which improved gravity data is required.

3.1 BACKGROUND

A Note on Units and Numbers. Gravity acceleration, or gravity, is the vector sum of the gravitation due to the mass of the Earth and the centripedal acceleration associated with the rotation of the Earth. Since gravity is an acceleration, the appropriate SI unit is m sec^{-2}. Since gravity is a vector, one can discuss gravity in specified directions and in terms of magnitude. This chapter emphasizes the magnitude of gravity. The unit still widely used for gravity studies is the "gal" which is 10^{-2} m sec^{-2}.   A sub-unit is the mgal which is 10^{-3} gal or 10^{-5} m sec^{-2}. The SI unit for density is kg m^{-3} which is equivalent to 10^{-3} g cm^{-3}.

The average gravity on the surface of the Earth at the equator is approximately 980 gal. Gravity varies from the equator to the pole because the Earth's shape is flattened towards the poles and because of the reduction of centripedal acceleration in going from the equator to the pole. Gravity also changes due to the variation in topography, lateral variations in internal mass, and to a much lesser extent by time variations associated with a variable rotation rate and changing internal mass distribution. Tidal variations in gravity are also caused by the attraction of the Sun, Moon, and planets. The variation of greatest interest for this discussion is that associated with internal density variations. The specific problem is known as the inverse problem, which is the determination of the distribution of masses inside the Earth from gravity observations. It is a key area of research in geophysics and geodesy. The observed variations of gravity due to mass inhomogeneity are of the order of 10^{-6} to 10^{-4} of the average value of gravity, so that a convenient unit for discussing the gravity variations is the milligal defined above.

Also of interest here is the linear gradient of gravity. The most commonly used unit is the Eötvös unit defined by 1 E = .1 mgal/km or 10^{-9} sec^{-2}. The vertical gradient of gravity at the surface of the Earth is about 3100 E.

Gravity and its Relatives. Descriptions of the Earth's gravity field are given in terms of the gravity potential, from which gravity can be obtained by computing the spatial derivative (see Table 1).   Normally, gravitational potential Φ and its derivatives are discussed after the removal of a reference potential Φ_{ref} associated with an ellipsoid whose shape (flattening) corresponds closely to the Earth either as it now exists (the "spheroid") or as it would be if it were a fluid affect only by gravitation and centripedal acceleration (the "hydrostatic figure"). The net values $\delta\Phi = \Phi - \Phi_{ref}$ can be used to calculate undulations in geoid height δN, gravity disturbances δg or anomalies Δg, and various gravity gradients $\delta\Gamma_{ij}$. The relationships among some of these quantities are shown in Table 1.

Table 4. Gravitational Relations

Definition	Example (point mass)
$\delta\Phi(r) = -\int \dfrac{G\,\Delta\rho(r_o)}{\lvert r-r_o\rvert}\,dr_o$	$\delta\Phi = -\dfrac{G\,\Delta m}{r}$
$\delta N = \dfrac{\delta\Phi}{g_o}$	$\delta N = \dfrac{\Delta m\,a^2}{r\,M_e}$
$\delta g = -\nabla\,\Phi$	$\delta g_r = -\dfrac{\partial\Phi}{\partial r} = -\dfrac{G\,\Delta m}{r^2}$
$\delta\Gamma = -\nabla\,(\nabla\,\Phi)$	$\delta\Gamma_{rr} = -\dfrac{\partial^2\Phi}{\partial r^2} = \dfrac{2G\,\Delta m}{r^3}$

G = Newton's gravitational constant; $\Delta\rho$ = density anomaly; r, r_o = position vectors; Δm = mass anomaly; g_o = average surface gravity; a = radius of the Earth; M_e = mass of the Earth; Φ = gravitational potential.

Spatial variations in these measures of the Earth's gravity field result from density contrasts distributed on and within the Earth. For density contrasts varying with horizontal length scales of thousands of kilometers, the density contrasts are commonly expressed in terms of spherical harmonics of degree ℓ and order m. For a given harmonic component of a density contrast, $\Delta\rho_{\ell m}$, distributed over a thickness h at a depth z below the Earth's surface, the geoid anomaly $\delta N_{\ell m}$, gravity disturbance $\delta g_{\ell m}$, and gravity gradient anomaly $\delta\Gamma_{\ell m}$ as evaluated at the Earth's surface (r=a) are given by:

$$\delta N_{\ell m} = \frac{4\pi Ga}{g_o(2\ell+1)}\left(\frac{a-z}{a}\right)^{\ell+1}\Delta\rho_{\ell m}\,h$$

$$\delta g_{\ell m} = \frac{2\pi(\ell+1)G}{\left(\ell+\tfrac{1}{2}\right)}\left(\frac{a-z}{a}\right)^{\ell+2}\Delta\rho_{\ell m}\,h$$

$$\delta\Gamma_{\ell m} = \frac{2\pi(\ell+1)(\ell+2)G}{\left(\ell+\tfrac{1}{2}\right)a}\left(\frac{a-z}{a}\right)^{\ell+3}\Delta\rho_{\ell m}\,h$$

The approximate wavelength λ of a spherical harmonic of degree ℓ is

$$\lambda \approx \frac{2\pi a}{\ell + 1/2}$$

When $\lambda \ll a$ ($\ell \gg 10$), the following asymptotic Cartesian relations are valid:

$$\delta N(\lambda) = \frac{G \lambda \Delta\rho(\lambda) h}{g_o} \exp(-2\pi z/\lambda)$$

$$\delta g(\lambda) = 2\pi G \Delta\rho(\lambda) h \exp(-2\pi z/\lambda)$$

$$\delta\Gamma(\lambda) = \frac{4\pi^2 G \Delta\rho(\lambda) h}{\lambda} \exp(-2\pi z/\lambda)$$

The above expressions quantify two important considerations which must be kept in mind in designing any program to measure the geopotential. The first is attenuation by the factor $\exp(-2\pi z/\lambda)$ of the gravity signal with distance from its source (or, equivalently, height of observation).   The amplitude is reduced by a factor of 0.1 for $z = 0.37 \lambda$, by a factor of 0.01 for $z = 0.74 \lambda$, and by a factor of 0.001 for $z = 1.1 \lambda$. At the elevation of a satellite 200-km high, surface gravity anomalies of 1000, 500, 200, 100 and 50 km wavelengths are attenuated by factors of 0.3, 0.08, 0.002, 3×10^{-6} and 10^{-11} respectively. Clearly, once $z > 0.5 \lambda$, most gravitational signal is lost. Later in this chapter, we quote the spatial resolution and accuracy necessary for gravity observations in order to address certain problems. We evaluate these requirements at the Earth's surface and consider the resolution to be one half the shortest wavelength λ sampled by the data. If the data are measured by satellite, the more severe accuracy requirements can be calculated from these relations given the satellite altitude and the desired resolution.

The second consideration is the fact that short-wavelength gravity signals are dominated by the largest, shallowest density contrast, namely the surface topography. For example, all of the following produce a 1 mgal gravity anomaly at the Earth's surface: a 10-m layer of crustal rock; a 1-km deflection of the Moho at depth $z = 0.7 \lambda$; a 4-km deflection of the Moho at $z = 0.5 \lambda$; a thermal anomaly of 100°C at the top of the mantle extending over a thickness of 80 km; a surface layer of water 25-m thick. The bottom line is that if we wish to interpret gravity data accurate to x mgals in terms of crustal structure, thermal anomalies in the mantle, or dynamic sea surface topography, we must at the same time have measured the topography of the Earth's surface to within 10x meters.

The Concept of "Isostasy". In the following sections we will frequently refer to the state of isostasy for mass anomalies on and within the Earth. Gravity anomalies conclusively demonstrate that the Earth does not rigidly support large loads over geologic time periods. Instead, mass excesses and deficits are mechanically coupled so as to compensate each other. Here we will adopt a particularly broad definition of isostasy, in that we include all static and dynamic mechanisms of producing and maintaining departures from perfect hydrostatic equilibrium.

Classical mechanisms of isostasy include the models of Airy and Pratt, which pointwise compensate topographic variations (or in general any laterally varying surface mass distribution) by changes in crustal thickness or crustal density, respectively, immediately beneath the load. Plateaus appear to be compensated by thickened crust, in the manner of Airy isostasy, while midocean ridges are thought to be compensated by density deficits arising from thermal expansion of hot mantle rocks, in the manner of Pratt isostasy. In tectonic areas where loads are emplaced on a plate with lateral strength, regional mechanisms of isostasy, such as that of the thin elastic plate, are more applicable. The gravity anomalies near midplate volcanoes, deep sea trenches, and continental thrust belts are more consistent with this sort of regional isostasy. The elastic plate model includes as end members Airy isostasy, as the elastic plate thickness approaches zero, and total uncompensation, as the elastic plate thickness approaches infinity. These local and regional compensation mechanisms are all "static" in the sense that the stress differences arising from the spatial offset between mutually compensating mass excesses and deficits are supported by the solid lithosphere, even though the mass anomalies may themselves be time dependent (i.e., due to conductive cooling of a low-density body of thermal origin).

Convective patterns within the mantle, whether driven by thermal or compositional buoyancy, produce dynamically maintained topography at the Earth's surface, at the core-mantle boundary, and at any other internal density interfaces. As a first approximation, the total mass displaced in any column by this boundary deformation is about equal to the mass anomaly due to density contrasts in the convecting column - a sort of dynamic isostasy. The mass anomalies caused by this dynamic compensation have an effect opposite in sign to the mass anomalies due to interior density contrasts. The net gravitational effect, including the effects of dynamic topography, depends upon how this dynamic compensation is distributed among the boundaries of the convecting system (Richards and Hager, 1984; Ricard et al., 1984). Due to the fall-off of gravitational interaction with distance, the gravitational attraction of the deformed bottom boundary is attenuated more at the surface of the Earth than the gravitational attraction of the interior density contrasts, which are in turn attenuated more than the effects of the deformed upper surface. If the mantle is chemically stratified, boundaries between chemically distinct layers will also support dynamic topography, with a resultant contribution to the net gravity field. Density contrasts near the 670-km seismic discontinuity would be nearly completely compensated by warping of that boundary if it is a chemical discontinuity, and result in negligible gravity anomalies. As discussed below, the gravitational signatures of deep subducted slabs provide powerful tests of the presence of chemical stratification in the mantle (Hager, 1984).

The net gravity anomaly at the Earth's surface, regardless of whether compensation is static or dynamic, is the result of near cancellation of large effects. The magnitude, shape, and sign of the result is a sensitive function of the compensation mechanism. Matching gravity observations thus provides a very diagnostic test of lithospheric rheology and mantle dynamics.

3.2 OCEANIC LITHOSPHERE

That satellite gravity field observations can yield major advances in our understanding of the lithosphere in general and the oceanic lithosphere in particular has been amply demonstrated. Specifically, altimeter data from Seasat and GEOS have led to a significant increase in our understanding for the thermo-mechanical structure and evolution of the oceanic lithosphere. In the past decade, these data have been applied, with much success, to models of lithospheric structure in all major oceanic provinces, from the mid-ocean ridges where the lithosphere is formed to trenches where it is consumed. In the following sections, we discuss some outstanding problems concerning the oceanic lithosphere which remain to be solved because presently-available altimetric geoids lack sufficient resolution, accuracy, and/or continuity at shorelines.

Mid-Ocean Ridges.

   "*The global mid-ocean ridge is perhaps the most striking single feature on the solid surface of our planet* (Fig. 9, heavy line). *Sections of ridge extend along the floors of the world's oceans to a length in excess of 50,000 km. The mid-ocean ridge dominates the Earth's volcanic flux and creates an average of 20 km³ of new oceanic crust each year. The processes of generation and cooling of oceanic lithosphere contribute two thirds of the heat lost annually from the Earth's interior.*"

   "*Recent discoveries of the widespread nature of volcanically-driven submarine hot springs and their attendant chemosynthetically-based animal communities underscore the fact that the seafloor/ridge crest environment represents one of the current frontiers in the exploration and understanding of our planet.*"

These two excerpts from the proceedings of a workshop entitled "*The Mid-Ocean Ridge: A Dynamic Global System* (RIDGE)" (Ocean Studies Board, 1988) illustrate the importance of investigating the mid-ocean ridge using all the tools that are available including highly accurate satellite altimeters and precise satellite navigation of research vessels.

   In the past, the principal use of gravity data for mid-ocean ridge studies has been in characterizing the state of isostasy of topographic features at various scales. Such investigations have led to fairly good first-order representation of the mechanisms compensating mid-ocean ridge relief. At the longest wavelengths (>2000 km) corresponding to the elevation difference between ridges and the deep sea abyssal plains, the topography is supported by thermal isostasy. The ridges stand higher because the rock below is hot and buoyant relative to that beneath the old ocean basins.

   While the large-scale structure of the cooling oceanic lithosphere is fairly well understood, the small-scale thermal, mechanical, chemical and hydrothermal processes that occur at mid-ocean ridge axes are not. This is because only a small portion of the mid-ocean ridge has been explored by research vessels. Satellite-born radar altimeters are now able to make high-accuracy (< 1 mgal) measurements of the marine gravity field on a global basis. Moreover, new shipboard gravimeters, using GPS navigation, can provide even higher resolution gravity surveys of small areas. These new gravity data when used in combination with seafloor topography data provide a valuable tool for studying lithospheric dynamics and mantle flow at spreading ridges. Examples of processes that can be addressed using these data include:

> Ridge segmentation and along-axis variations in ridge dynamics
> The strength and rigidity of lithosphere at zero age
> The overall rheology of the mantle beneath ridges
> Median valley formation and its relation to spreading rate and hot spots
> The state of stress on transform faults and fracture zones
> The origin and support of off-axis seafloor topography,
> The planform and width of mantle upwelling beneath ridges
> The relative importance of passive and dynamic flows

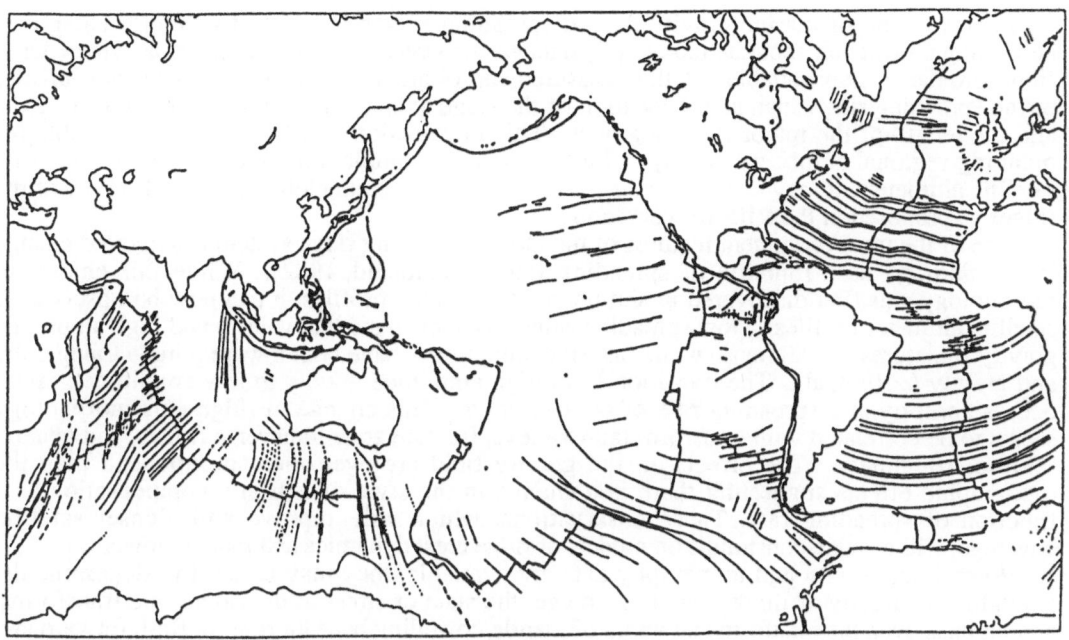

Fig. 9. The oceanic plate boundary consist of spreading ridges, transform faults and subduction zones (heavy line). The mid-ocean ridge extends for more than 50,000 km across the ocean basins and is segmented over a variety of length scales ranging from 20 - 1000 km by transform faults, overlapping spreading centers and minor departures from linearity. To date only a small portion of the mid-ocean ridge has been explored. Fracture zones (light lines) are linear scars in the seafloor produced by transform faulting. Topography along fracture zones consist of long ridges, troughs and scarps which separate regions of different depth. Because studying the contrast in the thermomechanical structure of the oceanic lithosphere. Moreover, fracture zone traced reveal seafloor spreading directions over geologic time. In this map, the locations of the ridges, transform faults and fracture zones were derived from a combination of satellite altimeter data, shipboard topography, gravity and magnetics data and earthquake epicenter data.

Interpretation of marine gravity data at spreading ridges falls into two major categories depending on whether accurate topography data are also available. In remote areas, where dense shipboard data are unavailable, satellite measurements of gravity can reveal overall characteristics of the spreading ridge such as the location of the ridge axis, the morphology of the ridge axis, and the length of the major ridge segment. This information will be extremely valuable for planning regional shipboard surveys. Moreover, with complete marine gravity coverage by satellite altimeters, it may be unnecessary to carry out a complete global shipboard survey of the ridges as proposed in the RIDGE workshop.

One of the most prominent features of the global spreading ridge system is a marked change in morphology as a function of spreading rate (Macdonald, 1982). While this change in morphology was first discovered in shipboard topographic profiles, it can now be observed in satellite altimeter profiles. Slow spreading ridges have deep median valleys and high amplitude gravity signatures (40-80 mgal) while fast spreading ridges have both low amplitude topography and gravity (~10 mgal). The transition from high amplitude gravity to low amplitude gravity occurs abruptly at a spreading rate of 60-70 mm/yr. This change in ridge-axis morphology seems to be correlated with transform fault valleys, fracture zone transverse ridges, and general seafloor roughness. The effects on the gravity field are even more pronounced than the topographic effects suggesting there is a change in the style of isostatic compensation as a function of spreading rate. These observations, which will improve with denser satellite coverage, will provide constraints on models of ridge crest dynamics and mantle upwelling.

When both gravity and topography data are available, they may be used to determine the strength and rigidity of lithosphere at zero age, the state of stress at the ridges and transforms, and perhaps even the planform and width of mantle upwelling beneath ridges. Both forward and inverse modeling techniques can be used for these types of studies. Models which predict a linear relationship between gravity and topography are most easily evaluated using the transfer function technique. The shape of the observed gravity/topography transfer function reveals the depth of compensation, the rigidity of the lithosphere, and in some cases, the presence or absence of a magma chamber. Preliminary studies suggest that the planform of mantle upwelling beneath the ridges is evident in the gravity data after the effects of topography are accounted for. In any case, there is more work to be done in modeling the gravity and topography signatures of spreading ridges because several thermal and mechanical processes are occurring in a narrow zone.

Recent availability of swath mapping equipment such as Seabeam and SeaMARC have revealed entire new classes of topographic features representing departures from linearity along the mid-ocean ridge system. Examples of these features include propagating rifts, overlapping spreading centers, 200-km undulations in axial relief, and minor departures from axial linearity. It is thought that many of these features are the topographic expression of variability in magma supply to the ridge crest as a function of space and time. Gravity information would provide key information concerning the origin and evolution of such features. Due to the fine-scale variability of mid-ocean ridge processes, we require a field accurate to a few mgals with resolution between 2 and 200 km.

Fracture Zones. One of the most intriguing observations to emerge from analysis of the altimetric geoid provided by Seasat is the failure of the simple plate model to describe the density structure of adjacent lithospheric plates of different ages across oceanic fracture zones. In only a few cases, such as the Mendocino Fracture Zone (Detrick, 1981), does the "step" in the geoid across the fracture zone conform to the predictions from conductive cooling of 125-km-thick thermal plates with the observed ages. For example, Fig. 10 shows the size of the geoid step across the Eltanin Fracture Zone as a function of the average age of the contiguous lithosphere. Theoretical thermal models derived from observations of heat flow and subsidence (Parsons and Sclater, 1977) predict a constant geoid step for young ages, with the step gradually decreasing with increasing plate age. Contrary to these predictions, the observed geoid step rapidly

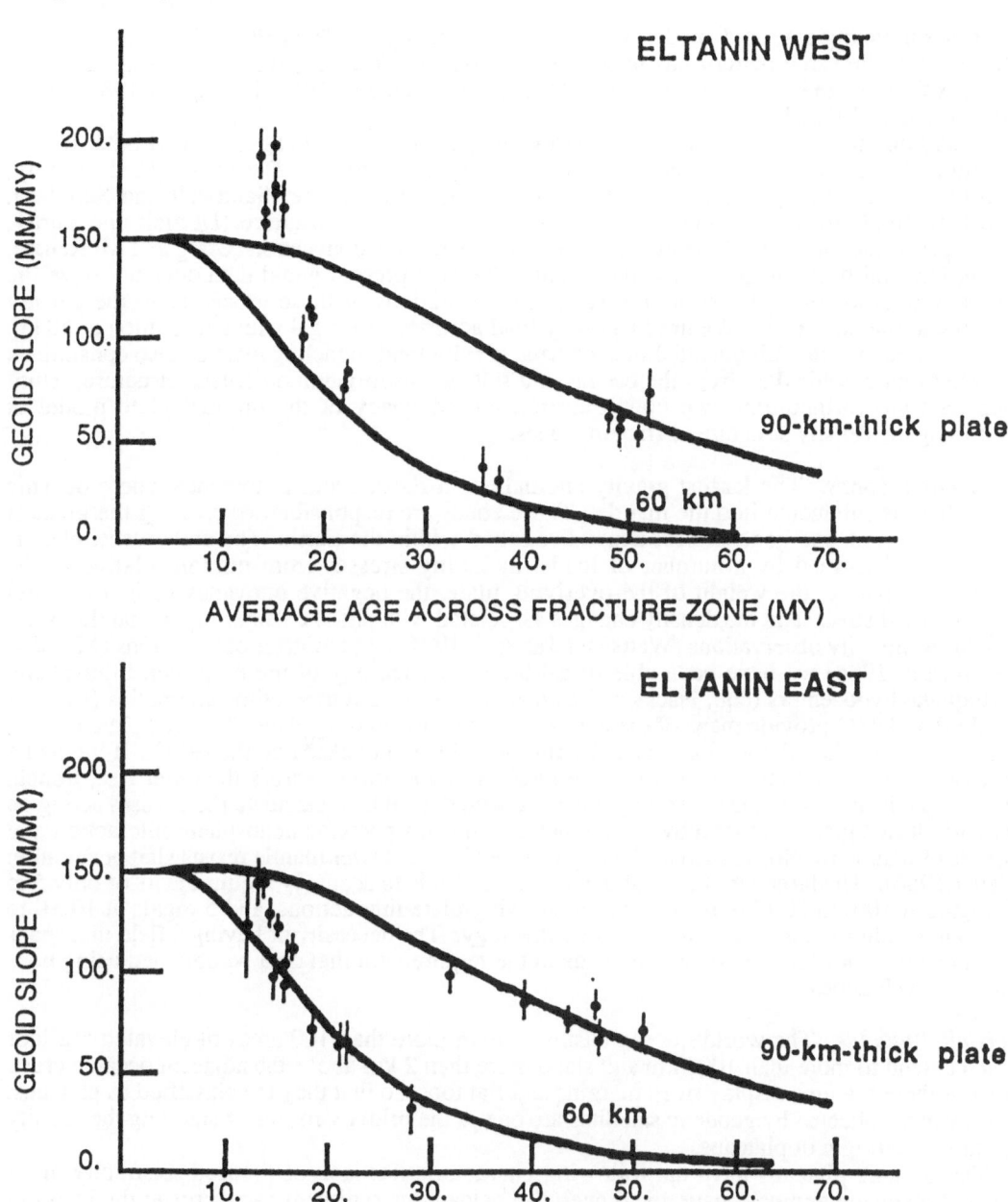

Fig. 10. Geoid slope estimates across the Eltanin Fracture Zone system. The circles represent the observed change in geoid across the fracture zone divided by the age contrast plotted as a function of the average of the ages on both sides of the fracture zone. The continuous curves give the expected relationship between geoid shape and average age for thermal plate models with thicknesses A of 90 km and 60 km. a) Data for the western (Antarctic) limb of the fracture zone. b) Data for the eastern (Pacific) limb. From Driscoll and Parsons (1988).

decreases at young ages and suddenly reappears at older ages. This behavior is not an isolated occurrence: the same pattern has been observed on the Udintsev, Ascension, and Falkland-Agulhas fracture zones (Cazenave, 1984; Driscoll and Parsons, 1988; Freedman, 1987; Marty and Cazenave, 1988a,b).

In addition to conductive cooling of thermal plates with differing ages, several other factors undoubtedly contribute to the geoid signature of fracture zones. Based on geoid, gravity, seismic, or topography studies, the effects of lithospheric flexure (Sandwell and Schubert, 1982), thermal stress (Parmentier and Haxby, 1986), crustal structure (Detrick and Purdy, 1980), peridotite intrusions (Fox et al., 1976), and small-scale convection (Craig and McKenzie, 1986), have all been suggested as significant. The best present geoid data does not have the accuracy or resolution to sort out the various contributions of these processes to the density structure at fracture zones. We need a gravity field accurate to 1 mgal with a resolution of 50 km or less. To realize the full potential of such data, gravity field modeling must be also constrained by better topographic data from the oceans and seismic information on crustal structure. Until such data are forthcoming, we must question the adequacy of the thermal plate model in describing the density structure of fracture zones.

Subduction Zones. The largest gravity anomalies on Earth occur at trenches where oceanic lithosphere is subducted into the mantle. These zones are responsible for creating the greatest thermal, seismic, and geochemical anomalies found within the Earth. The underthrust plate is flexed and deformed by a number of loads, including stresses from motion relative to the convecting mantle, the weight of the overlying plate, the negative buoyancy of its own cold mass, thermal stress, and the density changes associated with phase changes in the mantle. With sea-surface gravity observations (Watts and Talwani, 1974) and altimeter observations (McAdoo and Martin, 1984) we have been able to calibrate the rheology of the deformed lithosphere. Earthquake hypocenters (e.g., Isacks and Barazangi, 1977) and travel time anomalies (Creager and Jordan, 1984) provide maps of the geometry of the downgoing plate. Thermal plate models allow us to calculate the load associated with the cold slab (Toksöz et al., 1971). If we had a gravity or geoid map continuous from the undeformed seafloor, across the outer rise, trench, forearc, and island arc to the overriding plate, we would be able to calculate the stresses acting on the underthrust plate, and thereby learn much about lithosphere/asthenosphere interaction and aspects of mantle rheology, such as the degree to which the lower mantle resists slab penetration (Hager,1984). The large amplitude of the anomalies leads to accuracy requirements of only 5 to 10 mgals at 100- to 200-km resolution for studying plate interactions, and 5 mgals at 1000- to 3000-km resolution for investigating mantle rheology. The necessity of having a field that spans the transition from ocean to continent leads to the requirement that data be obtained using non-altimetric techniques .

Oceanic Plateaus. The world's ocean basins contain more than 100 areas of elevated seafloor which extend to more than 1000 km and stand more than 2 km above the adjacent oceanic crust. Most of these features display steep margins and flat tops, so that they are classified as plateaus. Gravity data collected by geodetic satellites are one of the primary means of studying the density structure and origin of plateaus.

The oceanic plateaus have the following characteristics: lack of focused seismicity, non-lineated magnetic anomaly patterns, (generally) calcareous sediment caps, crustal thickness in excess of 15 km, and topographically-correlated geoid anomalies (Carlson et al., 1980; Nur and Ben-Avraham, 1982; Sandwell and Renkin, 1988). In total area, oceanic plateaus cover more than 3% of the seafloor. Therefore, they must play a significant role in both the evolution of the ocean basins and the formation of collision-type margins (Vogt, 1973; Ben-Avraham et al., 1981). Nevertheless, the origin and subsurface structure of these features remain enigmatic. Are they continental fragments or oceanic in origin, formed by excess volcanism at or near mid-ocean ridges? The thick pelagic caps make direct sampling by dredging and drilling difficult.

Seismic refraction data are not always available, and in at least one case (Ontong-Java) the same set of travel times have been used to argue for both continental and oceanic origin. Since continental-type plateaus would have deeper isostatic roots than oceanic ones, gravity data provide important information as to the origin of plateaus. Satellite altimeter data have been used to estimate depths of compensation for a number of plateaus from the slope of geoid height versus topography (MacKenzie and Sandwell, 1986). For smaller plateaus, the accuracy of this procedure is limited by the accuracy and coverage of existing satellite altimeter data. A complete gravity/topography study requires a field accurate to 1 mgal at a resolution of 50 km and greater.

Mapping the Seafloor. It is startling that the topography of Mars (Carr et al., 1977) and Venus (Masursky et al., 1980) has been mapped at a higher resolution than the the topography of the Southern Ocean. There are still many 5° by 5° areas in the South Pacific, South Atlantic and Southern Indian Oceans that have not been explored by ships (see GEBCO charts 12-16, Canadian Hydrographic Service, 1984). Even with advanced swath mapping tools such as Seabeam, it will take many decades to survey significant portions of the seafloor in these remote areas. Satellite altimetry, which maps the topography of the equipotential sea surface (marine geoid), is an extremely useful tool for locating features in uncharted areas (Haxby, 1985; Sandwell, 1984). Details in the marine geoid reflect seafloor topography, especially at wavelengths shorter than 100 km. In addition to locating features, a complete two-dimensional mapping of the marine geoid could be used along with available shipboard bathymetric profiles to predict bathymetry in the uncharted areas.

The theoretical technique used to predict bathymetry from geoid height is based on the approximation that the oceanic lithosphere responds to applied topographic loads like a thin-elastic plate floating on a fluid asthenosphere. Under these conditions there is a linear-isotropic transfer function that maps topography into geoid height and vice versa. Thus if the geoid height and the transfer function are known then the topography can be calculated (see Dixon et al. 1983). The problem is that the topography must be known in order to determine the proper transfer function. An example of geoid height and bathymetry for a large area of the North Pacific surrounding the Emperor Seamounts is shown in Fig. 11. It is apparent that geoid height and bathymetry are well correlated at short wavelengths and poorly correlated at long wavelengths.

The ratio of geoid height to topography as a function of wavenumber is also shown in Fig. 11 along with the prediction of the thin elastic plate model (15-km-thick plate). An interesting feature of model transfer functions is that they are all similar for wavelengths less than about 200 km. This is because the short wavelength part of the transfer function depends only on the known density of the seafloor and the mean ocean depth. In contrast, for wavelengths greater than about 200 km, transfer functions are highly variable and may even be anisotropic. Thus, in theory, one could use a detailed two-dimensional geoid and a few shipboard topographic profiles to calibrate the longer wavelength part of the transfer function. Assuming the short-wavelength part of the transfer function depends only on the mean ocean depth, the topography can then be predicted from the detailed geoid.

The accuracy and reliability of the predicted topography will, of course, depend on the density of shipboard profiles available but more important, it will depend on the accuracy and resolution of the geoid. The limitations of the current satellite altimeter data set are apparent in Fig. 11 where the observed transfer function is systematically less than the model transfer function at short wavelengths (< 200 km). The problem is that the gridded geoid was derived from widely spaced Seasat altimeter profiles that do not resolve wavelengths less than about 200 km. If a high resolution geoid does become available then the accuracy of the predicted topography will be highly dependent on the accuracy of the geoid because the geoid topography ratio decreases rapidly at short wavelengths. For predicting seafloor topography the geoid should have both high resolution (~20 km wavelength) and high accuracy (~0.3 mm height precision or 1 µrad slope precision). Unfortunately, such dense satellite altimeter coverage is not yet available.

The Seasat radar altimeter, launched in June 1978, made precise measurements (Tapley et al., 1982) of geoid height (~ 0.1 m) over a large portion of the world's oceans (Fig. 12a). These data revealed many previously undiscovered features of the ocean basins such as fracture zones and seamounts (Haxby, 1985; Sandwell, 1984; Okal and Cazenave, 1985; Ruff and Cazenave, 1985) as well as large areas where the gravity field is lineated in the direction of absolute plate motion (Haxby and Weissel, 1986). Moreover, fracture zone lineations, derived from these data, improved our understanding of plate motions in the southern oceans (Shaw, 1987). Unfortunately, Seasat failed in early October of 1978 so it did not completely map the marine geoid.

The Geosat altimeter was launched by the U.S. Navy in March 1985 to map the marine geoid to a high spatial resolution on a global basis. Because of their military value, data collected during the first 18 months of the mission are classified. When this primary mission was completed, Geosat was placed into a new orbit with a ground track that repeats every 17 days (i.e., 244 revolutions, Fig. 12b) and overlays one of the 17-day Seasat ground tracks. The combination of ascending and descending profiles produces a typical Geosat ground track spacing of 70 km at the equator.

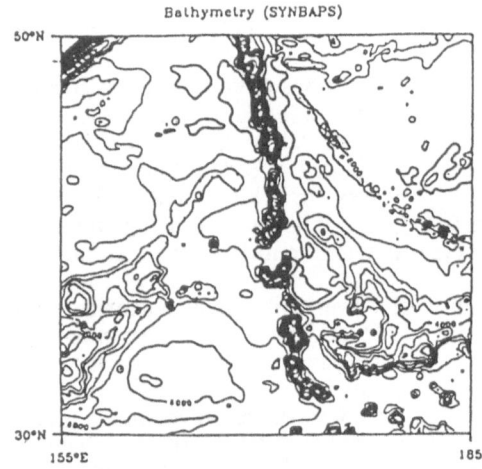

Bathymetry (SYNBAPS)

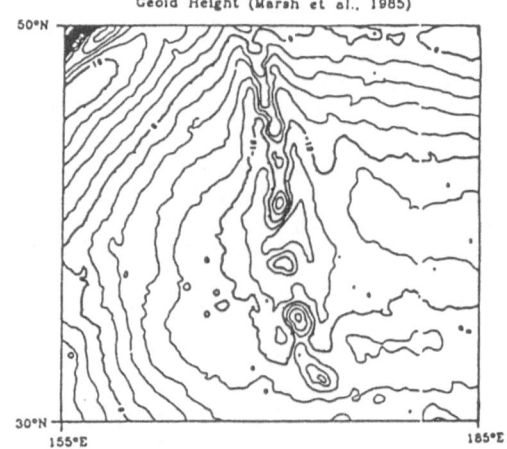

Geoid Height (Marsh et al., 1985)

Fig. 11. Bathymetry (upper left) and geoid height (upper right) for a large area in the North Pacific containing the Emperor Seamount Chain. The geoid height and bathymetry are well correlated at short wavelengths but poorly correlated at long wavelengths. The geoid/topography transfer function (right) reveals the thickness of the elastic lithosphere when the seamounts formed. The data transfer function is systematically less than the model transfer function for wavelengths less than 300 km. This poor agreement is caused by the inadequate density of satellite altimeter profiles used in constructing the geoid height map.

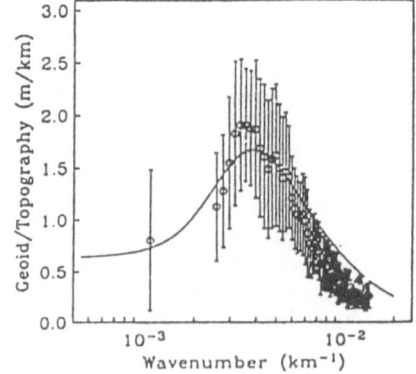

SEASAT Ground Tracks

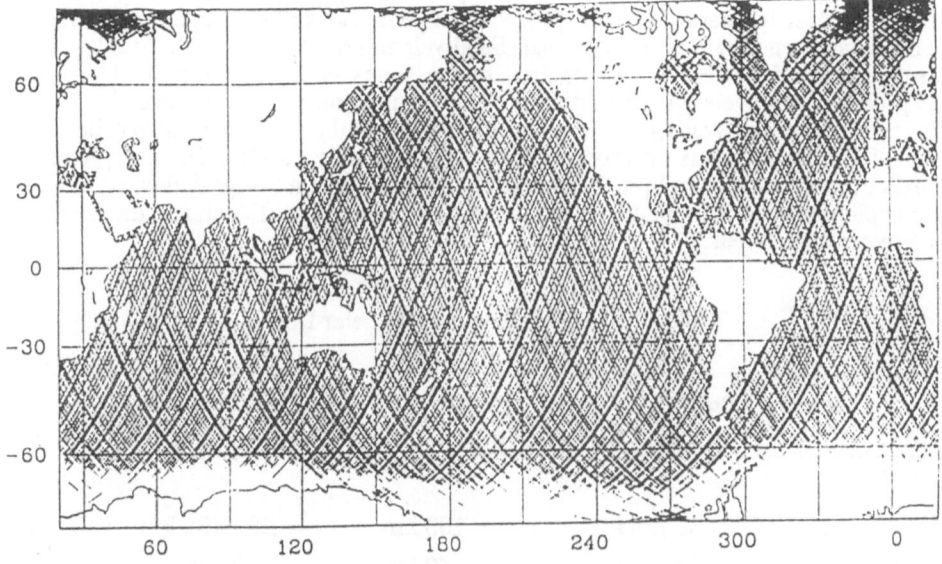

GEOSAT Ground Tracks

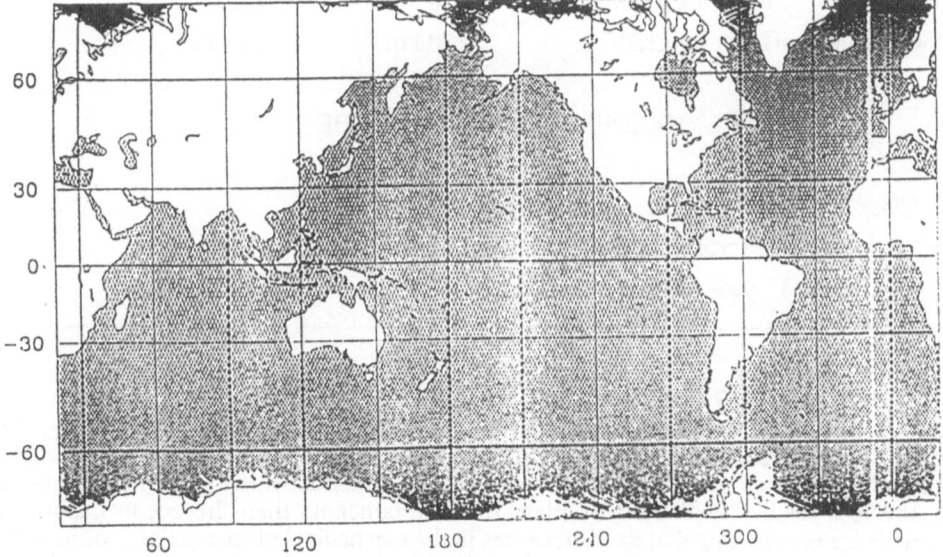

Fig. 12. (a) Seasat altimeter ground track for the entire mission (July-October of 1978). Data in the extreme Southern Ocean and Antarctic Margin were edited because of sea ice. The characteristic ground track spacing is 40 km so cross-track wavelengths of 80 km can be resolved. (b) Geosat altimeter ground track for first 10 repeat cycles. Coverage of the Antarctic margin and Southern Ocean is nearly complete. The characteristic ground track spacing is 80 km so 160 km cross-track wavelengths can be resolved. The along-track resolution of Geosat data is ~20 km.

A summary of the satellite altimeter data that will be available in the next 5 years is shown in Table 5. The precision of the satellite altimeters (1 μrad ≈ 1 mgal) has increased by a factor of about 50 over the past 10 years. By averaging many repeat cycles of Geosat altimeter data, <1 μrad precision has been obtained from Geosat (Sandwell and McAdoo, 1988). Similar precisions are expected for the ERS-1 altimeter and the Topex/Poseidon Mission. Along individual profiles the wavelength resolution of these instruments is now approaching 20 km. The major limitation of the available (and planned) data is that the spacing of the profiles is greater than 80 km which limits the cross track resolution to only 160 km. For complete, two-dimensional coverage with 1 μrad precision and 20 km resolution, a 5-year dedicated satellite altimeter mapping mission is required. A Geosat-class altimeter is sufficient and the repeat cycle should be greater than 1 year.

Table 5. Available Satellite Altimeter Data

Satellite	Precision (1μrad ≈ 1mGal)	Along-Track Resolution, λ	Cross-Track Resolution, λ
Geos-3	30 μrad	100 km	20 - 400 km
Seasat	10 μrad	50 km	80 km
Geosat (repeat)	<1 μrad	20 km	160 km
ERS-1 (repeat)	<1 μrad	20 km	?
Topex/Poseidon (repeat)	<1 μrad	20 km	314 km

NEEDED FOR 2-D COVERAGE

- Geosat-Class Altimeter (1 - 2 m radial orbit error)
- 6-Month Repeat Cycle (cross-track resolution ~ 15 km)
- 10-Repeat Cycles (5 - year mission)

3.3 THE CONTINENTAL LITHOSPHERE

The potential for accurate global gravity data to improve our understanding of lithospheric properties and processes is even more apparent for the continents, which are considerably more complex and less readily explained by plate tectonic concepts than the ocean basins. For example, evidence is mounting that the differences in the mechanical properties of continental and oceanic lithosphere are not simply explained by the presence of the thick, granitic continental crust, but rather requires thermal and/or compositional differences extending to depths of 200 km or more. At present, gravity data accurate to ±4 mgal at 100-km resolution are publicly available for only 22 percent of the Earth's land area, with political and geographical barriers preventing further acquisition by means of standard ground surveys. In order to comprehend the origin, evolution, and resource potential of that part of the planet which we inhabit, global gravity missions are a primary scientific priority.

Rifting and Continental Extension. The global distribution of continents today and the partitioning of their mineral and petroleum resources largely reflects the effects of continental rifting, and yet it is a process about which we understand very little. Does the location of rifting reflect the position of diverging currents in the mantle, a preexisting zone of weakness in the lithosphere, or both? Why do some rifting events fail after a short period of time while others succeed in leading to the formation of new ocean basins? How is the extension partitioned horizontally and vertically in the crust and lithosphere? Gravity data can bring important constraints on the problems concerning continental rifting in several ways.

On the most fundamental level, gravity anomalies have long been important in mapping the position and establishing the subsurface continuity of rifts. For example, before the 1300-km long Central North American Rift System (Ocola and Meyer, 1973) was identified as such, it was known as the "mid-continent gravity high" (King and Zietz, 1971). This feature appears prominently in the gravity map of the U.S. in the topographically featureless upper Midwest. Gravity studies played a crucial role in the studies of the Rio Grande rift in demonstrating the continuity of geologic structure along what was previously considered to be a string of disjoint and unrelated small basins (Bryan, 1938; Cordell, 1978). The strike of rifts and other intrusive bodies such as dikes and batholiths is a function of the orientation of tectonic strain axes at the time of emplacement. Thus mapping such features is one of the few means of determining paleostress directions.

Gravity anomalies over rifts are sensitive to the perturbed crustal structure from lithospheric stretching and any deep thermal anomalies responsible for doming and plate thinning. An outstanding problem in the study of extensional deformation is the disagreement among various measures of extension, such as heat flow, subsidence, and gravity anomalies, as to the total amount of lithospheric thinning in a single vertical column (Royden et al., 1983a,b; Wernicke, 1985). As compelling evidence for the discontinuous nature of extension in time as well as space (e.g. Wallace, 1984; Glazner and Bartley, 1984; Morgan et al., 1986) the need for models that go beyond one layer extension has been recognized (e.g. Royden et al., 1983a,b; Hellinger and Sclater, 1983). With particular reference to the Basin and Range, Wernicke (1985) proposes a simple shear model (Fig. 13) for the continental lithosphere in which motion along a low angle detachment allows lithospheric thinning in regions far removed from the surface zone of normal faulting. In general, this geometrical model is capable of explaining thermal uplift on the flanks of rifted regions where no crustal thinning has occurred, although the effects of small-scale convection induced by large lateral thermal gradients have also been invoked to explain the same observation (Steckler, 1985; Moretti and Froidevaux, 1985; Moretti and Chénet, 1987). Gravity anomalies have the potential to distinguish between these two explanations by providing bounds on the vertical and lateral extent of the low-density material providing flank uplift and by mapping out variations in flexural strength of the lithosphere caused by thermal reheating (Ebinger et al., 1988). Broad constraints on thermal structure would be obtained with gravity data accurate to 1 to 2 mgal with a resolution of 100 km. Specific information on variations in flexural rigidity, given the low elastic plate thicknesses to be expected, would require similar accuracy but a resolution of 20 km or better.

Sedimentary Basins and Passive Margins. The study of sediment deposits in continental basins and on passive continental margins has historically been the mainstay of Earth sciences. The accumulation and preservation of fossiliferous strata in sedimentary basins provided the first systematic basis for a geologic time scale. An inventory of the Earth's petroleum reserves begins with an inventory of sedimentary basin-fill volumes, and most geologists and geophysicists make their living thereby.

The principal research topics include: why do basins and margins subside? Why are the same basins periodically reactivated? Do sediment onlap/offlap patterns on passive margins reflect changes in eustatic sea level or temporal variations in lithospheric rheology? Gravity anomalies bear on these problems in several ways. For example, gravity maps of the Michigan Basin reveal a high-density body at the base of the sedimentary strata thought to correspond to a magmatic intrusion (Haxby et al., 1976). Cooling of this magma body may have supplied the driving force for basin subsidence. Additionally, gravity observations plus data on depths to distinct stratigraphic horizons yield estimates of the elastic thickness of the basin lithosphere as a function of time. The elastic thickness in turn constrains models of the long-term thermal evolution of the basin. Thus gravity observations supply key information on both the driving forces for basin subsidence and the history of how those forces affect the mechanical behavior of the lithosphere.

Global gravity data with ~50 km spatial resolution and an accuracy of 1 to 2 mgal, combined with bathymetry and radar ice-thickness data, would be required to answer these questions. Data at lower resolution and accuracy would allow the detection of sedimentary basins and their associated oil deposits hidden under ice sheets, allochthonous crystalline rock, and at offshore continental margins. In order to study passive margins, it is particularly vital that we obtain a gravity data set continuous across the coastlines.

Mountain Belts. Gravity observations have already played a major role recently in completely overturning the accepted notion that mountain belts on the Earth's surface are compensated by

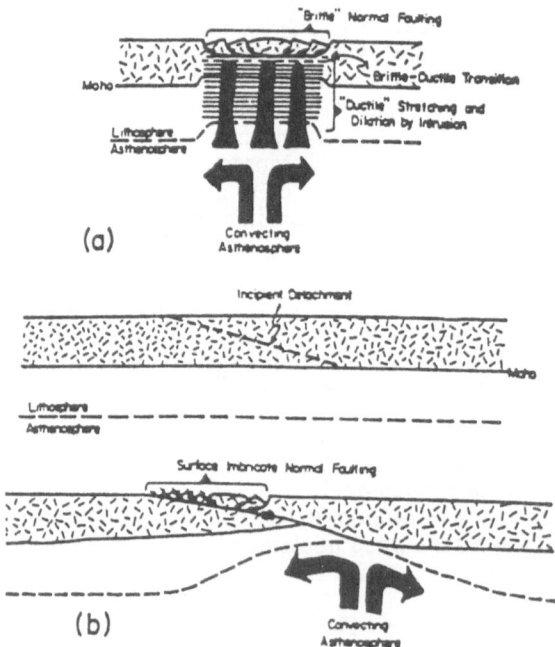

Fig. 13. End-member models of strain geometry in zones of lithospheric extension. (a) "Pure-shear" model, in which crust and mantle lithosphere are attenuated uniformly along any given vertical reference line. (b) "Simple-shear" model, in which relative extension of crust and mantle lithosphere along any given vertical line is non-uniform. From Wernicke (1985).

Gravity anomaly

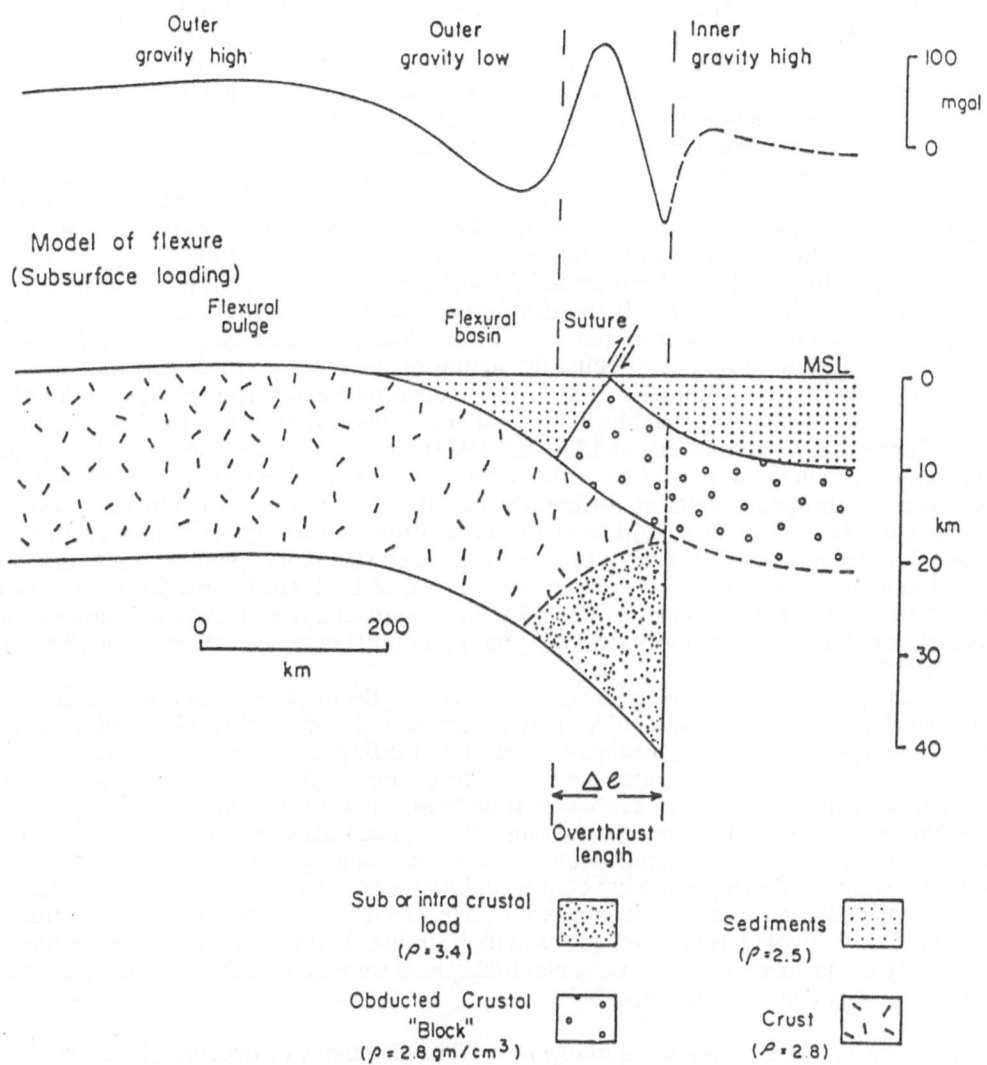

Fig. 14. Schematic model of the crustal structure and the predicted (Bouguer or free air) gravity anomaly over a completely eroded orogen. The flexure of the elastic plate on the left side is maintained by the weight of either an obducted block from the overriding plate or some intracrustal load. The gravity anomaly is characterized by a positive-negative couple in which the low in the foreland is due to flexural depression of the basement and the high in the hinterland is caused by the excess mass of the buried load. From Karner and Watts (1983).

simple crustal thickening through a form of Airy isostasy. Karner and Watts (1983) noted a consistent asymmetry in the Bouguer gravity field across the Alps and Appalachians. The Bouguer gravity low, which results from the low-density material at depth compensating the excess mass of the mountains, is consistently offset towards the foreland basin to the west of the Appalachians and to the north of the Alps, while a prominent gravity high, unassociated with any topographic feature and not predicted by Airy isostasy, appears in the hinterland on the opposite side of the orogens (Fig. 14). Karner and Watts (1983) demonstrated that this gravity pattern is consistent with a model in which the mountains are buttressed by a stiff elastic plate which has underthrust the mountains from the direction of the foreland in the process of continent-continent collision. The amplitude of the deflection of the elastic plate as revealed by the magnitude of the Bouguer gravity low requires loading by both the mountainous topography and by a buried high-density body in the hinterland, the mass presumably responsible for the Bouguer gravity high. This new model for the structure of mountain belts has thus established the validity of elastic flexure to describe the rheology of continental lithosphere and the existence of subsurface loads to maintain the deflection of foreland basins despite erosion of topographic loads.

Despite the importance we now place on buried loads in describing the conditions of mechanical equilibrium at mountain belts, the nature of these buried loads remains obscure. Buried continental margins (Stockmal et al., 1986), subsurface loads from cold slabs (Sheffels and McNutt, 1986), dense obducted blocks (Karner and Watts, 1983), and normal stress applied from flow in the mantle (Lyon-Caen and Molnar, 1983) have all been used to supply forces and bending moments to the lithosphere beneath mountain belts. Do all these factors contribute to the compensation of orogenic belts at different times in their geologic evolution, or do the peculiarities of plate collision lead to fundamentally different loading conditions at different locations? We require additional studies of thrust belts at all stages of evolution with a variety of pre-collision tectonic settings (e.g., presence or absence of back-arc basins, different ages of colliding plates, etc.). A wide range exists on Earth; unfortunately, we lack observations of the gravity field over many, particularly the very youngest collision zones, due to difficult terrain and/or political problems with access.

Gravity coverage over continental orogens at wavelengths of 50-100 km (i.e., less than the flexural wavelength of the lithosphere) with an accuracy of 2 to 5 mgal would allow us to test models of lithospheric rheology, mechanisms of plate loading, causes of vertical tectonics in orogens and the details of continental suturing. For example, McNutt et al. (1988) find that billion-year-old continental lithosphere where it underthrusts folded mountain belts in Europe, Asia, and the Americas can be either very strong (elastic plate thickness in excess of 100 km) or extremely weak (elastic plate thickness about 30 km) depending upon the steepness of decent of the underthrust plate. Steeply plunging continental plates are characterized by low values of elastic plate thickness, regardless of the age of the plate. They explain their result as the effect of massive brittle and ductile failure of the plates at high strains. In order to use this observation to better quantify the yield strength of continental lithosphere we require more precision and better resolution in gravity data over mountain belts.

Deep Structure of the Continental Lithosphere. The thickness of oceanic plates has been determined based on the cooling half space model. Generally speaking it varies from almost zero thickness at the mid-ocean ridges to about 100 km thickness beneath old oceanic basins. However, the thickness of continental lithosphere has not yet been agreed upon. The results of seismic studies on the thickness of continental lithosphere are controversial, with maximum thicknesses ranging from no more than 200 km (Anderson, 1979) to over 400 km (Jordan, 1979a). The flexural observations from foreland basins adjacent to mountain ranges point to an asymptotic thermal plate thickness for continental lithosphere of the order of 250 km or greater, at least twice that for oceanic lithosphere. The question remains as to how such a cold continental keel can be maintained against convective destabilization. One viable hypothesis for the deep structure of continents proposes a chemically-induced density reduction in the lower continental lithosphere that offsets the density increase from cooling (Jordan, 1979b). Regardless whether the bottom of the lithosphere is defined as a thermal boundary or a chemical boundary, density anomalies will exist at a depth between 100-400 km across the boundary of a "continental root." This horizontal density variation will give a surface gravity anomaly of about 1-5 mgal. The anticipated wavelength of the gravity anomaly will contain all wavelengths up to the length scale of the continent. Thus, an improved constraint on the thickness of the continental lithosphere can be derived based on improved surface gravity data and proper modeling of mantle thermal structure adjacent to the roots of continental lithospheres. Earth scientists do not have at present a precise global gravity field to search for the gravity signal from deep continental thermal structure.

3.4 MANTLE CONVECTION

The problem of mantle convection is fundamental to understanding the evolution of the Earth. The outgassing of the oceans and atmosphere, the differentiation of the crust, volcanism, and all tectonics - continental as well as oceanic - are ultimately dependent on energy sources within the mantle and core, and upon the transport of this energy and material by flow driven by thermal or compositional buoyancy. The oceanic crust and lithosphere are part of this convecting system: they make up its uppermost, cold thermal boundary layer. Their motion is associated with a flow pattern coming to the surface at the mid-ocean ridges or other rifting areas and returning to the interior at subduction zones. Most oceanic crust, as well as its associated lithosphere, is recycled to the interior. The continental lithosphere, which consists of the continental crust and sizeable pieces of sub-continental mantle, rides on top of the convective system. The velocities of the system are of the order of centimeters per year; the heat transport is an average of 0.08 W m^{-2}. These values, together with the thermal and rheological properties of rocks, indicate that the system must be more complicated than the smoothest flow necessary for the observed plate motions. Phenomena such as changes in the plate tectonic pattern on time scales of tens of millions of years, long-term episodicity of volcanism in tectonically complex areas such as western North America, exceptionally high heat flow on the continental side of subduction zones, and higher than predicted heat flow and topography in some parts of ocean basins, all suggest that there are secondary scales of mantle flow not directly connected to the precisely measured plate tectonic pattern. Observations which see through the lithosphere and into the mantle are needed. The gravitational field provides one such observable.

Seismic Tomography and Gravity Anomalies. Imaging of Earth's structure using seismic tomography is a discipline of geophysics currently undergoing explosive development. Spatial variations in seismic velocities have been obtained on a wide range of scales, from local (Sanders et al., 1988) to regional (Hearn and Clayton, 1986a,b; Humphreys et al., 1984) to continental (Grand and Helmberger, 1984) to plate (Grand, 1986) to global (Dziewonski, 1984; Woodhouse and Dziewonski, 1984; Nataf et al., 1984, 1986; Tanimoto, 1986; Hager and Clayton, 1989). One ultimate aim of geodynamics is to understand the physical processes causing these variations in seismic velocities. Are they the result of thermal differences associated with mantle convection, chemical differences caused by differentiation of the mantle and lithosphere, or some other process, such as anisotropy resulting from preferred alignment of crystals? Seismology by itself cannot discriminate among these hypotheses, whereas gravity can sense the magnitude and sign of the relationship between seismic velocity and density associated with each of these explanations.

For any assumed relationship between density and seismic velocity, the gravity field associated with tomographically-deduced velocity variations can be calculated. The gravity signal also depends on the mantle viscosity due to the dynamic boundary deformations produced by the density anomalies. Model solutions for a mantle with viscosity increasing significantly with depth can account for ninety percent of the gravity variations in wavelengths longer than 4000 km (Hager and Clayton, 1989).

As seismic tomography, now in its infancy, attains much greater detail in mapping the velocity anomalies of the mantle, there will develop a stronger need for an improved gravity field at shorter wavelengths to match the predictions. The appropriate accuracy needed for these comparisons depends upon the scale of the seismic model. Accuracies from 1 mgal at 100 km wavelengths to 0.01 mgal at 5000 km wavelengths are required in order to discriminate among models at a level of about 10 percent of the expected gravitational signal.

Vertical Scale of Mantle Convection. The vertical scale of the mechanical structure of mantle flow is a subject of current debate. If multi-layer convection exists, it is hypothesized that the 670 km seismic discontinuity will be the boundary between separate flow systems in the upper and lower mantle. Due to the upwelling and down-going currents associated with mantle flows, undulations or vertical displacements at this boundary will occur with a wide range of wavelengths (Christensen and Yuen, 1984). Due to the attenuating effects of distance, the ones with most signal will be in the range of several thousand km. The gravitational characteristics of a chemically stratified mantle are quite different than those of a mantle with uniform composition. Consequently, high resolution gravity data can be used to better delineate the competing hypotheses.

The depth of slab penetration can be studied particularly effectively with better gravity data. Based on seismic investigations of travel time (Creager and Jordan, 1984, 1986), subducting slabs are thought to be able to penetrate into the lower mantle. Since seismic anomalies are directly related to density variations, the existence of deep penetrating slabs can be examined with the gravity data derived from the proposed global measurements. The gravitational signature of a deep subducted slab is particularly sensitive to the presence or absence of a chemical discontinuity at 670 km depth. If a discontinuity is present, dynamic compensation of the slab will occur at that depth, resulting in smaller gravity anomalies for a given density contrast. The current long wavelength gravity field (> 4000 km) can be satisfied by either a model with normal slab density and mantle-wide flow or a model with high slab densities (caused by phase changes) and a chemical barrier to flow. The models can be discriminated using shorter wavelengths. Since subducting slabs lie beneath island arcs and typically span the ocean-continent transition, altimetric geoids are not sufficient to study this problem and gravity fields such as those obtained by a gravity-measuring satellite are required. The expected signal strength will be above 0.1 mgal with a typical wavelength of about 1000 km.

Small-scale Convection. The existence of small-scale convection (i.e., at horizontal scales much less than plate dimensions) beneath lithospheric plates has been predicted from the observed flattening of the depth-age and geoid-age relations for oceanic lithosphere, from laboratory experiments, and from theoretical convection calculations. For example, the fact that the slope of the geoid begins to flatten over older oceanic lithosphere (Parsons and Richter, 1981) has been one of the key observations in support of the thermal plate model (McKenzie, 1967) which does not allow conductive cooling to extend below about 125-km depth. Presumably heat is transported to the base of the lithosphere by some sort of small-scale convection. Theoretical calculations suggest that convective rolls can exist only if the upper mantle viscosity is extremely low (Yuen et al., 1981) and that they may have a typical horizontal wavelength of about 150 km with an amplitude of 5 mgal (Buck, 1985). Plume-style convection (Fleitout and Yuen, 1984) is an alternative mechanism.

One of the major discoveries of the Seasat altimeter mission is gravity undulations with the predicted wavelength, amplitude, and orientation in the Central Pacific (Haxby and Weissel, 1986). However, in the Indian Ocean, crossgrain features with the same wavelength but even larger amplitude (20-60 mgal) are thought to be due to buckling of the lithosphere in response to N-S compression of the Indian plate as it collides with Asia (Weissel et al., 1980; McAdoo and

Sandwell, 1985). Even lithospheric boudinage, a pinch and swell instability resulting from plate-wide tensile stresses (Froidevaux, 1986; Zuber et al., 1986), might be capable of producing some of the crossgrain lineations. Therefore, before crossgrain lineations are used to constrain models of small-scale convection, we must map them with greater accuracy and in more detail to determine their origin. Do these features contain information concerning asthenospheric viscosity, or are they indicative of lithospheric stress and rheology, or perhaps crustal structure? An improved gravity data set not only will be able to verify the existence or absence of such structures, it will also be able to delineate where such rolls begin and where they terminate as a function of plate age and spreading velocity. An accuracy of 1 mgal at 100-km resolution would allow these features to be traced to 20 percent of their amplitude.

Beneath the continents there is direct observational evidence from seismic tomography that small scale convection also occurs. For example, below the Transverse Ranges in Southern California a curtain of high-velocity material extending down to a depth of 250 km is evidence of convective downwelling of the cold thermal boundary layer at the base of the lithosphere (Humphreys et al., 1984) . This feature may explain the dynamics of the Big Bend of the San Andreas Fault. The gravity signature of this feature is calculated to be up to 15 mgal in amplitude with a wavelength of about 150 km. A model of the local gravity field (Sheffels and McNutt, 1986) indicates that this feature is, to a large extent, compensated from above by flexure of the overlying plates. In this particular instance, both gravity observations and velocity anomaly maps from seismic tomography were necessary in order to understand the interaction of the lithosphere with the asthenosphere. Once we have this system response well calibrated, it may be possible to identify regions of downwelling beneath continents from gravity and the surface geology alone.

Hot Spots and Thermal Swells. The observation of a regular progression in the age of the volcanism along several island chains led Morgan (1972) to propose the existence of "hot spots", long-lived melting anomalies in the mantle more or less fixed with respect to each other which produce a series of surface volcanoes as the lithosphere drifts over the mantle. The cause of hot spot volcanism is believed to be thermally and/or chemically buoyant plumes arising from the core-mantle boundary or perhaps an upper mantle-lower mantle interface (Morgan, 1972), although the number and present location of these plumes, their relative motion, their mode of interaction with the lithosphere, and their geochemical signature are still debated. Since hot spots constitute the only unambiguous sign of actively rising convection currents in the mantle, the study of these features is important for understanding the dynamics of the mantle.

The geoid and topography swells of intermediate wavelength (1000-2000 km) are now well mapped over the oceans by a combination of satellite altimetry and conventional sea surface oceanographic surveys. In the majority of cases, the swells are linked with active or recently active hot spot volcanism (Crough, 1978, 1983; Detrick et al., 1981,1986; Haxby and Turcotte, 1978; Fischer et al., 1986; McNutt and Shure, 1986; Sandwell and Renkin, 1988; Cazenave et al., 1988). What is unclear is whether these bathymetric swells represent the simple passive uplift of the lithosphere by dynamically maintained high temperatures in the underlying low-viscosity asthenosphere (Robinson and Parsons, 1988; Ceuleneer et al., 1988), thermal expansion and thinning of the lithosphere itself due to some more complicated plume-lithosphere interaction (Detrick and Crough, 1978; Menard and McNutt, 1982; Yuen and Fleitout, 1985; McNutt, 1987), or some combination of both effects (McNutt, 1988). If a thin, shallow, low-viscosity zone lies beneath the oceanic lithosphere, the geoid/topography ratios observed over swells can be adequately explained by thermally expanded mantle material either entirely within the lithosphere or entirely within the asthenosphere, and therefore such ratios are not diagnostic of the compensation mechanism. However, more precise gravity data would help constrain the depth distribution of the heat budget in swell in another way. Gravity anomalies with an accuracy of a few mgals and wavelength of 30-50 km have been the principal observational constraint used to measure the elastic plate thickness of the oceanic lithosphere supporting the individual hot spot volcanoes capping these swells (Watts, 1978). Because the base of the elastic plate corresponds to the isotherm marking the elastic/ductile transition for olivine at geologic strain rates, by comparing the elastic plate thickness on swells to that of similar-age lithosphere which has not experienced any thermal perturbation associated with its flexure (i.e., the outer rise seaward of subduction zones), we can calculate the amount of anomalous heat from the hot spot at mid-lithospheric depths (30-50 km) (McNutt, 1984). Such studies are presently impaired by the lack of adequate gravity and bathymetry data, particularly for hot spots south of the Equator.

Over the continents where the present gravity data are even more scarce, systematic studies of hot spot swells are completely lacking. Improved gravity data over continents are needed in order to obtain even a first order understanding of the structure of continental swells and how they might differ from the oceanic analogues.

A gravity data set with an accuracy of a few mgals at a spatial resolution of the order of 50-100 km would provide new constraints for known hot spot tracks and lead to the discovery of new hot spots. The small-amplitude, elongated features in the marine geoid attributed to small-scale convection or intraplate compressional deformation may in fact be tracks left by fixed plumes beneath moving plates. Accurate and high-resolution gravity data may therefore lead to the characterization of a new class of more subtle, hot-spot related activity.

Vertical Motion from Glacial Rebound (cm a^{-1})

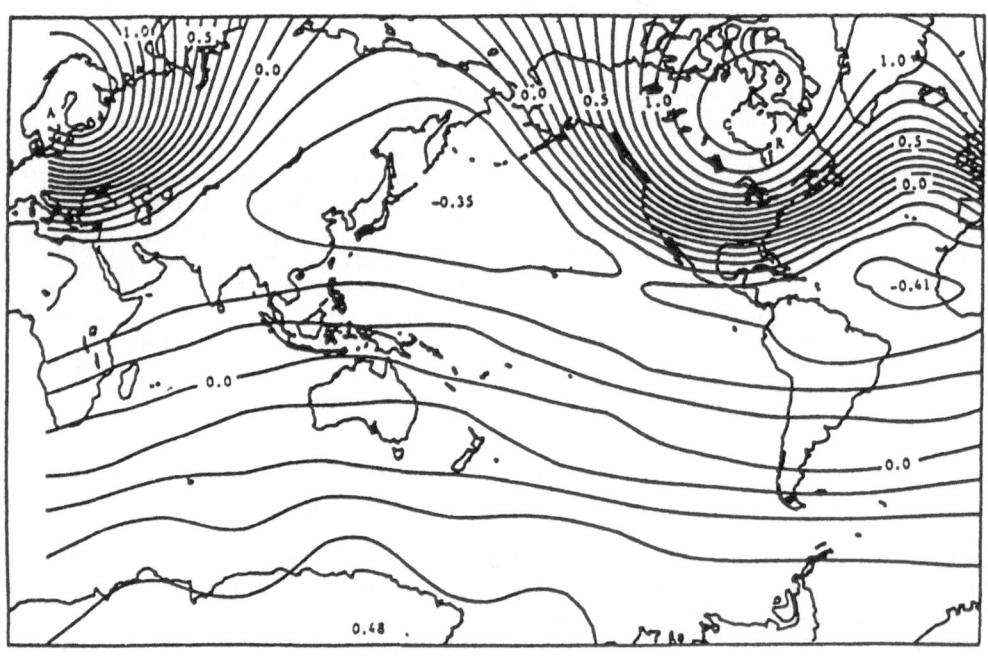

Fig. 15. Present-day rate of change in the radial position of the surface of the solid Earth calculated from simplified model of postglacial rebound. Units are cm/yr. From Wagner and McAdoo (1986).

3.5 TEMPORAL VARIATIONS IN GRAVITY

A less obvious but equally important application for high-precision satellite gradiometry and similar detailed, space-based mapping missions is the detection of time variations in the field. In most geophysical and geodetic studies, gravity is assumed to be static. Such an assumption is usually justifiable because expected changes in the field over typical observing lifetimes (~10^2 years) are relatively small. However, the precision of proposed satellite gravity measurements is such that heretofore undetected changes in gravity even over 6 months will be sensed. Furthermore, subsequent missions of similar design could provide additional time samplings of the field, thereby better resolving ongoing variations. Effects causing changes in gravity include ocean and solid Earth tides, postglacial rebound, secular melting of the ice caps, seasonal variations in ground water (e.g., snowpack) and great earthquakes. The magnitudes of these effects have been estimated in a recent study by Wagner and McAdoo (1986).

The characteristic periods for these effects range from minutes to thousands of years but to a large extent are global in nature, with the exception, perhaps, of the effects of great earthquakes. The cause of these changes in gravity are the orbital motion of the Sun and Moon, and changes in the physical state of the solid Earth, oceans, and atmosphere. We discuss below some effects that should give rise to temporal variations in gravity detectable with an extremely sensitive gravity gradiometer, for example. It is difficult to anticipate all such effects, and thus it is particularly in the frontier area of temporal variations in gravity that completely unexpected discoveries could result.

Postglacial Rebound. The most recent episode of Pleistocene deglaciation began roughly 18,000 years ago. The massive continental sheets of ice were largely melted 5000 years ago (Wu and Peltier, 1984). Although the melting has nearly ceased, the solid Earth continues to rebound in response to this deglaciation. This rebound continues today because of the high average viscosity (order 10^{22} Pa s) of the Earth's mantle. Concomitant changes in the gravity field also continue today.

Figure 15 shows the present day rebound rate. The largest rebound rates (~0.01 m/yr) occur in areas where ice caps were thickest. When the rebound is decomposed in spherical harmonics, the largest components occur at very low degree and order (≤ 6). Using the expression in Section 3.1 relating displacement of the solid Earth to gravity anomalies the maximum gravity change per year is 10^{-3} mgal. Such small variations in gravity associated with the second spherical harmonic have been measured as perturbations in the orbit of LAGEOS (Yoder et al., 1983) and been used to estimate the viscosity of the lower mantle. Higher harmonics of glacial rebound have not been observed from satellites although in several areas they are well constrained by shoreline emergence data (Wu and Peltier, 1983). Rebound of these higher harmonics contains information on radial variations in the viscosity of the upper mantle and asthenosphere, important for dynamic models of the Earth.

To understand the full extent and mechanism of the rebound, we require global models of the gravity field at wavelengths of 3000 km or longer and an accuracy of 10^{-3} mgal repeatedly measured at various epochs over a period of about a decade. Although changes in all the medium to long wavelength harmonics occur, those most easily detectable are restricted to the very low degree terms. There is reasonable expectation that variations up to degree and order four will be observable from low Earth orbit. Detailed knowledge of rebound will help solve the problem of the lateral and vertical variations of mantle viscosity.

Glacial Melting and Atmospheric Warming. One of the most important variations in gravity is due to the slow secular melting of glaciers (Meier, 1984). Climate studies show an accelerated warming of the atmosphere over the last century that is believed to be caused by increases in atmospheric carbon dioxide associated with burning of fossil fuels and deforestation (NASA

Advisory Council, Earth Systems Science Committee, 1987). Global warming may induce glacial melting resulting in sea level increases and coastal flooding.

Accurate satellite gradiometer missions could monitor the small changes in the global gravity field caused by the redistribution of mass from the glaciers to the oceans. The current estimate of yearly sea level rise, caused by melting of small glaciers, is 5×10^{-4} m. However, this number is very uncertain because only a few glaciers have been monitored over the last century (Meier, 1984). The yearly change in gravity over the oceans due to glacial melting is only 2×10^{-5} mgal. In glaciated areas, the yearly gravity change may be somewhat higher because the area is smaller.

Although these predicted temporal variations in gravity are extremely small, they are global and could perhaps be measured by a satellite gravity gradiometer which can render the highest resolution at long wavelengths by virtue of its long integration time. One method of increasing the signal associated with secular glacial melting is to measure over a longer period of time. This could be done with two or more missions spaced at five-year intervals.

Seasonal Variations in Groundwater and Excitation of the Chandler Wobble. The Chandler wobble is an apparent 427-day precession of the Earth's instantaneous pole of rotation about its axis of greatest moment of inertia. The Earth's anelasticity damps the Chandler wobble, such that in the absence of some excitation, the rotation pole and axis of figure would eventually coincide. The rate at which the wobble is damped depends upon the largely unknown viscosity structure of the mantle and core. Its Q (i.e., π times damping time divided by wobble period) could lie anywhere between 70 and 600. If the wobble is re-excited only occasionally, such as, for example, by mass movements associated with very great earthquakes, then its Q must be very large. If the forcing is more frequent, then Q must be small. A knowledge of the mechanism which excites the Chandler wobble would constrain its Q, which in turn would reveal the viscoelastic structure of the lower mantle and core.

Winter in the Northern Hemisphere is accompanied by a dramatic increase in continental water storage associated with the development of ice deposits and snowpacks (Hinnov and Wilson, 1987). Hinnov and Wilson (1987) and Wahr (1982) have suggested that this seasonal fluctuation in groundwater storage excites the Chandler wobble, implying a small Q. To test whether groundwater fluctuations excite the Chandler wobble and to determine its Q, better observations are needed. Due to a lack of hydrologic and meteorologic data, especially in Asia, the spatial and temporal variations in groundwater are not accurately determined. Typical annual variations in surface groundwater height over the continents are 0.1 m. During the seasons this water is transferred from the oceans to the continents and back to the oceans. Gravity variations associated with this mass transfer will be about 4×10^{-3} mgal. To determine the transfer function between the excitation and the observed wobble a gravity mission should last several years. A shorter mission (about 6 months) could be used to confirm that groundwater plays a major role in exciting the Chandler wobble.

Volcanoes and Earthquakes. Gravity data play an important role in understanding the physical processes associated with volcanic- and earthquake-related phenomena. For example, Rundle (1978) calculated the surface gravity changes arising from the expansion of a volumetric source in a half-space (a simple magma chamber), and found it to be -0.31 mgal/m. He also calculated the surface gravity changes to be expected from infinitely long thrust faults in an elastic medium, and found this to be equal to the Bouguer anomaly, -0.19 mgal/m. Other investigators who calculated the gravity changes from similar elastic half-space models using different methods included Walsh and Rice (1979) and Savage (1984). Observatories at Hawaii (Dzurisin et al., 1980) and Mount St. Helens (Jachens et al., 1981) have detected these gravitational changes caused by magmatic inflation events, which signal the possibility of impending eruptions. More recently, substantial surface gravity changes have been detected in Long Valley, California (Fig. 16), and these have been interpreted as a result of a magmatic inflation process there (Rundle and Whitcomb, 1986). Generally speaking, these gravity changes occur over regions of a few tens of kilometers in extent and have amplitudes at most of perhaps 0.5-1.0 mgal. These changes are therefore near the level of resolution of current gravity observations by space techniques. Similar changes in amplitude of gravity signals, but with substantially longer spatial wavelengths, are seen from major thrust faulting events. Upward continuation of the gravity signal from an earthquake to spacecraft altitude has been carried out by Wagner and McAdoo (1986). For the 1964 Alaska earthquake they show that a spacecraft passing over the region at 160-km altitude would have its velocity changed by approximately 15×10^{-6} m s^{-2} in a period of about 50 seconds as a result of the co-seismic vertical motion, indicating a change of about 0.1 to 1 mgal over an area of about 400 km^2. Changes of this magnitude, if detectable from near Earth orbit, might lead to the study of pre- and post-seismic behavior on a global scale.

3.6 SUMMARY AND RECOMMENDATIONS

In Fig. 17 we summarize in a very schematic way the resolution and accuracy we require in measurements of the gravity field in order to address the problems in solid Earth geophysics relating to long-term phenomena. The extremely high resolution requirements needed to test the next generation of models relating to the thermal and mechanical structure of oceanic lithosphere are so stringent that they can only be met with a dense global altimeter survey. The fact that gravity measurements must be acquired over continents to address the issues of mantle convection and continental tectonics dictates a non-altimetric form of potential field measurement, such as would be supplied by a gravity gradiometer in low orbit.

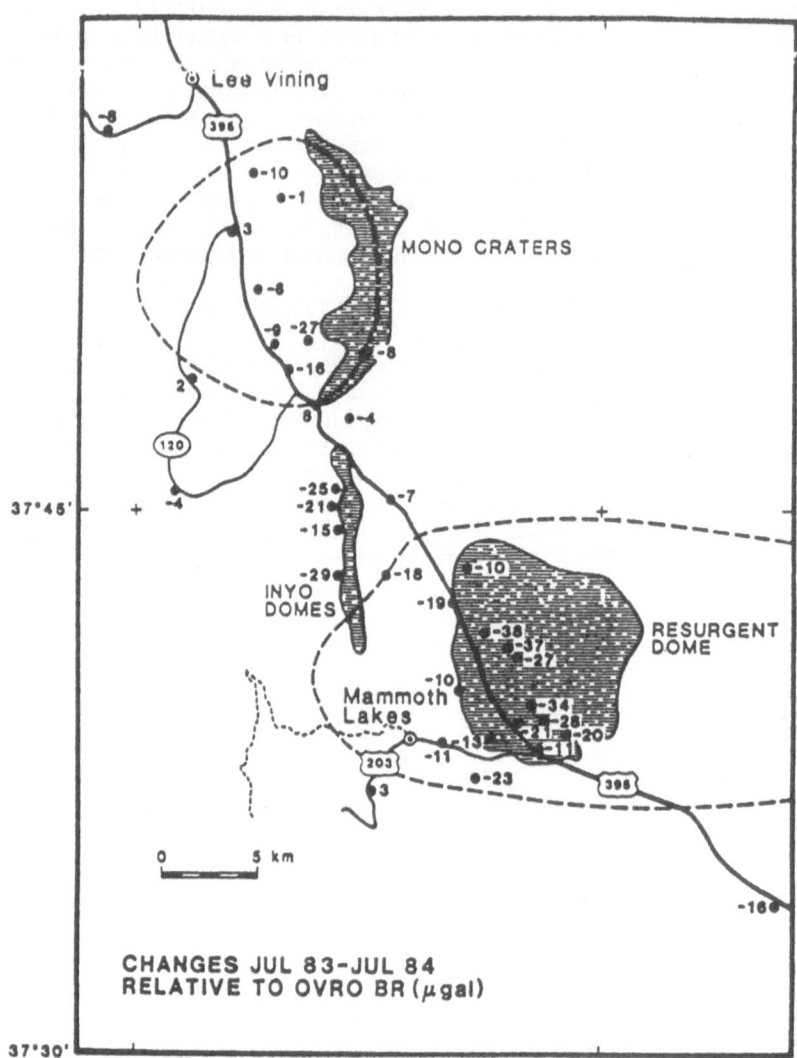

Fig. 16. Plots of the means of the gravity differences (μgals), referred to OVROBR (station 35), for the interval 1983-1984. From Rundle and Whitcomb (1986).

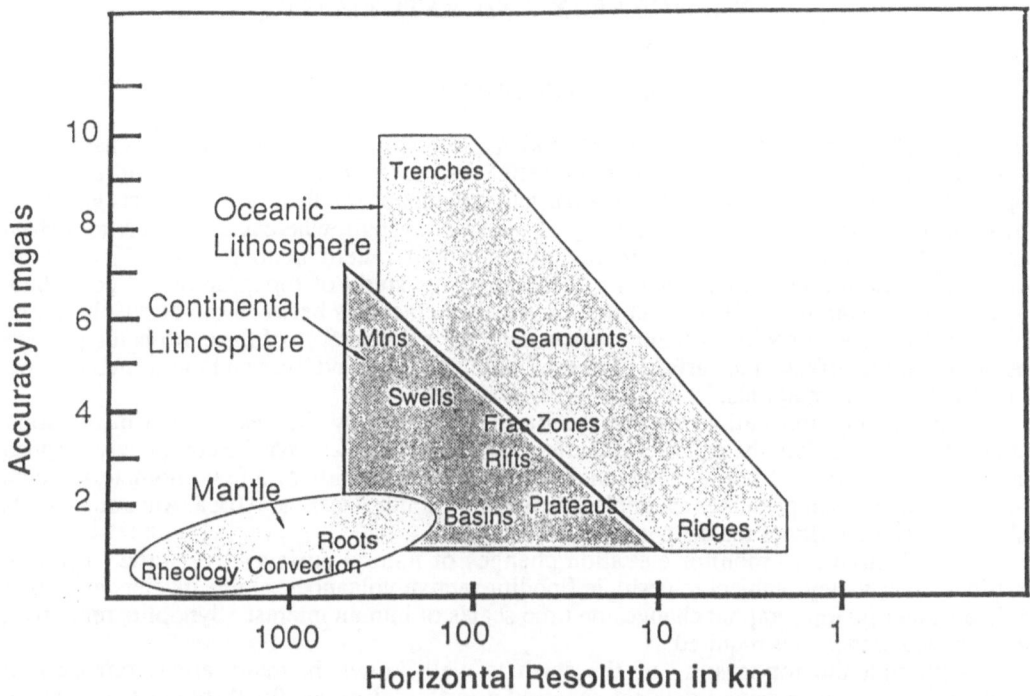

Fig. 17. Summary of requirements for gravity field accuracy and resolution in order to address various outstanding problems in solid Earth geophysics.

4. TOPOGRAPHY AND BATHYMETRY

4.1 INTRODUCTION

The solid surfaces of the terrestrial planets and satellites in the solar system are generally far from smooth. Variations in physiographic relief reflect thermal, mechanical and chemical properties plus the dynamic processes of a particular planet or satellite. Small objects generally provide information about the early history of the solar system, whereas the largest terrestrial planets (Venus and Earth) are more revealing about recent history. On objects with atmospheres (Earth, Venus and Mars) a comprehensive study of topography can contribute toward an understanding of chemical and physical interactions between the atmosphere and surface rocks. From a socio-economic perspective, the solid surface of the earth is the planet's most important interface, the surface closest to which all mankind lives and the source of most of our food and raw materials.

Topography and bathymetry of the Earth are presently the best known in the solar system. However, before the end of the century the surface of Mars may be completely mapped to a vertical resolution of a few meters and a horizontal resolution of a few kilometers. Such resolution exceeds that presently available over most of South America, Africa, Antarctica, parts of Asia and the majority of the deep oceans.

It is desirable to monitor elevation changes of natural surfaces over time. Eroding coastlines, river valleys subject to periodic flooding, active volcanoes, alpine glaciers and polar ice sheets undergo topographic changes on time scales of human interest. Synoptic, repetitive, high accuracy coverage is required.

Although the topography of the earth is well-known in many areas, our current knowledge represents a mixed data set acquired at different times, for different purposes, in different ways, presented in different formats, at different scales, on different projections, and tied to different reference levels. Using these data is complex and sometimes misleading; acquisition of a coherent topographic data set for the entire earth's surface in a readily accessible and usable form should constitute an immediate geodetic initiative. For the continents, such data are best obtained from space. The data should be in a form from which maps, images, perspective and stereo views can be readily produced, and they should be co-registerable with other data sets.

Topographic data have numerous applications in the solid earth sciences, including studies of active tectonics and interpretation of land, aircraft and space-based gravity data. Topographic data are also useful in hydrology, vegetation and ecosystem studies and are critical to monitoring ice caps and alpine glaciers, both potential indicators of global warming. This section will describe these applications, and conclude with a set of recommendations. Our key recommendation is that global, space-based elevation data in a uniform coordinate system over land and ice be obtained in the next decade with a resolution of the order of 100 m horizontal and 1 m vertical. Techniques for acquiring these data are described in the chapter on Instrumentation (Chapter 5).

4.2 LAND APPLICATIONS

Gravity Applications Topography and gravity data together are used to study internal density variations and compensation mechanisms in the earth. Topography is required to make terrain and Bouguer corrections to gravity measurements and can be combined directly with gravity for studies of lithospheric structure. To reduce topography-related errors in such studies, it is important that grid spacing (horizontal resolution) of the elevation measurements be finer than the gravity measurements, and the vertical accuracy should be commensurate with gravity accuracy ($\approx$ 0.3 mgal per meter of topography).

For land gravity surveys in high relief terrain, neighboring topography contributes an important signal, and detailed elevation data in the region of interest are required on grid spacings that may range from 1000 meters to as little as 100 meters. Ground-based gravity surveys can generally be obtained even in remote regions, but acquiring complementary elevation data, involving factors of 10 to 100 more data points, is often not feasible with ground-based techniques. Elevation data with horizontal and vertical resolutions of roughly 500 m and 1 m would probably satisfy most requirements for ground-based gravity studies, except in the highest relief terrains. The importance of such data for high precision ($\pm$1 cm) geoid undulation computations has recently been shown by Denker and Wenzel (1987).

For interpretations of space-based gravity data, spatial resolution requirements for topographic data are less stringent because the horizontal resolution of the gravity data decreases at orbital altitudes. One possible mission for the late 1990's is the Superconducting Gravity Gradiometer (SGG) which will deliver approximately 1 milligal accuracy and 50 km horizontal resolution. Topographic data adequate to support interpretation of such gravity data exist now or will soon exist for roughly two-thirds of the earth's surface. However, the high relief mountain belts (e.g., Tibet-Himalaya and the Andes) are important exceptions. Since these are precisely the areas of interest to many geologists and geophysicists, it is important that the necessary elevation data in these remote regions be acquired.

It is not sufficient to acquire elevation data on a 50 km grid to meet the SGG requirement. Rather, it is necessary to have a good estimate of the mean elevation within a 10 to 20 km resolution cell with a vertical accuracy of 5-10 meters. In areas where the topographic variance is high, measurements on a scale much finer than 10 km may be required to obtain a good estimate of the mean elevation within the 10 km resolution cell. For high relief mountain belts, measurements with a horizontal resolution of roughly 1-2 km would minimize the topographic error contribution for space-derived gravity data and ensure that elevation is not the limiting error source for interpretation of these data.

Tectonic Applications   Topographic data are fundamental to many geologic studies. In areas where erosion rates are low, topography primarily reflects tectonic and volcanic processes. Topographic data adequate to support most such studies are available in low relief, accessible or developed regions of the world. However, high relief mountain belts, important areas for studying continental evolution, are a major exception. Many tectonic problems in convergence zones such as the Himalayas and the Andes can clearly benefit from improved topographic data.

Isacks (1988) used regional topography to construct a tectonic model of the central Andes, linking Cenozoic convergence of the Nazca plate with the shape and amount of uplift of the Andean plateau. Because erosion rates are very low for much of this region and uplift is comparatively young (less than 10 Ma), topography can be directly interpreted in terms of tectonic processes. Some examples include:

> 1)   A regional unconformity is defined topographically, with Early Tertiary folded strata overlain by flat-lying Late Tertiary strata;
> 2)   A regional drainage divide accurately defines a major thrust front;
> 3)   Offsets of Holocene features along predominantly strike-slip faults can be used to constrain deformation rates.

The data used by Isacks (1988) was based on 1:1 M scale charts, digitized at 5 km intervals and having roughly 30 m vertical accuracy. This is similar to the best available global digital elevation data sets for the continents, although much more detailed data are available

regionally. It is technically feasible to obtain topographic data from space with at least an order
of magnitude improvement in both horizontal and vertical resolution compared to the Andean
data described above. Such data would have a major impact on our understanding of Neogene
tectonics in this and similar regions.

Erosional Effects   Erosion is a process that we know very little about, despite its
importance in such diverse subjects as geomorphology, isostatic rebound of mountains and
chemical cycling between oceans and continents. Quantitative models for erosional effects in
specific areas would allow estimates of tectonically important parameters from degraded
geomorphic features, for example dating of pre-historic earthquakes from fault scarp
morphology (e.g., Nash 1986). A better understanding of erosional processes also has
practical applications in areas such as river and near-shore navigation, waste disposal and
catastrophic events such as landslides.
        Erosion is likely to be complex spatially and temporally and involves both catastrophic
and steady state processes. A five-year mission would provide the opportunity to monitor
major catastrophic events, e.g., landslides and the effects of major floods.

Time-Dependent Phenomena (Dynamic Geomorphology)   Dynamic geomorphic
changes (e.g., those associated with catastrophic events such as landslides, floods, barrier
island reorganization during storms, earthquakes, debris flows and volcanism) have obvious
social implications. Their study also requires topographic data with high accuracy, high
resolution and in many cases, repeated coverage over an extended period. Comparable data
could be put together over a long time from piecemeal studies with aircraft, surface measurements and some space techniques (e.g., optical stereo), but the coherence, rapid coverage and
overall accuracy of a data set from direct measurement by a dedicated space-borne altimeter
could not be duplicated. Data acquired piecemeal also must ultimately be referred to a common
reference frame. The propagation of errors in such procedures greatly limits the value of these
partial data sets. Systematic acquisition of elevation data from an altimeter is highly desirable
for long-term monitoring of transient phenomena.

Hydrology and Ecosystem Studies   Elevation and related parameters such as slope and
aspect strongly influence surface hydrology and ecosystems. Topography influences intercepted radiation, precipitation and runoff, evaporation, snow ablation, soil moisture and
vegetation type and health. Energy and mass flux conditions determine dominant types of
vegetation and their succession in time on land surfaces, and these fluxes are affected by topography through combined influences of elevation, slope and roughness. Calculation of accurate
short (UV) and long (IR) wavelength energy exchanges in mountainous areas requires accurate
information on neighborhood topography, including distribution of local horizons for shading
and re-radiation estimates. Topographic parameters determine the exposure of a landscape to
weather and sunlight at a given latitude and thus determine its microclimate. By feedback
mechanisms, vegetation itself influences energy and mass fluxes, affecting not only the local
environment but also regional and global climate. Topography is therefore a key element in the
study of complex ecosystems.
        Data from MODIS (Moderate Resolution Imaging Spectrometer) and HIRIS (High
Resolution Imaging Spectrometer), proposed facility instruments for the Earth Observing
System (EOS), will provide high resolution spectral data over land. These data are important
for a variety of terrestrial studies including vegetation and lithology mapping, deforestation
studies, desertification, snow cover and intercepted radiation. However, accurate albedo
estimates depend strongly on slope and are important for interpreting spectral data properly.
Narrow-beam altimeters can provide accurate slope estimates and thus enhance the scientific
return from these instruments, particularly in high-relief terrain where slope affects are
important.

Quantitative modeling in hydrological and ecological sciences requires digital elevation data. Such data are available for many regions of North America, Western Europe and Australia, enabling numerous local scale studies, but they are scarce for other continents at moderate to high resolution. As our understanding of complex ecosystems has evolved, the importance of large-scale regional and global modeling has been recognized. Major improvement in digital topographic coverage is possible. Repeated coverage from a dedicated land altimeter may enable quantitative global monitoring of some vegetation characteristics through waveform analysis of the return pulse. No equivalent system exists at present.

Even existing high resolution digital elevation data have deficiencies. These are evident when we examine topographic effects on solar radiation, a major driver of topographic differences in vegetation, soil moisture, snowmelt, etc., and whose calculation requires accurate slope information. Local gradients and angles to the local horizon can be calculated from a digital elevation grid. This information can be combined with solar geometry to calculate the local incidence angle of solar radiation on the slope. In mountainous areas, noise in existing digital elevation data usually derived by digitizing contour lines from topographic maps is magnified by the differencing operations required to calculate gradients. It is often difficult to use the resulting gradient data for quantitative irradiation calculations. High quality, synoptic elevation data would improve existing data bases in remote areas. Horizontal resolutions of 100 m or better are desirable for the solar radiation calculations described above.

Quantitative modeling of nutrient transport and cycling is an important area of current ecosystem research. Human activities have greatly affected the cycle of some biologically active substances especially in aquatic ecosystems, for example, through nutrient leaching from agricultural areas. An important first step in modeling these fluxes is drainage basin analysis (Marks et al., 1984; Band 1986) Digital topographic data enable automatic calculation of watershed structure and to some extent, runoff. Topography strongly influences surface and subsurface fluxes of water, but in order to model these accurately, derivative quantities based upon elevation data (e.g., local gradient, upslope area) are required. Error propagation associated with parameters based on the first or second derivative of elevation dictate that original elevation data must be of very high quality for use in the quantitative models described above. Existing digital data bases have sometimes proved to be inadequate for such calculations because of uncertainties in the vertical component in the original contour maps.

Altimeters can be used to assess several important vegetation parameters. In areas of low to moderate vegetation density, echo power detected prior to the surface return may allow estimates of the difference between ground elevation and canopy top, depending on details of the backscatter mechanism and the nature of the vegetation. Regional vegetation height estimates would be a very useful quantity for determining total carbon storage in vegetation canopies and would give estimates of deforestation and other changes if measurements are repeated at intervals. It may also be possible to estimate vegetation density and other related parameters by quantifying certain characteristics in the return pulse. Repeat coverage and detailed ground calibration studies would be required for such measurements. Digital elevation maps can also be used to define barriers and corridors affecting species dispersal and speciation events and to predict biomass, timber site quality, potential vegetation cover and fire behavior over large areas. The need for accurate slope and aspect data is especially great in semi-arid mid-latitude regions where minor slope changes greatly affect local water availability and soil moisture.

4.3 ICE APPLICATIONS

Polar ice sheets are extremely important elements of the Earth's climate system, yet they are still largely unexplored. Antarctica alone contains 3×10^7 km^3 of ice, roughly 85% of the Earth's fresh water. Paleoclimate data indicate ice volume has varied greatly causing sea level changes and altering global climate. More must be learned about the current behavior of and dynamics within ice sheets to predict their effects on future climate. A wealth of information about ice sheets can be obtained from topographic data, and satellite-based sensors provide the continental-scale view required to collect a complete set of topographic data in this harsh environment.

Large ice sheets are nearly parabolic in cross section, almost flat in the center and increasing in slope toward the margins. Ice flow velocities are small in the center and increase toward the edges. Numerous amphitheater-like embayments exist around the margin where fast-moving outlet glaciers draw down the ice surface. In Antarctica, these outlet glaciers usually discharge into large floating ice shelves which eventually calve off as icebergs. Gravity drives ice flows along the direction of maximum surface slopes, therefore surface topography indicates flow direction. This relationship can be used to delineate the individual drainage basins which feed outlet glaciers. Adjacent outlet glaciers compete for inland ice, and changes in the location of ice divides between basins can occur.

Ice flow rates depend on surface slope, ice thickness and ice temperature. Surface slopes vary spatially more than either ice thickness or temperature, therefore variations in ice velocities are strongly correlated with spatial variations in surface slopes. Generally, large-scale surface slopes are small (50% of Antarctica has slopes less than 0.03%), however, local slope variations are frequently an order of magnitude greater. These variations affect ice velocities and flow directions.

Ice surface roughness correlates with the underlying bedrock roughness. Classic glacier theory predicts the correlation of surface and bed roughness decreases with ice thickness and is most effective for spectral components which are between three and four times ice thickness. This relationship applies to ice flowing by internal deformation. Ice principally sliding over the bed tends to move faster and has a smoother surface. Surface topography can thus be used to predict the character of ice flow.

Two factors limit the accuracy of surface elevations produced from existing altimeter data: poor tracking of the return pulse and wide beam angle. These limitations result from altimeters optimized for the smooth ocean surface. Tracking limitations for ice-sheet topography are overcome by retracking individual return waveforms (Martin et al., 1983) and improving the ability of on-board tracking circuitry to maintain lock on the return pulse (e.g., ERS-1). The remaining limitation to high-accuracy topography over ice sheets is the wide beam used by all previous radar altimeters.

Ice sheet surfaces are sufficiently rough that a wide-angle radar pulse does not first intersect the surface at nadir; the surface point closest to the satellite is often many kilometers from the nadir point. As with the case for land, an increase in the distance between the beam center and this closest point causes a broadening of the return pulse because the nadir echo comes back a correspondingly longer time after the first return. Experience with wide-beam altimeters has also confirmed that multiple, near-simultaneous returns from different points within the beam-limited footprint are common over ice sheets, further complicating interpretation of the altimeter signal. Only a fraction of these ambiguities and slope-induced errors can be removed from even the most accurate range data (Brenner et al., 1983).

While mapping ice-shelf margins is a technique which benefits from wide-angle beam (Thomas et al., 1983), most other scientific tasks are hampered by such elevation errors. Errors in altimeter-derived elevation surfaces can be reduced only by reducing the footprint size. Wide-angle altimetry will provide an important bounded measure of ice sheet volume change rate, but details of where change is occurring and the ability to compare this data with surface-based topographic surveys will be lacking. Narrow-beam altimeters are needed to map ice sheet surfaces accurately enough to allow comparison and combination with surface-based data to assess and monitor details of ice sheet mass balances and dynamics.

Ice sheet mass balances remain the single most important measurement of climatic significance and should be applied to the polar ice caps as well as the world's alpine glaciers, mainly in the high- and mid-latitudes. Both wide beam (pulse-limited) and narrow-beam (beam-limited) radar altimeters can contribute to mass balance studies of the polar ice caps, but narrow-beam systems are required for study of alpine glaciers. High accuracy laser altimeters may be the ultimate tool for studies of ice dynamics, particularly if good spatial coverage is possible, but may not be feasible in the near future. Future pulse-limited radar altimeter measurements can be directly compared with earlier data sets, extending the time-series of elevation measurements and improving our understanding of ice sheet mass balances. Narrow-beam radar altimeters are feasible now and would not contain the range interpretation ambiguities inherent in wide-beam altimeters. Narrow-beam altimetry (either radar or laser) would enable direct study of ice undulations, features important in the fuller understanding of ice dynamics.

Improved determination of topographic details of ice surfaces will also aid in the extraction of information from other data sets. Many features on visible and radar imagery of ice sheets result from elevation variations, but determinations of absolute elevations from such images are difficult. Altimetry provides an independent measure of elevation along specific ground tracks which can be interpolated to the remaining image. This technique holds promise because snow is homogeneous enough for reflectance properties to be constant. Thus, intensity variations seen in Landsat visible bands are strongly dependent on surface angles. The final elevation map promises to be more accurate than one produced from either data set alone. Mapping accuracy will be only as good as the accuracy of groundtrack elevations. Wide-beam radar altimetry would be unsuitable for this task; such an effort is possible with narrow-beam altimetry.

The smaller footprint of narrow-beam altimeters would enable extension of altimeter studies to smaller ice caps throughout the world. These tend to be thinner and therefore rougher than the larger Antarctic and Greenland ice sheets, rendering them unsuitable targets for wide-beam altimeters. While many small ice caps and glaciers exist in polar regions, some are found at lower latitudes and contain ablation areas indicating quick response to climate change. Although their overall mass is a fraction of larger ice sheets, their quick response-time could make them important contributors to sea level change (Meier, 1984), as well as useful indicators of global temperature change. Narrow-beam altimeters would allow inclusion of these smaller ice masses in a more complete inventory and monitoring of global ice.

Finally, there will continue to be areas of the larger ice sheets suitably flat to be studied with the wide-beam radar altimeters. This mode is useful over very flat ice shelves where raw data can be analyzed to locate crevasse fields and grounding lines where the ice shelf and ice sheet meet.

4.4 OCEAN FLOOR APPLICATIONS

Seafloor erosion rates are low compared to the continents and because all seafloor is geologically young (<200 Ma), bathymetry is primarily an expression of tectonic processes. For example, ridges and fracture zones delineate the axis and direction of seafloor spreading. Shipboard surveys to date have measured the large-scale (10^3-10^4 km) variations in bathymetry, but fine scale bathymetry (10-100 km) is known only in isolated regions. Large-scale variations (from shallow ridge crests, 2-5 km, to deep ocean basins, 5-6 km) are due to cooling and contraction of oceanic lithosphere as it moves away from a spreading axis. Departures from the normal depth-age relationship reveal areas where lithosphere has been reheated and/or crustal thickness varies. While these basin-scale variations are well described, the fine-scale bathymetry is not.

Fine scale bathymetry of a spreading ridge axis is sensitive to spreading rate, magmatic budget and distance from a transform fault. Preliminary bathymetric surveys show a high contrast between slow spreading ridges with their deep axial valleys (1-2 km) and fast spreading ridges with their narrow axial highs ($\approx$200 m). Similarly, major differences in morphology exist along transform faults and fracture zones. A comprehensive bathymetric survey of the ridges, as proposed in the NSF RIDGE program, will provide important information on complex thermal, mechanical and chemical processes at spreading ridges.

Fine scale intraplate features (i.e. features that formed away from the spreading ridge) are also mostly uncharted and therefore poorly understood. For example, it is estimated that only 10% of undersea volcanos (seamounts) have been detected and formation histories of only a fraction of these are known. Flexural moats and outer rises surrounding each undersea volcano reveal the thickness of the elastic lithosphere at the time the volcano loaded the seafloor. Other less well-understood intraplate features also must be charted and modeled, including regions of intraplate deformation, deep linear troughs and en-echelon volcanic ridges.

Finally, detailed bathymetric surveys of subduction zones and back-arc basins would improve our understanding of convergent plate boundaries. Flexural models of outer rises seaward of trench axes indicate that bending stresses are large enough to deform the entire lithosphere permanently ($\approx$500 MPa). Bending and compression also influence the topography of the fore-arc and accretionary prism. The downgoing slab may induce a secondary mantle flow causing back-arc spreading and tectonism. A more complete understanding of seafloor subduction will come from improved bathymetric and gravity surveys.

Bathymetry cannot be directly measured from a satellite. However there are at least two ways that the space program can contribute to charting bathymetry. The first involves navigation and communication. GPS navigation of research vessels is now required for the high accuracy swath mapping techniques such as SeaBeam and SeaMarc. During the next 20 years, unmanned ocean surface or subsurface vehicles may need to communicate with a central facility using a satellite link. The second use of satellites for seafloor mapping is described in the previous section on gravity. The basic approach is to take advantage of the correlation between the geoid or gravity field and bathymetry at short to medium wavelengths (e.g., Dixon *et al.*, 1983; Freedman and Parsons, 1986). High accuracy and resolution geoid and gravity field data, derived from pulse-limited radar altimetry, may be used to interpolate between existing shipboard bathymetry profiles. This technique relies on the assumption that the relationship between gravity and bathymetry is uniform over small areas ($\approx$200 km by 200 km). Because of upward continuation of the gravity field from the seafloor to the sea surface, the best achievable horizontal resolution of predicted topography will be approximately twice the average water depth. The estimated vertical accuracy of this technique is $\approx$100 m.

4.5 SUMMARY OF REQUIREMENTS

Users of topographic data have extremely diverse requirements. Recent detailed studies by several groups have attempted to define these requirements (e.g. Report of the Topographic Science Working Group, NASA,1988; Report on World Digital Database for Environmental Science-An IGU/ICA Project, 1987). Even within a specific discipline, investigators may have different needs for horizontal and vertical resolution, repeat interval, and region of interest. Fortunately there is considerable overlap in these requirements. For most land and polar applications, a data set of $\approx$100 m horizontal and $\approx$1 m vertical resolution is adequate. Global coverage is required. Repeat intervals range from approximately 6 months (where seasonal changes may be noticeable in very high resolution data) to 5 or even 10 years for more slowly changing land topography at moderate resolution. There is a very strong need to obtain accurate and uniform topographic data over the entire earth. This is the most glaring fault in existing data, which may nevertheless be very precise in selected areas. After a full global survey has been repeated once or a few times, it may be desirable to tailor the program to specific regions, perhaps at higher resolution or more frequent intervals. Continued pulse-limited altimetry at existing vertical resolution but with finer track spacing, is required for some ice and ocean bathymetry applications.

4.6 RECOMMENDATIONS

1. We recommend that global, space-based elevation data in a uniform coordinate system over land and ice with resolution of order 100 m horizontal and 1 m vertical be obtained in the next decade. Possible techniques include narrow-beam radar altimeters, scanning laser systems, microwave interferometers and advanced optical stereo systems.

2. We recommend continued pulse-limited altimeter measurements over the oceans, ice and land, to achieve a track spacing of 5 -10 km.

3. We support continuing efforts to collect, validate and unify existing but disparate regional topographic data over the continents and collection of detailed bathymetric data from ships.

REFERENCES

Anderson, D.L., 1979, *J. Geophys. Res.*, **84**, 7555.

Ben-Avraham, Z., A. Nur, D. Jones, and A. Cox, 1981, *Science,* **213**, 47.

Band, L.E., 1986,*Water Resources Res.,* **22**, 15.

Bird, P. and R.W. Rosenstock, 1984, *Geol. Soc. Am. Bull.,* **95**, 946.

Brenner, A.C., R.A. Bindschadler, R.H. Thomas, and H.J. Zwally, 1983, *J. Geophys. Res.,* **88**, 1617.

Bryan, K., 1938, in *Rio Grande Joint Investigations in the Upper Rio Grande Basin in Colorado, New Mexico and Texas,* National Resources Commission, Regional Planning, Part 6, Washington, D.C., 196.

Buck, W.R., 1985, *Nature,* **313**, 775.

Canadian Hydrographic Service, 1984, *General bathymetric Chart of the Oceanis (GEBCO)*, 5th ed., Ottawa, Ontario, Canada.

Carlson, R., N. Christensen, and R. Moore, 1980, *Earth Planet. Sci. Lett.,* **51**, 171.

Carr, M. H., R. Greeley, K.R. Blasius, J.E. Guest and J.B. Murray, 1977, *J. Geophys. Res.,* **82**, 3985.

Castle, R.O., J.P. Church, and M.R. Elliott, 1976, *Science,* **192**, 251.

Cazenave, A., 1984, *Nature,* **310**, 401.

Cazenave, A., K. Dominh, M. Rabinowicz, and G. Ceuleneer, 1988, *J. Geophys. Res.,* **93**, 8064.

Ceuleneer, G., M. Rabinowicz, M. Monnereau, A. Cazenave, and C. Rosemberg, 1988, *Earth Planet. Sci. Lett.,* **89**, 84.

Chase, C.G., 1972, *Geophys. J. R. Astr. Soc.,* **29**, 117.

Chase, C.G., 1978, *Earth Planet. Sci. Lett.,* **37**, 353.

Christensen, U.R., and D.A. Yuen, 1984, *J. Geophys. Res.,* **89**, 4389.

Christodoulidis, D.C., et al., 1985, *J. Geophys. Res.,* **90**, 9249.

Cohen, S.C., 1984, *J. Geophys. Res.,* **89**, 4538.

Cordell, L., 1978, *Geol. Soc. Amer. Bull.,* **89**, 1073.

Craig, C.H. and D.P. McKenzie, 1986, *Earth Planet. Sci. Lett.,* **78**, 420.

Creager, K. C., and T. H. Jordan, 1984, *J. Geophys. Res.,* **89**, 3031.

Creager, K. C., and T. H. Jordan, 1986, *J. Geophys. Res.,* **91**, 3573.

Crough, S.T., 1978, *Geophys. J. Roy. Astron. Soc.,* **55**, 451.

Crough, S.T., 1983, *Annu. Rev. Earth Planet. Sci.,* **11**, 165.

DeMets, C., R.G. Gordon, S. Stein, and D.F. Argus, 1987, *Geophys. Res. Lett.,* **14**, 911.

Denker, H., and H.G. Wenzel, 1987, *Bull. Geodesique,* **61**, 349.

Detrick, R.S., 1981, *J. Geophys. Res.,* **86**, 11751.

Detrick, R.S. and S.T. Crough, 1978, *J. Geophys. Res.,* **83**, 1236.

Detrick, R.S. and G.M. Purdy, 1980, *J. Geophys. Res.,* **85**, 3759.

Detrick, R.S., R. P. von Herzen, S.T. Crough, D. Epp, and U. Fehn, 1981, *Nature,* **292**, 142.

Detrick, R.S., R. P. von Herzen, B. Parsons, D. Sandwell, and M. Dougherty, 1986, *J. Geophys. Res.,* **91**, 3701.

Dixon, T.H., M. Naraghi, M.K. McNutt, and S.M. Smith, 1983, *J. Geophys. Res.,* **88**, 1563.

Driscoll, M.L. and B. Parsons, 1988, *Earth Planet. Sci. Lett.,* **88**, 289.

Dziewonski, A.M., 1984, *J. Geophys. Res.,* **89**, 5929.

Dzurisin, D., L.A. Anderson, G.P. Eaton, R.Y. Koyanagi, P.W. Lipman, J.P. Lockwood, R.T. Okamura, G.S. Puniwai, M.K. Sako, and K.E. Yamashita, 1980, *J. Volcan. Geotherm. Res.,* **7**, 241.

Ebinger, C., T.D. Bechtel, D.W. Forsyth, and C.O. Bowin, 1988, Effective elastic plate thickness beneath East African and Afar plateaus and dynamic compensation of the uplifts, preprint.

Elsasser, 1969, in *The Application of Modern Physics to the Earth and Planetary Interiors*, S.K. Runcorn, ed. 223.

England, P., G. Houseman, and L. Sonder, 1985, *J. Geophys. Res.,* **90**, 3551.

Fischer, K., M. McNutt, and L. Shure, 1986, *Nature,* **322**, 733.

Fleitout, L. and D. Yuen, 1984, *J. Geophys. Res.,* **89**, 9227.

Fox, P.J., E. Schreiber, H. Rowlett, and K. McCamy, 1976, *J. Geophys. Res.,* **81**, 4117.

Freedman, A.P., 1987, *Marine Geophysical Applications of Seasat Altimetry and the Lithospheric Structure of the South Atlantic Ocean,* Ph.D. Thesis, Department of Earth, Atmospheric, and Planetary Sciences, Massachusetts Institute of Technology.

Freedman, A. P., and B. Parson, 1986, *J. Geophys. Res.,* **91**, 8325.

Froidevaux, C., 1986, *J. Geophys. Res.,* **91**, 3625.

Glazner, A.F., and J. M. Bartley, 1984, *Tectonics,* **3**, 385.

Grand, S.P., 1986, Shear velocity structure of the mantle beneath the North American plate, Ph.D. Thesis, 228 pp., California Institute of Technology.

Grand, S.P., and D.V. Helmberger, 1984, *Geophys. J. Roy. Astr. Soc.,* **76**, 11465.

Hager, B.H., 1984, *J. Geophys. Res.,* **89**, 6003.

Hager, B.H., and R.W. Clayton, 1989, in *Mantle Convection,* W.R. Peltier, Ed., in press.

Haxby, W.F., 1985, *Gravity Field of the World's Oceans,* Lamont Doherty Geol. Obs., Palisades, New York.

Haxby, W.F. and D.L. Turcotte, 1978, *J. Geophys. Res.,* **83**, 5473.

Haxby, W.F., D.L. Turcotte, and J.M. Bird, 1976, *Tectonophysics,* **36**, 57.

Haxby, W.F., and J.K. Weissel, 1986, *J. Geophys. Res.,* **91**, 3507.

Hearn, T.M., and R.W. Clayton, 1986a, *Bull. Seismo. Soc. Am.,* **76**, 495.

Hearn, T.M., and R.W. Clayton, 1986b, *Bull Seismo. Soc. Am.,* **76**, 511.

Hellinger, S., and J.G. Sclater, 1983, *J. Geophys. Res.,* **88**, 8251.

Herring, T.A., et al., 1986, *J. Geophys. Res.,* **91**, 8341.

Hinnov, L.A., and C.R. Wilson, 1987, *Geophys. J. Roy. Astron. Soc.,* **88**, 437.

Holdahl, S.R., *J. Geophys. Res.,* **87**, 9374.

Humphreys, E., R.W. Clayton, and B.H. Hager, 1984, *Geophys. Res. Lett.,* **11**, 625.

Isacks, B.L., 1988, *J. Geophys. Res.,* **93**, 3211.

Isacks, B.L., and M. Barazangi, 1977, in *Island Arcs, Deep Sea Trenches, and Back-Arc Basins, Maurice Ewing Ser.,* vol 1, M. Talwani and W.C. Pitman III (eds.), 99, AGU, Washington, D.C.

Jachens, R.C., D.R. Spydell, G.S. Pitts, D. Dzurisin, and C.W. Roberts, 1981, in *The 1980 Eruptions of Mount St. Helens, Washington, U.S. Geological Survey Professional Paper # 1250,* 175.

Jachens, R.C. et al., 1983, Science, 219, 1215.

Jackson, D.D., A. Cheng, and C.C. Liu, 1983, Tectonophy. 97, 73.

Jordan, T.H., 1979a, Sci. Amer., 240, 92.

Jordan, T.H., 1979b, in *The Mantle Sample: Inclusions of Kimberlites and Other Volcanics,* F.R. Boyd and H.O.A. Meyer (eds.), AGU, Washington, D.C.

Jordan, T.H., and J.B. Minster, 1988a, in *The Impact of VLBI on Astrophysics and Geophysics,* (M.J. Reid and J.M. Moran, eds.) *Proc. IAU Symposium 129,* 341.

Jordan, T.H., and J.B. Minster, 1988b, *Scientific American, August,* 48.

Karner, G.D. and A.B. Watts, 1983, *J. Geophys. Res.,* **88**, 10449.

King, E.R., and I. Zietz, 1971, *Geol. Soc. Amer. Bull.,* **82**, 2187.

Le Pichon, X., 1968, *J. Geophys. Res.,* **73**, 3661.

Li, F.K. and R. Goldstein, 1988, Studies of multi-baseline spaceborne interferometric synthetic aperture radar. IEEE Trans. Geoscience and Remote Sensing, in press.

Lyon-Caen, H., and P. Molnar, 1983, *J. Geophys. Res.,* **88**, 8171.

Lyon-Caen, H., and P. Molnar, 1985, *Tectonics,* **4**, 513.

Macdonald, K.C., 1982, *Annu. Rev. Earth Planet. Sci.,* **10**, 155.

MacKenzie, K., and D. Sandwell, 1986, *EOS, Trans. Amer. Geophys. Union,* **67**, 1229.

Malinverno, A., and W.B.F. Ryan, 1986, *Tectonics,* **5**, 227.

Marks, D., J. Dozier, and J. Frew, 1984, *Geo. Processing,* **2**, 299.

Martin, T.V., H.J. Zwally, A.C. Brenner, and R.A. Bindschadler, 1983, *J. Geophys. Res.,* **88**, 1608.

Marty, J.C. and A. Cazenave, 1988a, *Geophys. J.,* **93**, 1.

Marty, J.C. and A. Cazenave, 1988b, Geophys. Res. Lett., 15, 593.

Masursky, H., E. Eliason, P.G. Ford, G.E. McGill, G.H. Pettengill, G.G. Schaber, and G. Schubert, 1980, *J. Geophys. Res.,* **85**, 8232.
McAdoo, D.C., and C.F. Martin, 1984, *J. Geophys. Res.,* **89**, 3201.
McAdoo, D.C., and D.T. Sandwell,1985, *J. Geophys. Res.,* **90**, 8563.
McKenzie, D.P., 1967, *J. Geophys. Res.,* **72**, 6261.
McKenzie, D.P., 1972, *Geophys. J. Roy. Astr. Soc.,* **30**, 109.
McKenzie, D.P., 1978, *Geophys. J. Roy. Astr. Soc.,* **55**, 217.
McNutt, M.K., 1984, *J. Geophys. Res.,* **89**, 11180.
McNutt, M.K., 1987, in *Seamounts, Islands, and Atolls,* Geophysical Monograph #43, B. Keating, P. Fryer, R. Batiza, and G.W. Boehlert (eds), AGU, Washington, D.C.
McNutt, M., 1988, *J. Geophys. Res.,* **93**, 2784.
McNutt, M.K., M. Diament, and M.G. Kogan, 1988, *J. Geophys. Res.,* **93**, 8825.
McNutt, M. and L. Shure, 1986, *J. Geophys. Res.,* **91**, 13,915.
Meier, M.F., 1984, *Science,* **226**, 1418.
Melosh, H.J., 1976, *J. Geophys. Res.,* **81**, 5621.
Melosh, H.J., 1977, *Pure Appl. Geophys.,***15**, 429.
Melosh, H. J., 1983, *Geophys. Res. Lett.,* **10**, 47.
Menard, H.W., and M.K. McNutt, 1982, *J. Geophys. Res.,* **87**, 8570.
Mercier, J.C., 1977, *Bull. Geol. Soc. Fr.,* **7**, 663.
Merifield *et al.,* 1983, *Tech. Rep. 83-3, Lamar-Merifield Geol. Inc.* Santa Monica, CA
Minster, J.B. and Jordan, T.H., 1978, *J. Geophys. Res.,* **83**, 5331.
Minster, J.B. and Jordan, T.H., 1984, *Pac. Sec. Soc. Econ. Paleontol. Mineral.,* **38**, 1.
Minster, J.B. and Jordan, T.H., 1987, *Geophys. Res.,* **92**, 4798.
Minster, J.B., T.H. Jordan, P. Molnar and E. Haines, 1974, *Geophys. J. Roy. Astr. Soc.,* **36**, 541.
Molnar, P. and P. Tapponnier, 1985, *Science,* **189**, 419.
Molnar, P. and P. Tapponnier, 1978, *Geophys. Res.,* **83**, 5361.
Molnar, P., B.C.Burchfiel, L. K'uangyi and Z. Ziyun, 1987, *Geology,* **15**, 249.
Moretti, I. and P.Y. Chenet, 1987, *Tectonophys.,* **133**, 229.
Moretti, I. and C. Froidevaux, 1985, in *Modeling the Thermal Evolution of Sedimentary Basins,* J. Burrus (ed.), Ed. Technip, Paris, 107.
Morgan,W.J., 1972, *Amer. Assoc. Petrol.Geol.Bull.,* **56**, 203.
Morgan, P., W.R. Seager, and M.P. Golombek, 1986, *J. Geophys. Res.,* **91**, 6263.
Nataf, H.-C., I. Nakanishi, and D.L. Anderson, 1984, *Geophys. Res. Lett.,* **11**, 109.
Nataf, H.-C., I. Nakanishi, and D.L. Anderson, 1986, *J. Geophys. Res.,* **91**, 7261.
Nash, D.B., 1986, In: Active Tectonics, National Academy Press, Washington, D.C., 181.
National Aeronautics and Space Administration Advisory Council, Earth Systems Science Committee, 1987, "Earth Systems Science: A Closer View", Washington, D.C.
Nur, A. and Z. Ben-Avraham, 1982, *J. Geophys. Res.,* **87**, 3644.
Ocean Studies Board, 1988,*The Mid-Ocean Ridge: A dynamic Global System,* National Academy Press, Washington, D.C.
Ocola, L.C. and R.P. Meyer, 1973, *J. Geophys. Res.,* **78**, 5173.
Okal, E. and A. Cazenave, 1985, *Earth. Planet. Sci. Lett.,* **72**, 99.
Parmentier, E.M. and W.F. Haxby, 1986, *J. Geophys. Res.,* **91**, 7193.
Parsons, B. and F. Ritcher, 1981, *Sea,* **7**, 73.
Parsons, B. and J.G. Sclater, 1977, *J. Geophys. Res.,* **82**, 803.
Ricard, Y., L. Fleitout, and C. Froidevaux, 1984, *Annales Geophysicae,* **2**, 267.
Richards, M.A. and B.H. Hager, 1984, *J. Geophys. Res.,* **89**, 5987.
Robinson, E.M. and B. Parsons, 1988, *J. Geophys. Res.,* **93**, 3144.
Royden, L., F. Horvath, A. Nagymarosy, and L. Stegena, 1983a, *Tectonics,* **2**, 91.
Royden, L., F. Horvath and J. Rumpler, 1983b, *Tectonics,* **2**, 63.
Royden, L.,1988, in *Geodetic studies and Crustal Dynamics,* U.S. Geodynamics Committee Progress Report, 3.1-3.10.
Ruff, L. and A. Cazenave, 1985, *Phys. Earth Planet. Int.,* **38**, 59.
Rundle, J.B., 1978, *Geophys. Res., Lett.,* **5**, 41.

Rundle, J.B., 1988a, *J. Geophys. Res.*, **93**, 6237.
Rundle, J.B., 1988b, *J. Geophys. Res.*, **93**, 6225.
Rundle, J.B. and D.D. Jackson, 1977, *Geophys. J. Roy. Astr. Soc.*, **49**, 575.
Rundle, J.B. and H. Kanamori, 1987, *J. Geophys. Res.*, **92**, 2606.
Rundle, J.B. and W. Thatcher, 1980, *Seis. Soc. Amer. Bull.*, **70**, 1869.
Rundle, J.B., and J.H. Whitcomb, 1986, *J. Geophys. Res.*, **91**, 12675.
Sanders, C.. P. Ho-Liu, D. Rinn, and H. Kanamori, 1988, *J. Geophys. Res.*, **93**, 3321.
Sandwell, D.T., 1984, Along-track deflection of the vertical from Seasat, GEBCO overlays, *NOAA Tech. Memo.*, NOS NGS-40.
Sandwell, D., 1984, *J. Geophys. Res.*, **89**, 1089.
Sandwell, D.T. and D.C. McAdoo, 1988, *J. Geophys. Res.*, **93**, 10389.
Sandwell, D.T. and M.L. Renkin, 1988, *J. Geophys. Res.*, **93**, 2775.
Sandwell, D. and G. Schubert, 1982, *J. Geophys. Res.*, **87**, 4657.
Sauber, J., K. McNally, J.C. Pechmann, and H. Kanamori, 1983, *J. Geophys. Res.*, **88**, 2213.
Savage, J.C., 1983, *Ann. Rev. Earth Planet. Sci.*, **11**, 11.
Savage, J.C., 1984, *J. Geophys. Res.*, **89**, 1945.
Savage, J.C., and G. Gu, 1985, *J. Geophys. Rers.*, **90**, 10301.
Savage, J.C., W.H. Prescott, and M. Lisowski, 1987, *J. Geophys. Res.*, **92**, 4785.
Scandone, 1979, *Bull. Soc. Geol. Ital.*, **98**, 27.
Shaw, P.R., 1987, *J. Geophys. Res.*, **92**, 9363.
Sheffels, B. and M. McNutt, 1986, *J. Geophys. Res.*, **91**, 6419.
Snay *et al.*, 1986, *Royal Soc. New Zealand Bull.*, **24**, 131-140.
Spiess, 1985, *IEEE Trans. on Geoscience and Remote Sensing GE-23(4)*, 502-510.
Steckler, M.S., 1985, *Nature*, **317**, 135.
Stein, R.S., 1987, *Rev. Geophys.*, **25**, 855.
Stockmal, G.S., C. Beaumont, and R. Boutilier, 1986, *Amer. Assoc. Petrol. Geol. Bull.*, **70**, 181.
Tanimoto, T., 1986, *Geophys. J. Roy. Astron. Soc.*, **84**, 49.
Tapley, B.D., G.H. Born, and M.E. Parke, 1982, *J, Geophys. Res.*, **87**, 3179.
Thatcher, W., 1979, *J. Geophys. Res.*, **84**, 2351.
Thatcher, W., 1983, *Nature*, **299**, 12.
Thomas, R.H., T.V. Martin, and H.J. Zwally, 1983, *Annals Of Glaciology*, **4**, 283.
Toksoz, M.N., J.W. Minear, and B.R. Julian, 1971, *J. Geophys. Res.*, **76**, 1113.
Tse, S.T. and J.R. Rice, 1986, *J. Geophys. Res.*, **91**, 9452.
Turcotte, D.L., J.Y. Liu, and F.H. Kulhawy, 1984, *J. Geophys. Res.*, **89**, 5801.
Vogt, P., 1973, *Nature*, **241**, 189.
Wagner, C.A. and D.C. McAdoo, 1986, *J. Geophys. Res.*, **91**, 8373.
Wahr, J.M., 1982, *Geophys. J. Roy. Astro. Soc.*, **70**, 349.
Wallace, R.E., 1984, *J. Geophys. Res.*, **89**, 5763.
Walsh, J.B. and J.R. Rice, 1979, *J. Geophys. Res.*, **84**, 165.
Watts, A.B., 1978, *J. Geophys. Res.*, **83**, 5989.
Watts, A.B. and M. Talwani, 1974, *Geophys. J. R. Astron. Soc.*, **36**, 57.
Weissel, J.K., R.N. Anderson, and C.A. Geller, 1980, *Nature*, **287**, 284.
Weldon, R. and E. Humphreys, 1985, *Tectonics*, **5**, 33.
Wernicke, B., 1985, *Can. J. Earth Sci.*, **22**, 108.
Wilson, P., 1987, *Geojournal*, **14.2**, 143.
Woodhouse, J.H. and A.M. Dziewonski, 1984, *J. Geophys. Res.*, **89**, 5953.
Wu, P. and W. R. Peltier, 1983, *Geophys. J. Roy. Astro. Soc.*, **74**, 377.
Wu, P. and W. R. Peltier, 1984, *Geophys. J. Roy. Astro. Soc.*, **76**, 753.
Wyatt, F., K. Beckstrom, and J. Berger, 1982, *Bull. Seismo. Soc. Amer.*, **72**, 1707.
Yoder, C.F., J. G. Williams, J.O. Dickey, B.E. Schutx, R.J. Eanes, and B.D. Tapley, 1983, *Nature*, **303**, 757.
Yuen, D. and L. Fleitout, 1985, *Nature*, **313**, 125.
Yuen, D.A., W.R. Peltier, and G. Schubert, 1981, *Geophys. J. Roy. Astro. Soc.*, **65**, 171.

Zebker, H. and R. Goldstein, 1986, *Geophys. Res.*, **91**, 4993.
Zhang, 1987, *Rate, amount, and style of late Cenozoic deformation of sourthern Ningxia northeastern margin of Tibetan Plateau*, Ph.D. Thesis, Mass. Inst. of Tech., Cambridge, Mass.
Zuber, M.T., E.M. Parmentier, and R.C. Fletcher, 1986, *J. Geophys. Res.*, **91**, 4826.

Chapter 4

INTERACTION WITH OTHER DISCIPLINES AND PROGRAMS

A. Anderson, F. Gilbert, H. Grassl, E. Harrison, R. Hide, S. Houry, K. Lambeck,
M. Lefebvre, J. Melosh, A. Morelli, R. Sabadini, D. Smith, C.-K Tai, G. Visconti,
S. Wilson, C. Yoder, and V. Zlotnicki

1. INTRODUCTION

The structure of planet Earth and the dynamics of its constituent parts have received the intellectual attention of natural philosophers since ancient times. Only in the present century has the quantitative, physics-oriented approach led to a deeper and more profound understanding of the subject, and only in the past two decades has the role of satellite data become both well established and centrally important.

As we strive to learn more about geodesy from satellite data we are compelled to take into account that the data contain the effects of the atmosphere, noise and bias to the geodesist but signal to the meteorologist. Thus, it is not surprising that branches of geophysics other than geodesy are directly involved in the proper interpretation of satellite data. The relation between atmospheric angular momentum and the length of day, for example, compels geophysicists to understand the fluid dynamics of the earth's cores and atmosphere, and the coupling between them. Coupling occurs throughout the disciplines of geophysics, including geodesy, and is the theme of this chapter.

Beyond the Earth we realize that geophysical techniques, particularly those based on artificial satellites, allow us to study the other planets in the solar system.

In less than one year after the meeting in Erice that led to this report the planet Neptune will have been visited by a spacecraft. Interesting as such "flybys" may be, future studies of the other planets would benefit considerably from satellite data, one of the subjects treated in this chapter.

Careful design of interplanetary space probes can convert the solar system into a huge laboratory for experimental physics. Gravity is the subject of primary interest. The most enigmatic force in Nature, it is the focus of experiments on all scales, from antiproton experiments through geophysical investigations of the Newtonian constant to the large scale projects proposed and discussed in this chapter. The interaction between satellite geodesy and fundamental physics is intriguing and represents the breadth and depth of geophysics.

2. GEODYNAMICS

2.1 INTRODUCTION

Geophysical observations provide the principal source of information on the interior structure of the Earth. They include the travel times, amplitudes, and frequencies of seismic waves, the flux of heat through the surface, and magnetic field parameters. Geodetic observations of the external gravity field, of the shape of the planet and its time dependencies, and of the rotational motions, provide further constraints. Together with geological observations (including geochemistry and geochronology) these geophysical and geodetic observations provide the key to understanding the structure, dynamics, composition and evolution of the Earth's hydrosphere, crust, mantle, and core.

Central problems to which geodetic observations contribute are of two types. The first is where observations provide a measure of the response of the Earth to known forces (e.g., tidal deformation or post-glacial rebound). The second is where the observations are used to infer the forces themselves (e.g., the forces driving plate tectonics from surface strain observations). In neither case can the geodetic observations alone provide unique answers and the interpretation of these data requires complementary geophysical, meteorological, oceanographic and geological data. In particular, many geodetic signals pertaining to the solid Earth contain signals originating from the hydrosphere and atmosphere and these must be measured by independent techniques in order to understand the solid Earth signatures.

Plate tectonics has provided the major impetus to Earth Sciences in the past two decades and the kinematics of global plate motions are now fairly well understood. Major developments in the future will be directed towards understanding the dynamics of the mantle and the mantle-lithosphere interaction. Geodetic observations can provide important constraints on this problem by providing information on the planet's rheology through a wide range of measurements including the gravity field, planetary rotation, tides and crustal motion. One area of particular current interest is the dynamics of the lower mantle and nature of, and processes occurring at, the core-mantle boundary. A number of the geodetic observations complement the geophysical and geological evidence and arguments. Another important problem is the interactions at the solid Earth-hydrosphere boundary and here also the geodetic observations contribute through the study of tides and planetary rotation and through observations of mean sea level and sea level changes.

Most of the past progress in the Earth Sciences has come from the examination of long series of observations that were collected because of advances made in technology and engineering. Examples include the measurement of heat flow in the ocean crust, marine bathymetry and the use of satellites to measure the Earth's gravity field. It is important that observation programs of fundamental properties of the Earth be continued. In the geodetic context this includes satellite altimeter missions such as TOPEX and the high-spatial resolution Earth gravity field studies, missions to map the Earth's topography, and the magnetic field and its time dependence, and long-term observing programs of passive LAGEOS and STARLETTE type satellites using laser ranging. It includes long term VLBI observing programs for monitoring global motions and deformations of the planet and GPS- or GLRS-type missions for crustal deformation monitoring on regional and local scales. In parallel, observing programs in seismology and other geophysical (including oceanographic, glaciological and meteorological) and geological quantities are essential, for without this the geodetic observations are only of limited value. Interactions between the geodetic and other Earth Science disciplines are essential in planning and executing the parallel programs of observation.

Some specific examples of the interactions between the geodetic observations and other Earth Science disciplines are given below (§2.3). The examples are not exclusive and many others, equally valid, can be found. Some specific requirements for ensuring that the geodetic data can now be collected by space techniques are compatible with the overall scientific goals of the study of the dynamic Earth are given in §2.2. These requirements also are more indicative than complete. The nature of science is such that major progress often occurs where it is least anticipated.

2.2 SUMMARY OF PRINCIPAL POINTS MADE BY SUB-PANEL

(1) Further LAGEOS and STARLETTE-type missions are required in different inclination orbits, to study, interalia, the time-dependent behavior of the Earth's gravity field. These time dependencies include the tidal deformations of the Earth and long-period and secular changes in the long wavelength components of the Earth's gravity field.

(2) Meteorological data must be collected to support the geodetic missions both for making appropriate corrections for the propagation characteristics of the satellite tracking data and for monitoring the mass movements of the hydrosphere and atmosphere.

(3) There is a need for continued improvements in ocean-tide and ocean-circulation models. It is particularly important to establish ocean-tide models free from geophysical inputs (assumed Earth responses to tidal forces and surface loading) in order to separate dissipation of tidal energy in the oceans from that in the solid Earth. The most appropriate way of achieving this is by using both tide gauge data, including deep ocean seafloor records, and satellite altimetry observations.

(4) There is an important need to upgrade the world network of sea-level monitoring stations to establish periodic and secular trends in sea level and to establish the global patterns of these changes. Such measurements should be made in conjunction with observations of changes in geocentric heights of the stations.

(5) Satellite altimeter missions are essential to achieve the above objectives. It will be important that successive missions permit long-term fluctuations in sea-surface topography to be monitored. All such satellite missions will need to be tracked with the highest possible accuracy. It is important that the orbital parameters of these missions are such as to give global ocean coverage.

(6) GPS-type missions for rapid and precise positioning (cm-level) are important for monitoring crustal motions and deformation in tectonically active regions. There is a need for relatively inexpensive equipment that can operate remotely in a wide range of environments. High accuracy, relative height measurements are particularly important. Such studies will contribute to understanding the deformations of the crust along plate margins as well as the study of the stress-strain cycles of large earthquakes.

(7) GLRS-type missions (geodynamics laser ranging system) with mm relative positioning precisions and of long operational lifetimes are strongly supported for future crustal deformation studies.

(8) High accuracy, sub-microgal, absolute and relative gravity meters need to be deployed at selected sites in support of vertical motion studies. Such monitoring should be conducted routinely at principal satellite stations and at VLBI sites.

(9) High spatial resolution gravity field studies are essential to support oceanographic altimeter missions. They also contribute significantly to studies of continental tectonics.

(10) There is an important need to measure the Earth's topography with high resolution. This includes both continental and oceanic areas. Without such observations the full tectonic potential of the gravity missions cannot be achieved.

(11) Seismological studies of the radial and lateral variations in seismic velocities form a very important complement to the gravity, topography and magnetic observations, and the establishment of global digitally-recording, broadband seismic networks is strongly supported.

(12) There is a need to monitor the Earth's magnetic field, both for separating the core and crustal components of the total field and for mapping the time dependent behavior of the field. It will be particularly important to be able to separate the time-dependent contributions from the internal and external sources. Such observations contribute to the understanding of the Earth's dynamo, core motions, core-mantle coupling processes and lithospheric tectonics.

(13) There is a need to establish and maintain significant modeling programs for the routine interpretation of the data collected by these various programs. Innovative research based on these analyses as well as on the original data to complement these larger analysis programs should be encouraged.

2.3. EXAMPLES OF INTERDISCIPLINARY GEODETIC STUDIES

Temporal variations in Earth's gravity field. High-accuracy laser tracking of the LAGEOS and STARLETTE satellites has revealed signatures in the orbits that have been interpreted in terms of temporal changes in the Earth's gravity field. These include (i) periodic variations due to solid-Earth and ocean tides, (ii) seasonal variations associated with periodic redistribution of mass in the oceans and atmosphere, and (iii) quasi-secular changes driven by post-glacial rebound, and the present-day melting of mountain glaciers. A fourth possible change, not yet identified, are step-like changes in the Earth's inertia tensor that are associated with major earthquakes.

Tidal variations provide a direct test of ocean models derived either from tide gauge data or from satellite altimetry. In particular, these tidal variations provide a direct measure of the dissipation of tidal energy in the Earth and, by combining these results with independent ocean model estimates of dissipation, they provide estimates of dissipation or Q within the solid Earth. If very accurate ocean tide models can be derived from satellite altimeter data, it may also become possible to detect lateral variations in the Earth's response to the tidal force. Such observations contribute to understanding the structure and dynamics of the Earth. Accurate Earth-nutation data also places constraints on the solid Earth's Q.

These objectives can only be achieved if ocean-tide models for the principal tidal frequencies can be established with accuracies of better than 5% for the long wavelengths in the ocean tide spectrum. The design of altimeter missions such as TOPEX, should be commensurate with this requirement. Complementary observations to these studies include surface tidal gravimetry observations and radial and horizontal station displacements of VLBI and satellite laser ranging sites. From such observations it becomes possible to probe the lateral structure in the $_0S_2$ mode to complement seismic free oscillation studies.

Seasonal variations in the Earth's gravity field provide important global constraints on models of ocean and atmospheric circulation and on models of the Earth's water budget. When combined with accurate observations in the seasonal variations of the ocean height and air pressure such observations may help in determining the thermal contributions to the ocean height variations as distinct from contributions arising from water added into the ocean.

Secular variations in gravity measure the combined effect of the rebound of the Earth's crust to the removal of the Late Pleistocene ice sheets and any present day glacial melting. The actual rates of change will depend on the spatial and temporal distribution of these ice sheets and of the Earth's response to the change in surface loading. Any rate of change of tesseral coefficients or the geopotential will be primarily the result of northern hemispheric contributions whereas any change in the zonal coefficients, particularly in the odd harmonics, will be strongly influenced by any Antarctic contributions. Thus a combined data set of secular change in the potential coefficients should provide an important constraint on the poorly known Antarctic melting histories.

Discontinuous time-dependent changes in the gravity field have not yet been observed, mainly because there have been no major ($m \geq 8$) earthquakes since the launch of LAGEOS. Earthquakes such as the 1964 Alaska earthquake, have the potential of producing major changes in the low-degree gravity field coefficients the observation of which provides constraints on models of elastic seismic dislocation, pre- and post-seismic relaxation and of the Earth's fluid core spin-up time constant.

The harmonic spectrum of the time dependence of gravity parameters that can be derived from the LAGEOS and STARLETTE orbits is limited and satellites in different orbits are required. In designing these orbits, attention must be given to how such new satellites improve the accuracy of the time-variable gravity field. Near resonant orbits should also be considered as a means to probe the time dependence of the tesseral field. A realistic goal should be one that maps the slowly changing field to harmonic degree and order 3 and for zonal harmonics up to degree about 6 or 8 with an accuracy of a few parts in 10^{-11} yr^{-1}. Observation programs of satellite laser ranging should have a lifetime of the order of at least 2 decades to separate out long-period tidal oscillations from secular trends.

The Earth's deep interior. It is now widely recognized that significant dynamical processes may be occurring in all parts of the Earth, including the lower mantle and at the core-mantle interface. Geodetic observations make several contributions to the understanding of these processes. These include the Earth's gravity field and rotation.

Departures from axial symmetry in the thermal and mechanical boundary conditions, imposed on motions in the fluid core of the Earth by convection in the highly viscous lower mantle, may produce a number of observational consequences at the Earth's surface. For example, the density anomalies associated with deep mantle convection and a bumpy core-mantle interface would contribute to the long wavelength gravity field observed at the surface. Fluctuations in the fluid flow past the mantle produce torques at the core-mantle boundary which may result in variations in the length of day on the decade time scale and in polar motion. These torques are of two types: (1) tangential stresses produced by viscous forces in a thin boundary layer just below the interface, and the Lorentz forces associated with the electric currents in the lower mantle, and (2) normal stresses produced by dynamical pressure forces acting on the equatorial bulge and irregular topography of the core-mantle boundary.

By combining Earth rotation, geomagnetic, seismological and gravity data, it is possible to estimate the magnitude of this topographic coupling. These observations also contribute to models of the structure, dynamics and composition of the outer core and lower mantle, including the boundary layers on both sides of the interface.

To make use of the rotation data it is necessary to remove irregularities in rotation produced by meteorological and oceanographic variations in mass distribution for these make dominant contributions to the observed changes in length-of-day and polar motion. Daily components of all three components of the wind surface torques or of the atmospheric angular momentum are essential, including winds up to stratospheric heights. Information on seasonal variations in ocean circulation are desirable. Observations that part of the time-dependent part of the magnetic field that is of internal origin are also required.

Secular accelerations of the Earth and C_{20}. Astronomers have long been aware that the Earth's diurnal rotation has not been constant through time but that the rotational velocity has decreased slowly. Likewise, astronomers have long recognized that the Moon has been subjected to an acceleration in its orbit of about -20 arcsec/cy^2. These accelerations are small but if they have persisted throughout the history of the Earth-Moon system their consequences are dramatic for they imply that the Moon has been much closer to the Earth in the past than it is now. These accelerations are a consequence of the torque exerted by the Moon, and to a lesser degree by the Sun, on the tidally deformed and elastically imperfect Earth.

Evidence for the secular acceleration of the Earth can only be obtained from very long series of historical records of the motions of the Earth-Moon-Sun because of the existence of decade and century scale variations in rotations that are generally attributed to differential rotations of the core and mantle. New evidence for the lunar acceleration in its orbit comes from lunar laser ranging from relatively short records and significant improvements in accuracy can be expected. The tidal part of the secular acceleration of the Earth can therefore be computed with considerable accuracy and the difference between this estimate and the observed estimate constitutes the nontidal acceleration of the Earth. Estimates of this quantity have indicated a positive nontidal acceleration of the Earth but the amounts are generally unreliable. These accelerations have been attributed to several mechanisms, including a secular change in G, a consequence of ongoing isostatic adjustment of the crust and mantle to Late Pleistocene glacial unloading, to electromagnetic core-mantle coupling and to growth of the core. The nontidal acceleration would therefore provide a significant constraint on cosmological theories or on geophysical models, provided that one could establish that one or other mechanism was appropriate. As in many problems of the rotations of the Earth, a number of mechanisms are likely to operate together.

Some separation of these mechanisms is nevertheless possible. If glacial rebound is responsible for the acceleration then there should be a concomitant secular polar motion. All solutions of the equations of polar motion driven by realistic models of the deglaciation of the ice sheet predict a shift of the mean rotation pole that is remarkably close to the observed shift since about 1900. Electromagnetic torques could possess an equatorial component and hence contribute to the polar wander on this time scale but it would be fortuitous if this response aligned itself with that of glacial rebound. The agreement therefore suggests that the primary excitation of the secular polar wander is the Late Pleistocene deglaciation and it should be possible to predict the corresponding secular change in nontidal acceleration.

Another significant observation is the acceleration $\ddot{\Omega}$ in the node of satellite orbits because glacial rebound will also contribute to this whereas electromagnetic torques do not. Because $\dot{\Omega}$ is proportional to C_{20} it becomes possible to estimate the nontidal acceleration ω_{NT} and recent analyses suggest that the results are consistent.

Author	$\ddot{\Omega}$ $10^{-23} s^{-2}$	$\dot{C}_{20}$ $10^{-19} s^{-1}$	$\dot{\omega}_{NT}$ $(10^{-22} s^{-2})$
Yoder *et al.*, 1983	-7.2 ± 0.6	11.1 ± 1.0	1.6 ± 0.4
Rubincam, 1984	-5.3 ± 1.2	9.2 ± 1.8	1.2 ± 0.26
Stephenson & Morrison, 1984			1.7^*

*From Lambeck 1988, based on the Stephenson & Morrison (1984) estimate of $\dot{\omega}$ for the past 10^3 years.

This also suggests that magnetic torques may not be important on this time scale. Any secular change in G contributes to ω but not to the secular polar motion so that it also becomes possible to separate these two factors. While the observational precisions are still inadequate, observations of the secular changes in length-of-day, polar motion and the rate of precession of the node of satellite orbits have the important potential of being able to discriminate between quite different excitation mechanisms. Of importance is that the resulting mantle viscosity estimates complement the studies of glacially induced sea level change that are generally more sensitive to upper mantle than to lower mantle viscosity.

Sea level change. Analysis of long-term tide gauge records indicate that global sea level has been rising during the past century although estimates of the eustatic rate, taken here to mean the global average value, have differed significantly, ranging from 1 to 3 mm/year. There are several reasons for these different estimates. (1) Sea level changes, irrespective of their causes, are nonuniform over the Earth because the slowly changing sea surface remains an equipotential at all times. Hence, estimates of the secular rate vary with the region. (2) Sea level changes occur over a wide range of time scales and estimates of the secular trend are strongly dependent on how the other fluctuations have been eliminated and on the record length. (3) Vertical tectonics along many parts of the coastlines of the world are significant and this may contaminate the eustatic sea level rise estimates because the tide gauge observations only yield sea level positions relative to the Earth's crust. The result of an attempt at separating tectonic from eustatic contributions and at examining regional variations in the latter term, is illustrated in Fig. 1. The contribution (which also exhibits regional variation) from the melting of the present mountain glaciers as modeled by Meier (1984) has been removed. This contributes about 0.46 mm/year to the observed eustatic amount of 13 mm/year. Part of this regional variation illustrated in Fig. 1 can be attributed to the warming of the upper 100–200 m of the ocean surface then the required water warming is about 0.02–0.04°C/year, consistent with changes observed in, for example, Atlantic mean surface temperatures.

What results such as those in Fig. 1 indicate is that monitoring of secular change in sea-level cannot be accomplished successfully with a few tide gauges only. Nor can meaningful results be achieved from records as short as a decade or two. Whether these changes can be measured with satellite methods, by a combination of very high-accuracy laser ranging and satellite altimetry, is debatable and any monitoring program will be one of the most

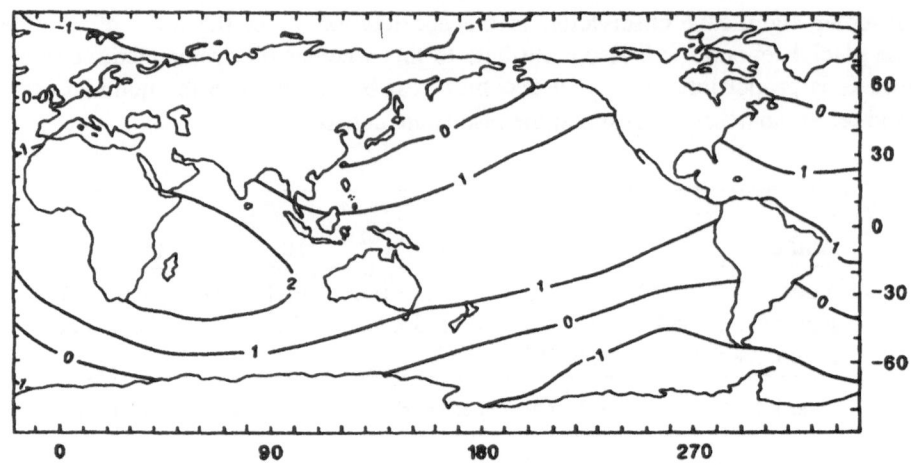

Fig. 1. Global sea-level changes, corrected for mountain glacier melting. The contour interval is 1 mm/yr. The corrected eustatic rise is 0.7 mm/yr (from Nakiboglu, Lambeck and Hekimoglu, 1988).

challenging and interdisciplinary goals of the geodynamics program. Fortunately, the sea level rise is an integrated effect and it may be possible to obtain less direct estimates by monitoring some of the suspected causes of the global change. Thus, supporting oceanographic data will be required, including the monitoring of surface temperatures, salinity and wind stress, as will the monitoring of the ice volumes in the polar ice sheets and mountain glaciers. Satellite-based monitoring of sea surface temperatures and of ice volumes will be an important complementary aspect of the geodynamics satellite program.

3. EARTH STRUCTURE

The first and basic step in the evolution of our knowledge of the dynamics inside the earth has to deal with a reliable determination of the *structure*. Only a detailed image allows us to infer the modes of the processes in act, in order to define them in a quantitative way. Many problems today remain open. In particular, the mechanisms of interacting (chemical, mechanical, magnetic) between different structural units are far from being fully understood. A detailed snapshot of the zones of transition between these units (mantle, outer core, inner core) represents a fundamental constraint for any dynamical theory. It is important, therefore, to recognize our need for improving the quality of the pictures.

In the last few years, renewed interest and new data yielded considerable progress in the study of lateral variations of the deep structure of the Earth. After decades of one-dimensional models, in which properties vary only with depth, geophysicists think now in terms of three-dimensional models. Seismology provides images of variations of elastic wave speed in the mantle. These images derive from the analysis of observations spread over a very broad range of frequencies, including travel times, body waveforms, and normal modes of oscillation. Such velocity models can be translated into density models, and then used as an input datum in studies of mantle convection. Very encouraging results have been obtained which yield substantial agreement with surface plate motions and long-wavelength geoid anomalies. However, several elements are still missing, such as a detailed description of the effect of subducted slabs. The depth of penetration of slabs represent a fundamental piece of information for defining the convection style of the mantle. Slabs are currently imaged, on one side, by means of higher-resolution seismological techniques, and, on the other side, by means of jointly fitting geoid anomalies, plate velocity data and convection patterns.

As a consequence of the increased resolution achieve in the latest studies, laterally heterogeneous models of the mantle will presently show their importance in modeling the response of the earth to surface loading (post-glacial uplift, tides). The translation of seismic velocity into a viscosity model has to rely on assumptions about material properties of mantle constituents. This is related with the estimation of the mineral composition of the mantle, and therefore mineral physics is also involved.

The transition between mantle and core is perhaps the most critical region for our understanding of the dynamics active inside the earth. Core-mantle coupling (viscous, thermal, magnetic) has consequences on many different processes, like mantle convection, outer fluid motion—hence magnetic field generation—and earth rotation. The structure affects different observables. The inversion of seismic data yields models which do not satisfy the constraints which can be inferred by means of current theories and parameter estimates. Also, not all the different seismological observations properly fit into a unique picture. This means that our current view cannot be regarded as complete, and actions should be taken in order to develop a comprehensive picture.

In view of this, and considering unsatisfactory a situation in which every discipline has good models for its own data, but fails to explain the other observations, we believe that a recommendation for the immediate future of the research in earth structure can be formulated as follows:

All available pieces of information regarding the internal structure of the earth should be gathered and a concerted effort should be moved in an interdisciplinary way in order to derive a coherent picture. They include: gravity, magnetic, rotational, and seismic data. Theoretical interpretation of: (i) large-scale geoid anomalies and plate velocities, (ii) rotational data, and (iii) magnetic field observations in terms of: (i) mantle convection, (ii) core-mantle coupling, and (iii) core motions should be considered in an objective way, and uncertainties and assumptions be properly stated. Efforts toward models able to fit all available data—within their relative uncertainties—should be thrusted.

Higher detail will be possible in the future by virtue of improvements in instrumentation quality and deployment, as well as better computational techniques. In order for this higher detail to have real physical meaning, we need to find the coherent way to interpret observations. This necessarily passes through the cooperation among all of the disciplines involved.

4. OCEAN PHYSICS

4.1 INTRODUCTION

The ocean has traditionally been undersampled. Since the Williamstown report, however, major advances have occurred in the development and utilization of satellite-borne sensors, in situ instruments, and numerical models of the ocean circulation.

Table 1. Ocean-Related Satellite Missions

Ocean Parameters	Mission	Launch	Status
temperature	NOAA-n	continuing	operational
sea level, wave	GEOSAT	3/85	operational
temperature, color, sea ice	MOS-1,-1B	2/87,90	oper., app.
wind, sea ice	DMSP-n	continuing	operational
sea level, temperature, wind vector, wave, sea ice	ERS-1,-2	9/90,93	app., prop.
sea level, wave	TOPEX/POSEIDON	12/91	approved
color	LANDSAT-6	91	approved
sea ice, wave	JERS-1	92	approved
color	SPOT-4	93	proposed
temperature, color, wind vector	ADEOS(w/NSCAT)	95	proposed
sea ice, wave	RADARSAT	94	proposed
SUPPORTING INFORMATION			
insitu data transmission and instrument positioning	NOAA-n/Argos	continuing	operational
ship positioning	GPS	continuing	operational
high acc. satellite tracking	SPOT-2,-3	88,89	approved
high acc. satellite tracking	TOPEX/POSEIDON	6/92	approved
high acc. satellite tracking	ERS-1,-2	9/90,93	app., prop.

Advances in satellite altimetry, scatterometry, radiometry, and tracking have demonstrated that the sea surface slope, vector winds, color, temperature, wave height, and sea-ice cover can be sampled from orbiting satellites (see Table 1) with the accuracies, space-time resolution, and global coverage needed to solve a variety of open questions in Oceanography. Satellite sensors also track our ships, buoys, and drifters, and transmit the data from these and tide gauges to land stations in near-real time.

Advances in in-situ instrumentation (to measure temperature, salinity, pressure, velocity, and the concentration of tracers) have also occurred, and have shown the potential to greatly increase our knowledge of the ocean interior.

Advances in theoretical understanding and computer power have resulted in numerical models of the ocean circulation, driven by realistic winds, and including assimilation of satellite and in-situ data, that will lead to useful upper ocean predictions within the next decade.

Taken together, the observations from the above satellites and associated in-situ instruments represent a capability to address how the ocean is forced by the atmosphere, how it responds, how it advects heat and nutrients, how phytoplankton productivity responds, and how the sea ice changes.

Physical Oceanography is moving to establish new global programs to exploit these technological developments and to advance our understanding of both ocean dynamics and of how the global ocean interacts with the atmosphere and the rest of the Earth System. By contrast, all past oceanographic experiments were carried out in limited areas, because ships are about 1400 times slower in sampling the ocean than oceanographic satellites are, and because some ocean areas provide limited accessibility. These global experiments will also define the sensors and strategies needed to monitor the oceans both for operational utilization and for research leading to estimate the magnitude of climate changes. The World Ocean Circulation Experiment (WOCE) and the Tropical Ocean-Global Atmosphere experiment (TOGA) are key components of the World Climate Research Programme. For more information on these two efforts, the reader is referred to WOCE International Planning Office (1988), US WOCE Science Steering Committee (1988), and WMO/IOC Inter-Governmental TOGA Board (1988). A timely collection of papers outlining the present status of physical oceanography—including satellite techniques, in-situ techniques, and numerical modelling—can be found in Knox (1987) and the companion articles in that issue.

WOCE and TOGA have arisen from recognizing that the Earth climate system changes with time and that we do not have appropriate data to describe these variations and understand their physical mechanisms, and that in-situ technology has advanced to the point that basin-scale observation programs are feasible. WOCE is designed to provide the best possible global look at the ocean circulation to understand its present state. TOGA will observe and hopefully explain the interannual variations in the ocean and atmosphere that collectively are known as the El Nino-Southern Oscillation phenomenon, a shift in global weather and tropical ocean circulation patterns that affects many parts of the globe.

Among the space techniques useful in Physical Oceanography, satellite tracking, altimetry, and scatterometry are probably those closest to the interests of Geodesy. Satellite tracking and altimetry provide sea level and the gravity field, classical objects of geodetic study yielding the oceanographically important slope of sea surface relative to the geoid. Scatterometry provides winds which affect the Earth's rotation (another classical object of geodetic study) and are the driving force of ocean motions at most scales.

4.2 SEA LEVEL

From all the observables of use in Physical Oceanography, we now concentrate on the one more closely related to Geodesy—sea level. Our later recommendations are geared at extracting maximum oceanographic information from altimetry and tide gages, two techniques to measure sea level.

Studies of sea level time series reveal a variety of oceanic phenomena, including waves, tides, mesoscale eddies, the large scale circulation, the heating and cooling of the oceans, or the El Nino-Southern Oscillation (e.g., Wunsch, 1972; Wyrtki, 1984; Cheney *et al.*, 1983). Secular trends in sea level (e.g., Barnett, 1983) and some tidal effects can reflect either oceanographic or geophysical phenomena and were briefly discussed in the Geodynamics section of this chapter.

Tide gages (e.g., Wyrtki *et al.*, 1988) sample at a few locations over the Earth for long times with high resolution in time. Satellite altimeters (e.g., Fu *et al.*, 1988; Douglas *et al.*, 1987) provide unprecedented global coverage in a matter of days. Clearly, the two measurement systems are complementary.

Both techniques also benefit from geodetic advancements: positioning either the satellite or the tide gage is a geodetic concern; the gravity field estimate needed for both the satellite orbit computation and to estimate absolute surface currents (by the departure of sea level slope from the geoid) is a classic geodetic concern. Also, because both systems are complementary, it is important that they be positioned with comparable accuracies, in the same geodetic reference system, an issue discussed at length in the chapter on reference systems, and underlying current plans to deploy new sea level stations containing tide gages (Diamante *et al.*, 1987).

4.3 ALTIMETRY

Using altimetry from Seasat (1978) and Geosat (1985), procedures were developed to estimate sea level variability, marine geoid, accurate orbits, and various altimeter corrections. We are confident that we understand the issues that need to be addressed to meet our future sea level data needs. This experience was already used in the design of TOPEX/POSEIDON (further discussed in the next paragraph). Seasat, in 90 days, provided global statistics on ocean mesoscale variability, detected strong currents and showed the potential to detect the longer wavelengths of the circulation. Geosat, in almost 2 years of oceanographic mission and still healthy, is supplying a blurred picture of the annual cycle and equatorial waves, somewhat tarnished by incomplete atmospheric corrections (due to the lack of a companion radiometer to measure atmospheric water vapor and a second radar frequency to measure ionospheric effects) and by a 3-m orbit error. Existing technology has produced data important for our discipline, and has shown us which improvements are needed to turn it into an operational tool to monitor the ocean.

The TOPEX/POSEIDON mission, to be launched in mid-1992, is expected to provide the most accurate sea level data yet obtained from space. It will use a dual frequency altimeter to correct for ionospheric path delays, a radiometer to correct for delays due to water vapor in the troposphere, relatively dense orbit tracking, a carefully chosen orbit to minimize tidal aliasing, and well understood algorithms. ERS-1 will include a single frequency altimeter, and a radiometer. The ERS-1 orbit will be chosen to satisfy the constraints of several instruments in addition to the altimeter, but it will reach a higher latitude than Topex/Poseidon, providing complementary coverage. From these advances, we expect to resolve the time averaged circulation, interannual variability, and tidal variations out to the longest wavelengths and periods allowed by the mission duration and current technology. The satellite missions will profit from the in-situ oceanographic and meteorological observations of WOCE and TOGA, and will benefit them with their rapid global coverage. Satellite and in-situ data will be combined in hydrodynamic models of the general ocean circulation and in models of specific processes, for example tides.

Inaccurate knowledge of the Earth's gravity field causes the two limiting errors in altimetry: orbit error and marine geoid error.

The 14-cm total accuracy (single pass) estimated for the Topex/Poseidon measurements of sea level are dominated by the orbit error, a consequence of both insufficient knowledge of the gravity field and the density of tracking stations. Interannual variability at intermediate and longer scales will most likely be seen with T/P as currently configured, assuming the total error in the surfaces built from the data can be brought down to a few cm by clever processing. The marine geoid model is presently known to 2 cm at basin scales (5000 km), and worse than 70 cm at wavelengths shorter than 1000 km. This restricts calculations of the time averaged circulation to the components at the longest length scales.

We recognize that it is a difficult technological feat to measure the gravity field so as to satisfy a ''1 cm rss at n < 800'' long-term goal stated in the recommendations below. The wide variety of time and space scales in the ocean, their nonlinear interactions, mass conservation, and complementary data allow us in principle to use accurate information in any wavenumber band to constrain the circulation at all wavelengths. As a guide to the accuracies required, at spherical harmonic degrees lower than 20 the variance in the mean circulation is about 30–50 times smaller than the variance in the gravity field. At shorter wavelengths a spherical harmonic representation is less useful due to regional variations, but the signals we try to recover cause slopes of 30 cm over 100 km.

With this background, we now proceed to a set of recommendations to maximize the return to physical oceanographic research of current and emerging geodetic advances.

4.4 RECOMMENDATIONS

Recommendation 1—Near Term. Utilize the approved future missions, ERS-1, TOPEX/POSEIDON, and the proposed ADEOS/NSCAT, together with in-situ data and models, to estimate and model the 3-D large-scale seasonal and interannual variability of the oceans in the first period of WOCE (1990–95); to study its response to wind forcing; to estimate those wavelengths of the time-averaged circulation that orbit and marine geoid accuracies permit; to improve models of tides. Locate the satellite tracking stations (laser, DORIS, and GPS) and tide gauges with respect to a common Earth reference frame to a vertical accuracy approaching one cm.

Recommendation 2—Near Term. Immediately commence design and tradeoff studies to define the methods and space systems that will allow long-term monitoring of the oceans (WOCE second objective) with altimetry and scatterometry after 1995. This follow-on has to be defined within the next two years to ensure continuity, due to the long lead times required to put instruments in space. A continuing monitoring capability with uniform accuracy is essential to build the long, uninterrupted time series on which climate change predictions are based.

Recommendation 3—Medium Term. Measure the gravity field with the accuracy and resolution needed to:

(a) recompute the TOPEX/POSEIDON orbit a-posteriori leaving a radial orbit error whose gravity component is of the order of 1 cm (S. Klosko, 1988, pers. comm., stated a 1 cm gravity component of orbit error as an achievable goal).

(b) provide a marine geoid that is much closer to the goal of 25 km resolution, 1 cm uniform accuracy (rss over all degrees <800) than we have today.

(b') if (b) cannot be achieved, any improvements in the marine geoid at wavelengths larger than $2\pi \times 30$ km (the first baroclinic Rossby radius, which varies with latitude and stratification; see Gill, 1982), are useful to physical oceanographic research if the computation of the geoid model does not cause a correlation with oceanographic signals, and its error is below the oceanographic signals sought. Even regional geoid improvements are useful, but a global geoid improvement with uniform accuracy is most useful.

A dedicated satellite gravity mission is probably the best way to fulfill these goals, because of the need for uniform accuracy over the oceans and for the orbit calculation.

Recommendation 4—Long Term. We encourage technology developments that lead to the following long-term goals, while recognizing that advances that fall somewhat short of these goals may also be extremely useful:

(a) positions of tide gage and satellite tracking stations in the same reference system, with a vertical accuracy in the mm range. Local atmospheric and tectonic conditions at the stations must be monitored: atmospheric pressure, wind, etc, as well as long-term local gravity changes and vertical motions of the ground.

(b) orbits of altimetric satellites with no more than 1 cm rms error, single pass, without orbit discontinuities, relative to the reference frame mentioned in (4a). This differs from recommendation (3a) in that the total orbit error, not just its gravitational component, is decreased.

(c) altimetric measurements of sea level covering the ocean every 7 days at 25 km resolution to resolve energetic mesoscale features. This may achieved either by multiple simultaneous missions or by multibeam altimeters.

5. ATMOSPHERE AND CLIMATE

5.1 INTRODUCTION

In November 1987 a workshop was held in Erice on "Interactions of the solid planet with the atmosphere and climate" dealing with most of the topics which overlap with this particular chapter of the Geodesy Workshop. In this section we shall refer mainly to the conclusions of that workshop dealing with the interactions of the present atmosphere and climate with the solid Earth and the oceans. There are several other topics discussed at the conference that will not be reported here, namely on the chemical evolution of planetary atmospheres and the role of planetary impacts. On the other hand, ocean circulation, which has been discussed during this Conference, has a well known influence on the regulation of the CO_2 level in the atmosphere, and, although barely touched at Erice last year, will be mentioned in what follows.

Space Geodesy will soon reach such a high precision in the measurement of distance between two locations on earth that the intervening atmospheric delays in travel time will become the main source of error, and, thus, will determine accuracy. On the other hand, space geodesy is needed for a multitude of environmental disciplines including atmospheric physics in two different ways. First, it helps in the location of any pixel of images from space missions through precise orbital elements from satellite tracking, and, second, it delivers distance measurements as a prerequisite for the understanding of many climate processes; sea level and inland ice cap height being two obvious examples for parameters to be measured.

If the atmosphere changes travel time, and, through dispersion, phases of electromagnetic waves with different wavelengths, there should exist, in principle, a possibility to derive from space geodetic data such atmospheric (and surface) parameters as temperature, pressure and humidity.

The following subsections outline the basic features for atmospheric correction, space geodesy's value for the other disciplines, and the combination of space geodesy instruments for geodetic atmospheric and surface parameters from which the recommendations in the final section are derived.

5.2 CORRECTION OF THE ATMOSPHERE

The refractive index n of air as a measure of refraction and the deviation of the propagation velocity of an electromagnetic wave from that in vacuum is for all wavelengths a function of the total number of air molecules; thus temperature T and pressure p and especially for radiowaves, also a function of the number of polar molecules thus depending on absolute humidity. If n (≈ 1.000282 for $p = 1013$ hPa, $T = 288.15$ K and 60% relative humidity at 0.69μm wavelength) is needed to an accuracy of 10^{-6} tolerable pressure p, temperature T and water vapor pressure e errors are $\Delta p = \pm 3.6$ hPa, $\Delta T = \pm 1.0°$C, $\Delta e = \pm 25.6$ hPa within the optical domain. Within the microwave range we find $\Delta p = \pm 3.7$ hPa, $\Delta T = \pm 0.8°$C and $\Delta e = \pm 0.23$ hPa. Humidity, therefore, is the most critical parameter in the microwave space geodesy. Past and current instruments measuring water vapor include the Scanning Multichannel Microwave Radiometer (Seasat) and the Special Sensor Microwave/Imager (DMSP8).

The wave traveling time increase translates into a large distance, reaching more than two meters (~2.30 m) for the dry air contribution and fluctuating (only for microwaves) between a few centimeters and half a meter for the "wet" contribution, if both ranges are given for zenith observations. Present simple models of refractivity, $N = (n-1) \cdot 10^6$, versus height compared to meteorological soundings leave, at given surface observations of pressure, temperature and humidity, a root-mean-square error of a few centimeters (at least 2 cm) due to water vapor fluctuations but also to errors in temperature profiles for the dry air contribution. The bias for the humidity contribution is strong for the tropical areas and may reach 4 cm at 5.5 cm rms error certainly also partly due to radiosonde errors.

A presently possible strong improvement would be the use of models accepting global meteorological analyses on main pressure surfaces available from meteorological forecasting centers. However, if accuracies below the cm level are needed a direct measurement of temperature and humidity, best in the form of profiles, along the line of sight of a geodetic instrument would be the optimum choice.

5.3 GEODETIC INSTRUMENTS FOR ATMOSPHERIC PARAMETERS

The masking of all raw data by the effects of the atmosphere either through refraction, absorption, or scattering is not only a reason for correction and the cause of growing relative contributions to errors (see §5.2) but should also be exploited for the derivation of atmospheric parameters.

Refraction effects of systems like GPS and GLOMASS equipped with sensors like PRARE/DORIS could be used to derive stratospheric temperature profiles with sufficiently dense horizontal resolution in the height interval between tropopause and 35 km using all satellite occultations. Present sounding systems onboard operational meteorological satellites show rms errors above 2 K for the stratosphere; thus the GPS/PRARE/DORIS has to show its superiority.

All laser tracking stations could record the backscattering from aerosol layers in order to establish an aerosol backscattering climatology not available at present.

The parameters most urgently needed in meteorology and climatology, especially over oceans, are precipitation amount and rate. Scanning multi-frequency altimeters are candidates for this purpose, since their signals are attenuated frequency dependent by rain drops of different size and their footprint is small enough to resolve convective precipitation.

5.4 ANGULAR MOMENTUM BUDGET OF THE EARTH-ATMOSPHERE

The measurements of the length of day and the position of the rotation pole of the Earth have revealed that part of the short term changes in the frequency range of 0.01 cycles per day to 1 cycle per day are due to changes in angular momentum of the atmosphere (Hide, 1984). Thus geodetic data together with global meteorological wind data open the possibility of separating the core-mantle boundary dynamics contribution from the atmospheric one. Length of day time series in turn would allow a hindcasting of atmospheric circulation anomalies like El Niño.

A comparison has been made by Rosen (1989) between the estimate for the total angular momentum of the atmosphere obtained with the European Center for Medium Range Weather Forecast (ECMWF) and with the National Meteorological Center (NMC) models. He found that the NMC gives a higher estimate with a rms difference of 5.68×10^{24} Kgm^2 s^{-1} which would translate in l.o.d. of about 0.095 ms. The difference between the atmospheric and geodetic series is 0.13 ms and the mean uncertainty in l.o.d. is 0.045 ms so that the error in the atmospheric analysis of M is comparable to the one obtained from independent estimates. Subsequent analysis of the two data sets indicates that most of the overestimate in the NMC may be due to the absence in that model of the stratospheric levels. This example, however, is useful to clarify that most of the comparison between geodetic and atmospheric data is through the calculation of the atmospheric momentum with winds data on a grid with a resolution of about 300 km which is large enough to neglect the smaller scales. Tibaldi (1989) showed that such scales are of primary importance to explain the recent progress in weather forecasting. In the near future it is expected that satellite wind measurement could improve the accuracy from the present one of 5 m/s to about 2 m/s. This would probably affect the torque evaluation.

Rosen has also shown a loose correlation between the averaged standard deviation in the filtered l.o.d. and the global temperature regime (Salstein and Rosen, 1986). The same authors have shown that ENSO signature is clearly recognizable in the l.o.d. data set.

Madden (1989) has also shown a loose correlation between the averaged standard deviation in the filtered l.o.d. and the global temperature regime (Salstein and Rosen, 1986). The same authors have shown that ENSO signature is clearly recognizable in the l.o.d. data set.

Madden (1989) has studied the effect of 40–50 days tropical oscillation on the l.o.d. record and shown that this could be an important mechanism for the exchange of momentum between the Earth and the atmosphere. The geodetic data should also be studied as an independent validation for theories on atmospheric blocking.

5.5 ATMOSPHERIC LOADING

Atmospheric disturbances cause displacement of the solid Earth. In some instances these could be used to learn more about earth/atmosphere coupling processes. Wahr (1989) has shown that meteorological disturbances could cause displacements in surface of the oceans up to 1 cm. This result could be of some importance especially in the interpretation of gravity observations and gauge measurements of the ocean level. Dickman (1989) has shown the importance of considering the ocean response to atmospheric pressure and wind stress in evaluating the exchange of angular momentum between the atmosphere and the ocean.

5.6 LONG RANGE CLIMATIC AND ATMOSPHERIC EVOLUTION

Berner *et al.* (1983) have proposed an explanation for the warmer cretaceous based on a larger content of atmospheric carbon dioxide possibly due to a faster rate of seafloor creation with respect to the present epoch. Gaffin (1989) has developed a simple model which relates eustatic sea level changes in the last 10^7–10^8 years to seafloor creation rate changes and also found an anticorrelation with the rate of geomagnetic reversal. These studies are relevant for a precise estimation of seafloor spreading and plate motions and a possible reconstruction of the same data in the past.

5.7 OCEAN CIRCULATION AND ATMOSPHERIC CARBON DIOXIDE

The ocean is the main buffer for CO_2 in the atmosphere with surface water removing the gas and maintaining an equal pressure between the ocean and the atmosphere. The characteristic time of this circulation is only 10–20 years so that when water with a specific content of CO_2 goes back to the surface it will dissolve more gas that has increased in the meanwhile due to the burning of fossil fuel and deforestation. The organic compounds formed at the surface will settle to the bottom of the ocean where, in presence of oxygen, it will liberate nutrients (phosphates and nitrates) and again carbon (Broecker and Peng, 1982). The deep circulation of the ocean will then transfer carbon from the surface to the bottom together with oxygen. Sources of this deep water are also sources of carbon dioxide and changes in the ocean deep circulation would influence the climate both through the carbon dioxide content and the latent heat release.

5.8 CONCLUSION AND RECOMMENDATIONS

The problem of atmospheric masking, the value of geodetic data for other disciplines and the desirable combination of sensors for geodetic and atmospheric parameters have led to the following recommendation divided into four parts. Although part 1 is no real space geodesy problem, it is nevertheless kept, since most imaging missions would be of strongly increased value if geodetic information is improved. The other parts are a straight consequence of the foregoing discussion.

Recognizing (1) the need for more precise geodetic data for all imaging missions in meteorology and climatology, (2) the growing importance of the correction of atmospheric effects on space-borne geodetic measurements, (3) the strong links between motions of the Earth's pole and angular momentum of the atmosphere, and (4) the relation between paleo-climatic ice volume and glacial rebound of the entire Earth's crust, it is recommended that

(1) Adequate geodetic information should be added to operational and preoperational satellite data from meteorology, ocean and climate missions (this information should include time, orbital elements and satellite attitude allowing the location of any pixel from the imaging radiometer with the highest spatial resolution to a small fraction of a pixel).

(2) Planned geodetic space missions should be exploited for atmospheric and surface parameters, examples being (i) land surface roughness (emphasis on vegetated surfaces) laser altimeters), (ii) precipitation rate from scanning multi-frequency, radar altimeters, (iii) stratospheric temperature profiles from GPS/PRARE/DORIS satellites and ground stations, and (iv) aerosol particle backscattering from (multi-wavelength) satellite laser tracking.

(3) Information on atmospheric temperature profiles and water vapor content should be combined with geodetic raw data in order to reach the 1 mm accuracy goal for any distance between stations on Earth.

(4) Meteorological and climatological data should be provided for angular momentum patterns of the atmosphere in order to separate the core-mantle dynamics contribution to polar motions and crustal dynamics from the atmospheric one, both for short-term weather fluctuations and (paleo-)climatic states or time series. However, improvements of the initialization data of general atmospheric circulation models (which presently provide the angular momentum data) are required. Global coverage of wind vectors could be provided by satellite scatterometers for surface winds and a laser atmospheric wind satellite for profiles. A direct involvement of the World Meteorological Organization (WMO) would be helpful also considering the feedback Earth rotation data might have in explaining meteorological or climatological phenomena.

(5) A more realistic data base is needed to compare the geodetic data on angular momentum with those evaluated with meteorological data. A promising approach is provided by the next generation of satellite that will measure winds from space on a global basis with better spatial resolution and precision.

(6) Evaluation of the effects of atmospheric disturbances on geodetic data will grow in importance as the gravity measurements are further refined.

(7) Data and modeling on plate motions could give important input for long-range climatic evolution.

(8) More information is needed about deep ocean circulation styles as related to the problems of carbon dioxide content of the atmosphere.

6. EXTENSION TO THE PLANETS

6.1 INTRODUCTION

During the last two decades many important advances in our knowledge and understanding of the Earth have been made through the application of space geodetic and geophysical methods. These advances include the measurement of present-day tectonic plate kinematics, the measurement and interpretation of variations in the Earth's gravity field and tides, the measurement of the topography of the world's oceans and our understanding of its circulation, and the construction of models of the Earth's core and crustal magnetic fields.

In addition, major new technologies have been developed that have made the scientific advances possible, including the introduction of centimeter level tracking systems for spacecraft and very long baseline interferometry of extra-galactic radio sources, centimeter level laser and microwave altimeters, gravity gradiometers, and sophisticated software systems for determining orbits, processing data, and extracting the geodetic and geophysical results.

In order to further develop our understanding of the basic mechanisms and processes involved it is necessary (and technically feasible) to address the corresponding geophysical and geodetic problems as they occur on other planetary bodies in our solar system. These geophysical processes require that certain (space) geodetic measurements be made over appropriate time scales and with certain accuracies. These measurements include, but are not limited to: the gravity field of the planet and its variation with time due to external and internal forces, the rotation of the planet on its axis and the existence of any variations with time, the topography of the planet on the scale commensurate with the resolution of the gravity field and/or our knowledge of the geological processes, and the relative motions of blocks or segments of the planets crust arising from tectonic forces or realignments of material in the planets interior. In addition, direct geophysical measurements of magnetism, heat flow and seismology are needed. With these geodetic and geophysical data sets inferences can be made about several fundamental planetary processes and parameters, including, the elastic thickness of the lithosphere, the planet's mass, the state of isostasy, the nature of the planets rotation and possible connection with its atmosphere, the existence of present-day tectonic processes, and the possible existence of a fluid core, etc, etc. These processes may then be compared with those occurring on the Earth.

Table 2. Geodetic and Geophysical Experiments on Future Planetary Missions

Year of Launch	Mission	Destination	Geodesy/Geophysical Investigation
1988	Phobos 1&2	Phobos	Size, shape, gravity of Phobos. Transponder left on Phobos for orbit determination
1989	Magellan	Venus	Imaging radar, altimeter (30m, 20–30 km spot), Gravity experiment, limited to low and medium latitudes, eccentric orbit (250 km–??)
1992	Mars Observer	Mars	Altimeter (~ 1 m, 100 m spot) Gravity (~ 50 by 50) orbit, polar, 360 km–410 km Magnetics
1994	Mars 94	Mars	Radar (sounder) Magnetics Gravity (?)
1998(?)	Lunar Observer	Moon	Gravity(?) Altimetry (?) 100 km polar orbit
2000(?)	Sample Return Mission (US & USSR)	Mars	Unknown

6.2 PLANETARY MISSIONS AND PRESENT KNOWLEDGE

Several planetary missions are planned for the next several years; Table 2 summarizes these missions and briefly describes their geodetic experiments. None of these missions has geodesy or geophysics as their primary objective but some are undertaking gravity and topography measurements, albeit in a nonoptimal mode; some missions will also make magnetic field measurements. In order to improve our understanding of the processes already mentioned it will be necessary to move toward a full global scale program of observations and measurements on at least one planetary body, and preferably more, where the conditions appear to differ from those on Earth.

Table 3. Present Geodetic Knowledge of the Moon and Terrestrial Planets

Planet	Quantity	Knowledge
Mercury	Rotation variation	unknown
	$\sigma(GM)$	1 part in 10^4
	$\sigma(J_2)$	30%
	Mean radius	+/– 1 km
	Topography	unknown
	Flattening	unknown
Venus	Rotation variation	unknown
	Rotation period	234 days +/– 0.1 days
	"Chandler" wobble:	
	amplitude	200 km (?)
	period	10^5 years (?)
	$\sigma(GM)$	1 part in 10^6
	Gravity field	7×7, limited coverage
	Mean radius	+/– 0.1 km
	σ (Topography)	$\pm$ 100 m, 50 km block range $\leq$ 1 km
	Flattening	very small
	Atmospheric angular momentum	10^{-3} of solid body
Mars	Rotation variations	l.o.d. ~ 1 msec, nothing since Viking
	$\sigma(GM)$	1 part 10^5
	Gravity	18×18, patchy, not near poles
	Mean radius	+/– 0.1 km
	Topography	+/– 1 km, near equator
		+/– 2 km, poles
	Atmospheric motions	25% of atmosphere moves from pole to equator annually, change in gravity ($J2$) observable
	Magnetic field (core)	undetected
Moon	Rotation variations	nothing observed
	$\sigma(GM)$	1 part 10^6
	Gravity	16×16 near side poor, far side
	Radius	0.03 km
	Topography	100 m near side—not near poles (Earth-based radar) poor, far side

Table 3 summarizes our present geodetic knowledge of the moon and terrestrial planets and indicates the considerable variation in our depth of knowledge of these bodies.

It is apparent that any attempt at understanding these kinds of measurements in a systematic way will involve a long term program and the basis of any program of geodetic and geophysical exploration of the terrestrial planets should begin with a primary description of the fundamental parameters that describe the body. These should include, but not necessarily be limited to:

- mass
- gravity field
- topography
- core magnetic field
- crustal magnetic field
- rotation rate and any variations
- polar motion
- geodetic control (a few features established in a center of mass reference frame)

and if obtained globally on a uniform basis would enable some assessment of many of the processes described earlier. Further, the geodetic and geophysical requirements must in many cases be combined with more common planetology investigations, such as imagery and spectroscopic analyses.

Table 3 shows that our present knowledge of Mercury is very limited and that none of the parameters listed above are available for Mercury, and that only after Magellan will we have a preliminary model of the static gravity field and topography for Venus, and only after Mars Observer will the gravity field, topography, and magnetics be adequate for initial studies of Mars. For the Moon, nearly all our present knowledge is limited to the near side.

For none of the terrestrial planets do we have significant information about its rotation and no information about any time dependent changes in the gravity field arising from tidal distortions of the planet or mass redistributions due to atmospheric motions, as is believed to take place on Mars. The former is important as it can be used to provide information about tidal Love Numbers and the latter is important as it can be used to provide estimates for viscosity of the planet's interior. These methods have been successfully used for Earth by analysis of the observations of Lageos and other spacecraft over many years. Thus a spacecraft in orbit about a planet can, over a long period, provide important information about the planets interior as well as its gravity field.

6.3 A PROGRAM CONCEPT

The first step in such a program could therefore be the establishment of a geodetic spacecraft in orbit about the planet. In its simplest form this spacecraft would be in a close geodetic type orbit and carry only a radio beacon for very precise tracking. This system could provide the low degree and order gravity field, the planet's mass, and any time variation in the gravity, including tides and J2 dot. Inclusion of a magnetometer could be the next level of complication, followed by an altimeter. Other possible instrumentation includes a gravity gradiometer, and deployment of sub-satellites that permit closer approach to the surface. Whether several instruments could be combined on one spacecraft would depend on other factors and specific objectives. To obtain details of the planet's rotation will require visiting the surface of the planet and amounts to step two in any program because of its obvious increase in difficulty. With one or two beacons placed appropriately, estimates of rotation rate would be derivable and with three or more the complete rotation of the planet, including any polar motion, should be observable. At the same time the detection of tectonic motions through the change in relative positions of the beacons becomes possible. However, having reached the surface of the planet there are many other important geophysical measurements that should be made, including, seismology, heat flow etc. Knowledge of the positions of these "beacons" would be a requirement at probably the decimeter level or better, requiring some special space geodetic method, such as a GPS type measurement, or VLBI, to to applied or established on and around the planet. These locations could then become the components of a control net for the planet in a center of mass coordinate system.

6.4 RECOMMENDATION

The next long-range plan for geodynamics should include a planetary component and that we should:
1.   initiate a program of geodetic and geophysical exploration of the moon and terrestrial planets (in cooperation with existing programs in planetology);
2.   determine the following parameters for each planetary body: mass, gravity (dynamic and static), topography, magnetism, rotation rate and variations, polar motion, and establish a geodetic control grid on the surface.

The first step in the program could be the placing of a geodetic spacecraft into a stable orbit around one of the planets. A suitable choice would be the moon or Mars. The precise tracking of such a mission alone could provide detailed gravity information, including time variations, the planets mass and possibly details about the atmosphere. The mission could logically be extended to conduct topographic and/or magnetic field measurements.

The second step in the program could be the establishment of one to three instrumented sites on the surface that would enable the measurement of polar motion, planetary rotation, tectonic motions, and the measurement of heat flow, seismology, etc.

7. FUNDAMENTAL PHYSICS

7.1 INTRODUCTION

An important interdisciplinary field for Space Geodesy is the tie between areas of Astrophysics and Geophysics. In this century theoretical physics has created concepts which have provided us with dramatic changes in our perception of reality. Experiments in Space Geodesy have provided important confirmation of many of these theoretical physical concepts. There are a number of new and important continuing experiments in this field which need to be carried out in the coming decades. Currently there are three main areas of experimental physics in which the techniques of Space Geodesy can make significant contributions to fundamental physics. These are in the areas of (1) the search for gravitational waves, (2) the measurement of first- and second-order post Newtonian effects (PPN parameters in modern theories of gravitation), and (3) the measurement of feeble short-range forces associated with various higher-order theories in physics.

7.2 GRAVITATIONAL WAVES

Interplanetary spacecraft tracking is now used routinely to search for very low frequency gravitational waves. Interplanetary plasma, ionosphere and troposphere effects are currently limiting the detection capability of these experiments.

Improvement in correcting for or eliminating these effects must be pursued. This means the utilization of higher microwave frequencies for interplanetary spacecraft communication links, with the possibility of using laser carrier frequencies in the future. Development of these technologies should be pursued.

Alternative methods of detecting gravitational waves in space should be investigated. This includes the tracking of interplanetary spacecraft from space based observatories, the operation of tracking observatories on the moon, and in particular the building of dedicated multiarm interferometers for detection of gravitational waves. These interferometers should have the potential to measure spatial strain to 1 part in 10^{18} or greater.

7.3 FIRST- AND SECOND-ORDER POST-NEWTONIAN EFFECTS

The relativity gyroscope experiment GPB is now in an advanced stage of development. The predicted effect of the gravitomagnetic effect for this experiment is about 42 marcs per year and the accuracy to which it is to be measured is about 0.5 marcs. Geodesists should take account of all requirement and possible auxiliary data collection necessary to guarantee a successful outcome to this experiment. In particular this experiment represents an example of interdisciplinary cooperation which should be followed more often in future experiments.

In the field of the theory of gravitation it is possible to parameterize effects which are predicted by post-Newtonian theories of gravitation, such as those in General Relativity. These are called post-Newtonian (PPN) parameters as they can be made to show a deviation from a Newtonian reference frame.

Several of these parameters have been determined through spacecraft tracking experiments and currently set some of the best limits to these parameters. The gamma or time delay parameter and beta or perihelion shift parameter are currently known to a few parts in 40^{3}. It is feasible to measure these parameters to a few parts in 10^{6} by requiring more precise tracking of spacecraft, including drag free design and the technology to carry very precise frequency standards that operate in space for long durations. A number of solar system missions including Solar Probe, Icarus lander and Mercury orbiter are representative for these studies.

The most stringent limits to a possible change in the gravitational constant G have been set by precise ranging to Mars using the Viking planetary lander. These measurements are being extended in a cooperative experiment between the US and USSR using the recently launched Phobos spacecraft. Similar experiments should be encouraged in the future to allow even more precise limits to possible changes in this fundamental unit of gravitation to be determined.

7.4 FEEBLE SHORT RANGE FORCES

The inverse square law of gravity has most recently come under intense theoretical and experimental scrutiny. In some modern theories of physics the existence of particles of small but
nonzero mass (axions in QCD, dilations in superstrong theory, hyperphotons, gravitons, etc.)
can lead to Yukawa type contributions to the potential between bodies.

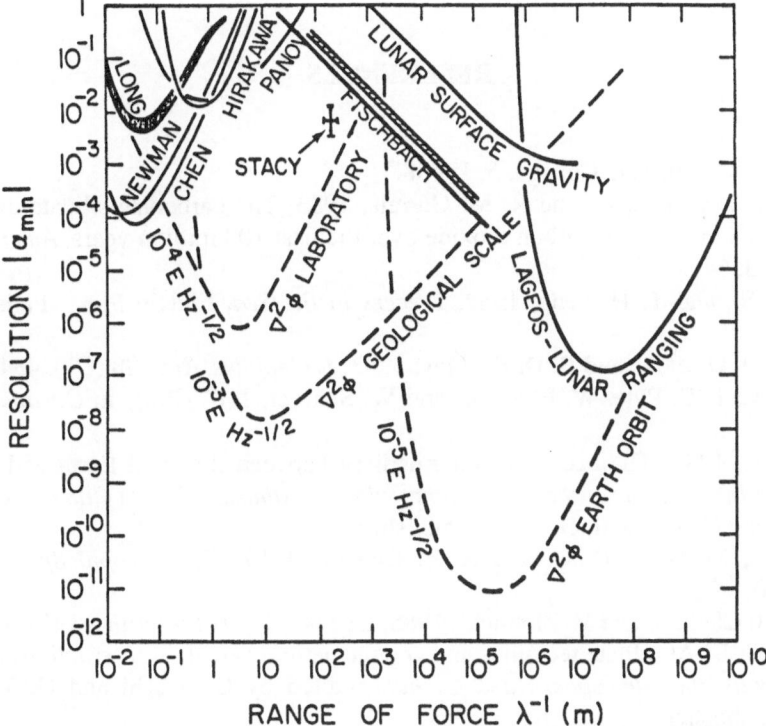

Fig. 2. The figure shows the current limits (solid lines) and projected limits (dashed lines) that can
be placed on the anomalous potential, where

$$\phi = -\frac{GM}{r}\left(1+\alpha e^{-r/\lambda}\right).$$

For ranges in excess of about 300 km the strength of such forces are constrained by
planetary and satellite orbit observations. In the range 1 meter to 300 km the constraints are
weaker. Space experiments designed to measure these potentials, such as precise gravity gradiometry, should be considered and may play an important role to discriminate between various fundamental theories of physics. Fig. 2 indicates the current limitations that can be set to
these anomalous potential interactions. Where appropriate future space geodetic experiments
should include planning for possible measurement of feeble short-range forces.

7.5. SUMMARY

A natural role for space geodesy exists in the bridging of experimental gaps between the fields of Astrophysics and Geophysics. Fundamental physical effects can be successfully measured and theoretical model parameters can be studied using techniques developed for space geodetic experiments. There is an opportunity for Space Geodesy to make important fundamental contributions to physics and this should be exploited in the planning of future missions.

REFERENCES

Barnett T.P., 1983, Climatic Change, 5, 15–38.

Berner, R. A., A. C. Lasaga, and R. M. Garrels, 1983, The carbonate-silicate cycle and its effects on atmospheric carbon dioxide over the past 100 million years, *Amer. Jour. Science,* **283**, 641.

Broecker, W. S., and T. H. Peng, 1982, *Trecers in the Sea*, Elalgio Press, Palisades, New York.

Cheney, R. E., J. G.Marsh and B. D. Beckley, 1983, *J. Geophys. Res.,* **88** (C7), 4343–4354.

Diamante J. M., T. E. Pyle, W. E. Carter and W. Scherer, 1987, Prog. in *Oceanography* **18**, 1–21.

Dickman, S. R., 1989, The oceans as intermediary between the solid Earth and the atmosphere, in *Interactions of the Solid Planet with the Atmosphere and Climate*, edited by E. Boschi and G. Visconti, G. Galilei Publisher.

Douglas B. C., McAdoo, D. C. and R. E. Cheney, 1987, *Rev. Geoph. Space Phys.* **25**, 875–880.

Fu, L.-L., D. B. Chelton, and V. Zlotnicki, 1988, *Oceanography Magazine,* **1** (2) (in press).

Gaffin, S., 1989, BLAG, Plate tectonics and geomagnetic reversal, in *Interactions of the Solid Planet with the Atmosphere and Climate*, edited by E. Boschi and G. Visconti, G. Galilei Publisher.

Gill A. E., 1982, *Atmosphere-Ocean Dynamics*, Academic Press, 662 pp.

Hide, R., 1984, Rotation of the atmospheres of the Earth and planets, *Phil. Trans. R. Soc. Lond.* A, **313**, 107–121.

Knox, R. A., 1987, *Rev. Geophysics,* **25** (2) 181.

Lambeck, K., 1988, *Geophysical Geodesy: The Slow Deformations of the Earth*, Oxford University Press, 718 pp.

Madden, R. A., 1989, Relationship between the 40–50 day oscillation in the tropics and the length of day, in *Interactions of the Solid Planet with the Atmosphere and Climate*, edited by E. Boschi and G. Visconti, G. Galilei Publisher.

Meier, M. F., 1984, Contribution of small glaciers to global sea level, *Science,* **226**, 1418.

Nakiboglu, S. M., K. Lambeck, and S. Hekimoglu, 1988, Present secular changes in global sea-level, *J. Geophys. Res.*, submitted.

Rosen, R. D., 1989, Recent progress in the study of the Earth-Atmosphere angular momentum budget, in *Interactions of the Solid Planet with the Atmosphere and Climate*, edited by E. Boschi and G. Visconti, G. Galilei Publisher.

Rubincam, D. P., 1984, Postglacial rebound observed by LAGEOS and the effective viscosity of the mantle, *J. Geophys. Res., 89*, 1077–1087.

Salstein, D. A., and R. D. Rosen, 1986, Earth rotation as a proxy for interannual variability in atmospheric circulation, 1860–present, *J. Climate Appl. Meteor., 25*, 1870.

Stephenson, F. F., and L. V. Morrison, 1984, Long-term changes in the rotation of the Earth: 700 BC to AD 1980, *Phil. Trans. Roy. Soc. Lond.,* **A313**, 47–70.

Tibaldi, S., 1989, Impact of orography on global scale weather forecasting, in *Interactions of the Solid Planet with the Atmosphere and Climate*, edited by E. Boschi and G. Visconti, G. Galilei Publisher.

Wahr, J., 1989, Some effects of the atmosphere on the Earth's rotation and on crustal deformation, in *Interactions of the Solid Planet with the Atmosphere and Climate*, edited by E. Boschi and G. Visconti, G. Galilei Publisher.

WOCE Science Steering Comm., 1988, US WOCE Planning Office, Dept. of Oceanography, Texas A&M Univ., College Station, TX. USA.

WOCE International Planning Office, 1988, WMO/WCRP-11 and 12, Joint Planning Staff for the WCRP, WMO, Case Postale 5, CH-1211 Geneva 20, Switzerland.

WMO/IOC Inter-Governmental TOGA Board, 1988, WMO/TD No. 216., WMO, Case Postale 5, CH-1211 Geneva 20, Switzerland.

Wunsch C., 1972, *Rev. Geoph. Space Phys., 10*, 1–49.

Wyrtki, K., 1984, *J. Geophys. Res., 89*, 10419–10424.

Wyrtki, K., K. Constantine, E. J. Kilonsky, G. Mitchum, B. Miyamoto, T. Murthy, S. Nakahara, and P. Caldwell, 1988, The Pacific Island sea level network, JIMAR report 88-0137, University of Hawaii, Honolulu, HI 96822.

Yoder, C. F., J. G. Williams, J. O. Dickey, B. E. Chutz, R. J. Eanes, and B. D. Tapley, 1983, Secular variation of earth gravitational harmonic J2 coefficient from LAGEOS in nontidal acceleration of earth rotation, *Nature, 303*, 757–762.

Chapter 5

INSTRUMENTATION

1. INTRODUCTION

William G. Melbourne and Christoph Reigber

Over the past two decades the technological advances in space-based instrumentation have led to remarkable improvements in our scientific knowledge of solid Earth and oceans processes. In geodetic systems for example, we have realized two orders of magnitude improvement in system accuracy since Williamstown. Through satellite orbit perturbation analysis comparable improvements have been achieved in the long wavelength spectral components of our gravity models. Satellite orbit determination accuracy has improved from dekameters to decimeters, and even to centimeters for specialized satellites such as LAGEOS. Knowledge of ocean topography and the variability of ocean circulation was virtually non-existent 20 years ago. Today, through advances in satellite radar altimetry, we understand a great deal more about ocean circulation. Similarly, magnetometry and other remote sensing systems have advanced dramatically. Even so, there remain a number of technological challenges for space-based instrumentation and flight missions that arise from current scientific requirements for higher accuracy, broader applicability, and greater temporal and spatial resolutions.

This chapter deals principally with microwave-based geodetic systems, laser ranging systems, gravitational and magnetic field measurement systems, and topographic mapping systems. It addresses the status and accuracies of current systems, their limiting error sources, and the future trends in performance of these current systems. In some instances, the chapter also takes a long view toward systems planned for the later part of the coming decade such as Gravity Probe-B, ARISTOTELES, GLRS on Eos, and a gravity mission flying a 3-axis superconducting gravity gradiometer.

The emphasis in this chapter is on the technology aspects of these systems, particularly with regard to accuracy and applications. No attempt has been made to establish priorities or costs for these systems; it is difficult to do the former without having done the latter. The forum and time constraints of the Erice meeting made it difficult to align the instrumentation priorities with the scientific priorities emerging from the companion panels. Similarly, there was little time to establish standards for cost assessment and for assembling cost technical information so that realistic cost-performance comparisons among competing and/or complementary technologies and systems could have been made. As a result, there are a number of unanswered questions regarding priorities relating to the development of key technologies and the deployment of systems. Nevertheless, the technical information in this chapter should serve future panels deliberating on priorities and on program plans for development and deployment of these systems.

2. SCIENTIFIC AND MEASUREMENT REQUIREMENTS

William G. Melbourne and Christoph Reigber

The scientific requirements articulated in earlier chapters should profoundly affect plans for development of technology and for deployment of measurement systems. These requirements, as they pertain to space-based instrumentation, can briefly be summarized as follows:

1. Relative velocities of crustal blocks at 1 mm/yr accuracies are required; but this requirement goes a great deal further. It emphasizes the need to determine departures, including transients, of current tectonic motions from long-term averages based on models and geological data; it requires high accuracy strain distribution measurements in plate boundary deformation zones with unprecedented spatial and temporal resolutions ranging over 10^0-10^3 km and from hours up to years.

2. Positioning accuracies on land approaching 1 mm are called for.

3. Earth rotation vector accuracies of 0.1 mas, measured down to sub-diurnal periods, are required.

4. A global gravity model with accuracies in the 1 mgal region and 100 km resolution is required.

5. Determination of the vector geomagnetic field with 1 nT accuracy and 100 km spatial resolution is required. Measurements at ~200 km and at ~600 km are required to separate the crustal and main field.

6. Complete global land and ice topography surveys are required in a uniform reference system with 1 m vertical accuracy and 100 m horizontal resolution with temporal resolutions ranging from several months to decades.

7. Global ocean topography is sought at 1 cm accuracy with spatial resolution down to 10 km and temporal resolutions of 1 week spanning several years; comparable accuracies are required for the spectral components of the ocean geoid to derive ocean circulation models from the topography. For altimetry missions this requirement also translates into 1 cm orbit determination accuracies in the radial components of the satellite orbit.

8. Sea floor position accuracies of 1 cm are required to support deformation studies at spreading centers and subduction zones.

The above summary of the scientific requirements should provide the basis to define most of the enabling technology that will need to be developed and the required programs, both ground-based and space-based, for deploying existing, improved and/or totally new measurement systems. There are additional scientific requirements pertaining to such problems as time variable phenomena in gravity and tides, Love numbers, reference frame alignment, etc.; but these objectives can mostly if not entirely be achieved with instrumentation developed to meet the primary set discussed above.

Geodetic Systems. It is likely that both microwave and laser geodetic systems will achieve instrumental accuracies approaching 1 mm within the 1990's if appropriate resources are applied. Whether environmental errors, e.g., tropospheric delay, seasonal ground effects, etc., can concomitantly be reduced to 1 mm remains an open question. However, the greatest impact on the future development and deployment strategies of geodetic systems arises from the new scientific requirements resulting from the successes of the past two decades. These requirements now call for less emphasis on detecting plate motion, particularly on plates where knowledge of relative motion is well-constrained from models. Instead, they emphasize measurements of differential motion with very high resolution within plate boundary deformation zones over widths of 100 to 1000 km; they also emphasize the detection of temporal variability including directions orthogonal to the long-term plate motion vectors. The extensiveness of the measurement programs inferred from these requirements, if widely conducted, would lead to enormous costs with present systems. Hence, the costs of high accuracy geodetic systems will have to be greatly reduced, by upwards of an order of magnitude, to carry them out practicably. In addition to accuracy, a major challenge will be the development of the enabling technology that will result in major reductions in capital costs, and equally important, in the costs associated with the operations of these systems and in the data management and analysis tasks.

These and related requirements have strong implications for the future course of geodetic systems. The strategies for their development and deployment are fairly evident and can be addressed by posing the following questions:

a)   To best exploit their respective strengths, how should the mix of geodetic systems, satellite lasers, VLBI, and satellite microwave systems, e.g., GPS and GLONASS, be developed to improve accuracy and costs?

b)   On a relative basis how many systems of each type should be acquired?

c)   How should they be used?

The scientific requirements on geodetic systems can be placed in the following applications categories:

a)   Arrays for deformation studies,

b)   Networks for reference frames (including constraining rigid plate motions and monitoring Earth rotation variability),

c)   Tracking systems for satellite dynamics and satellite-based remote sensing applications, and

d)   In-situ applications, e.g., sea floor deformation and tides.

For array applications the GPS, possibly augmented by other navigation and spaceborne tracking systems, is almost certain to be the technique of choice throughout most of the 1990's. Inasmuch as the accuracies of GPS derived baselines are now comparable to those obtained by SLR and VLBI (and they are likely to improve in the future), the prospects for greatly reduced capital and operating costs with the GPS far outweighs other considerations. Indeed, equipment costs, already more than an order of magnitude lower than SLR and VLBI costs, are likely through large scale VLSI technology to drop another order of magnitude by 1992. This will enable unattended, continuously operating and remotely monitored arrays of GPS receivers that are networked together and that are highly automated, much like existing seismic arrays today. The prospect of GLRS on an EOS platform may alter this stategy in the late 1990's because of potential low ground deployment costs; the complementary uses of these two systems should be studied further.

For reference frame applications, SLR and VLBI appear to be the primary techniques. For Earth rotation monitoring the GPS is likely to fill a complementary role that would alter the mix of these three techniques for this application. VLBI appears to be most capable of detecting short period variability. Although the number of systems required for reference frame applications is moderate, the highest performance is required. This means that for SLR and VLBI systems more emphasis should be placed on the development of 1 mm system accuracy (even at the risk of increased costs) and less on procurements of large numbers of these systems.

For satellite tracking applications such as precision orbit determination for satellite geodesy and ocean circulation, GPS and SLR are likely to remain the primary techniques through the 1990's. However, for gravity recovery microwave tracking techniques, such as satellite-to-satellite tracking (SST), and other field measurement systems such as gravity gradiometry, may supersede this approach by the end of the decade for these particular applications.

Sea floor geodesy will be developed using a combination of ocean platform-based acoustic ranging for differential postioning of sea floor monuments and GPS to tie the ocean platform to existing reference frames

Field Measurement Systems. *Gravity Field.* The development of systems to achieve the desired detailed and accurate gravity field (100 km spatial resolution and 1 mgal accuracy) will likely proceed in an evolutionary way. Starting from the best current gravity models for the long and medium wavelength features (e.g., GEM-T2) and short wavelengths features (e.g., OSU 86 F), steady but increasingly logarithmic improvement will be made with (i) SLR tracking of forthcoming LAGEOS type satellites at lower altitudes and appropriate inclinations, (ii) microwave tracking of NOVA, SPOT, TOPEX, and ERS-1 satellites and (iii) with more accurate and more complete gravity data in the continental and oceanic regions.

TOPEX will carry a GPS receiver and would mark the first use of the GPS in an SST mode to recover gravity. This SST experiment will be followed by the Gravity Probe-B, which will also carry a GPS receiver, will fly in a considerably lower orbit (600 km) and will be drag free. The GPS SST tracking from this mission could significantly improve the medium wavelength features of the field.

ARISTOTELES could be the first dedicated gravity mission, providing the 100 km resolution of the field. Planned to be flown around 1995 at 200 km altitude and being equipped with a room temperature gradiometer, a GPS receiver and a magnetometer, but not being drag-free, the accuracy of the recovered field would be somewhere in the range of 3-5 mgals. Although less sensitive, it would be a very valuable precursor to the ultimate step planned near the end of the 1990's: a 3-axis superconducting gravity gradiometer mission (SGGM) in a drag-free environment and with measurement accuracy of 0.0001 $E/Hz^{1/2}$. This mission would provide similar resolution, with accuracy of roughly 1 mgal.

Magnetic Field. A similar stepwise procedure might be realizable for the improved determination of the various components of the Earth's magnetic field: (i) long-term monitoring of the main field and its temporal variations with the MAGNOLIA/MFE mission, (ii) investigation of the external fields with the NASA EMF mission, and (iii) global high resolution crustal anomaly field recovery with a few nT accuracy, with ESA's low orbiting ARISTOTELES mission, if the satellite is equipped with a scalar and vector magnetometer. Tethered systems could provide improved accuracy and resolution.

Topographic Mapping Systems. Topographic mapping accuracy and spatial resolution requirements differ considerably for various research and application tasks. For most land and polar applications, a surface elevation data set of about 100 m horizontal resolution and 1 m vertical accuracy is adequate. This data set should cover the entire Earth. Possible techniques by which such resolutions might be achievable are: narrow beam radar altimeters, scanning laser systems, microwave interferometers, and advanced optical stereo systems.

3. LASER RANGING TECHNIQUES

Steven C. Cohen and Michael R. Pearlman

This section deals with Satellite Laser Ranging (SLR), Lunar Laser Ranging (LLR), and Spaceborne Laser Ranging (SBLR); it reviews the current status of each, the current limiting factors, work in progress, and progress anticipated over the next decade.

The Williamstown Conference in 1969 projected a requirement on laser ranging of ± 2 cm. This was presented as range accuracy or systematic error on a pass-by-pass basis. The community saw the need for doing better, but with range accuracies then about 1 meter, they tried to put their requirement on a practicable basis. Unfortunately, with ± 2 cm range accuracy, measurements of crustal movement would require long integration times, and high frequency, low amplitude structure would be impossible to measure.

Over the past several years, lasers and other techniques have surpassed the Williamstown requirement, having reached measurement accuracies of 1 cm, and still offering promise of considerably better performance. Based on this, and the evolving need to see smaller structure over shorter observation times, the Laser Ranging community is now striving for measurement accuracies of 1 mm with occupation times of a week or less (cf. Chapter 3).

3.1 SATELLITE AND LUNAR LASER RANGING

Current Status. Satellite and Lunar Laser Ranging use short pulse lasers to make range measurements from the ground to retroreflectors on artificial satellites or the moon (Degnan et al., 1985). The quantity of interest is time-of-flight corrected for ranging system internal delay (calibration), atmospheric refraction (delay), retroreflector offset to the spacecraft center-of-mass (or lunar reflector position), and network epoch synchronization. SLR measurements are referred to the Earth center-of-mass, and geodetic and geophysical quantities are measured by aggregating data over a number of satellite passes. At the moment, satellite targets include one high satellite, LAGEOS (5900 km), and a complex of low satellites.

LLR measurements are referred to the inertial frame of the lunar/planetary ephemeris system (Dickey et al., 1984). Lunar and geophysical quantities are measured by aggregating data from one or more of the four arrays over a number of lunar transits.

Worldwide there are approximately 30 fixed SLR stations, of which 20 have high performance and routine or scheduled operation. A list of the stations providing data during the last year is included in Table 1. In addition, there are systems under development in Japan, PRC, USSR, Saudi Arabia, Italy, and India. At the moment, density of coverage varies with geography: there are many stations located in Europe, the U.S., and the Orient; there is no coverage in Africa, Indochina, and the Indian Ocean. Many groups tend to keep their fixed SLR systems "close to home." There is also considerable variation in the technology from system to system: some are at the state-of-the-art, using mode-locked lasers, multi-channel plate tubes, pulse shape/size sensitive discriminators, 20 ps resolution electronics, and either internal or very close range target calibration. The very oldest vintage systems are still using Q-switched lasers, conventional photomultipliers, threshold detection, 1 ns resolution electronics, and billboard target calibration. Fortunately, some of these older systems such as those in Arequipa, Matera, and Bar Giyyora are now in the process of being upgraded or replaced.

Ranging performance for typical SLR systems is shown in Table 2. Although it is common to quantify ranging machine performance in terms of single shot range noise, the most relevant quantity is accuracy, taken here as systematic excursions seen on a pass-by-pass basis. In practice, these error signatures are observed either through careful ground-based measurements designed to isolate individual error components, such as range variation with signal strength or azimuth (see Fig. 1), or through side-by-side collocation or ranging by two systems (see Fig. 2).

The recovery of station positions or other geophysical parameter involves orbital computations in which data is averaged over a number of passes and, therefore, sky geometry. The newer ranging systems with 1 cm quality data are producing station positions of 2 cm accuracy with about 100 passes of data (Christodoulidis et al., 1985; Tapley et al., 1985). Older systems, even those with accuracies of 3-6 cm, are producing station positions only a factor of two worse owing to the strength of data averaging.

Table 1 Fixed Satellite Laser Ranging Stations Providing Data During the Last Year

AREQUIPA; PERU
BAR GIYYORA; ISRAEL (MOBLAS 2)
BOROWIECZ; POLAND
CHANG CHUN; PRC
GRASSE; FRANCE
GRAZ; AUSTRIA
GREENBELT; MD (MOBLAS 7)
HELWAN; EGYPT
HERSTMONCEUX; UK
MATERA; ITALY
MAZATLAN; MEXICO (MOBLAS 6)
METSAHOVI; FINLAND
MONUMENT PEAK; CA (MOBLAS 4)
MT. FOWLKES; TX
MT. HALEAKALA; HI
ORRORAL VALLEY; AUSTRALIA
POTSDAM; GDR
QUINCY; CA (MOBLAS 8)
RIGA; USSR
SANTIAGO DE CUBA; CUBA
SHANGHAI; PRC
SIMOSATO; JAPAN
WETTZELL; FRG
YARRAGADE; AUSTRALIA (MOBLAS 5)
ZIMMERWALD; SWITZERLAND

Table 2 SLR System Performance

	Current Technology	*Older Vintage Technology*
SINGLE SHOT NOISE	0.7-3.0 cm	10-15 cm
NORMAL POINT NOISE	0.2-0.4 cm	5 cm
RANGING MACHINE ACCURACY (SYSTEMATIC ERRORS/PASS)	0.5-1.0 cm	3- 6 cm
RANGE MODELING ERRORS (REFRACTION, C/M CORRECTION)	0.5 cm	0.5 cm
STATION POSITION RECOVERY (100 PASSES/3 MOS.)	2 cm	4 cm
BASELINE CHANGE (MONTHLY SOLUTION)	±FEW MM/YR	±SEVERAL MM/YR
POLAR MOTION	3-6 cm	
LENGTH OF DAY	0.02 ms	

There are presently seven mobile SLR systems:

System	*Owned/Operated*
MTLRS-1	IFAG/FRG
MTLRS-2	DUT/Holland
TLRS 1-4	NASA/USA
HTLRS	Hydrographic Office/Japan

Additional systems are under development in Italy and the USSR. Typically, mobile systems occupy sites for periods of 6-12 weeks and then require 1-10 days for relocation, depending upon distance and logistics. Since the mobile laser systems are fairly new, they tend to be of recent technology and at the higher end of the performance spectrum.   Sites currently available to accommodate mobile SLR are listed in Table 3.

There are five Lunar Laser Ranging (LLR) stations: Mt. Fowlkes (Texas), Mt. Haleakala (Hawaii), and Grasse (France) which are in routine operation; and the Crimea (USSR) and Orroral Valley (Australia) which are in engineering status. In addition, stations are under development in Japan and the Federal Republic of Germany and should be operational within a year. The current operating network is geographically very lean, the lack of an operational Southern Hemisphere station being the most critical shortcoming.

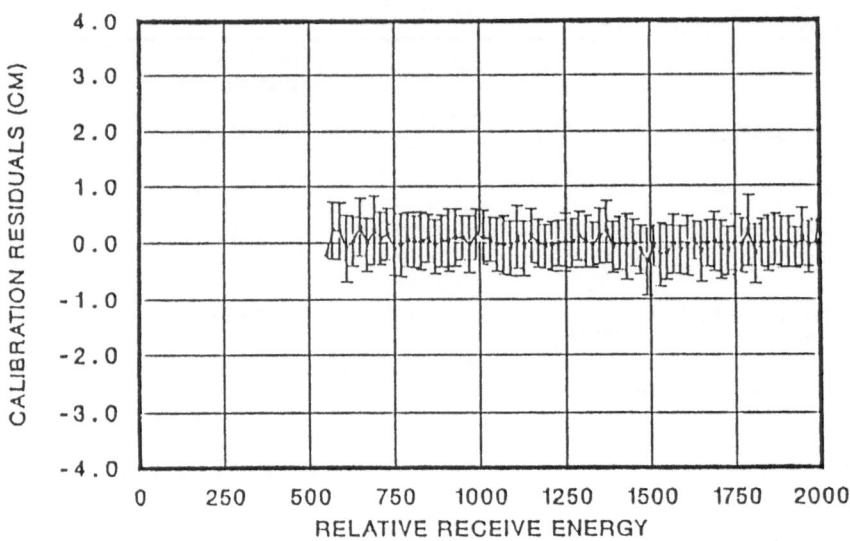

Fig. 1 Receive Energy Dependence; Moblas 8 LAGEOS 3/25/88 AT 12:09 GMT

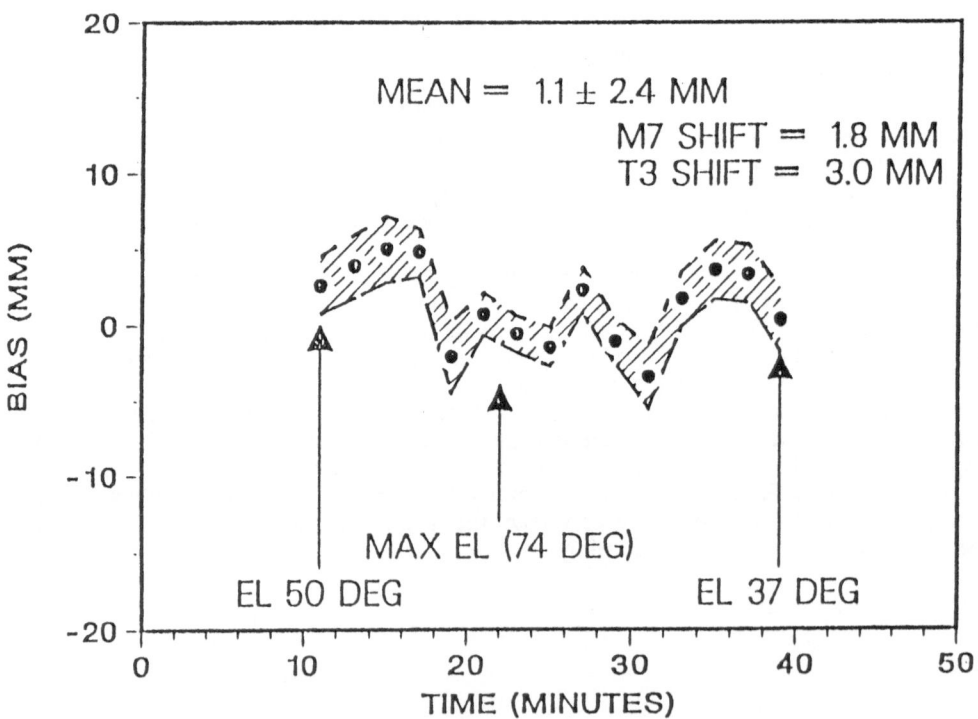

Fig. 2 TLRS-3 minus MOBLAS-7; LAGEOS 11/01/87 3:29

Table 3 Mobile SLR Sites
_____

WEGENER/MEDLAS Program

SWITZERLAND	MONTE GENEROSO
ITALY	MATERA
	LAMPEDUSA
	PUNTA SA MENTA
	BASOVIZZA
	MEDICINA
GREECE	ROUMELI
	KARITSA
	XRISOKELLARIA
	ASKITES
	KATTAVIA
	DIONYSOS
TURKEY	YOZGAT
	DIYARBAKIR
	YIGILCA
	MELENGICLIK

Crustal Dynamics Program

WESTFORD, MA
RICHMOND, FL
BEAR LAKE, UT
MOJAVE, CA
FLAGSTAFF, AZ
PLATTEVILLE, CO
VERNAL, CA
YUMA, AZ
OWENS VALLEY, CA

CERRO TOLOLO, CHILE
SANTIAGO, CHILE
AREQUIPA, PERU
CABO SAN LUCAS, MEXICO
EASTER ISLAND
TAHITI

Japanese Marine Geodetic Control
ISHIGAKI-SHIMA
TITI-SHIMA
_____

The LLR stations have very significant differences in design, but all are using recent technology. An overview of the LLR System Performance is shown in Table 4.

Table 4 Lunar Laser Ranging System Performance

SINGLE SHOT NOISE	2-6 cm
NORMAL POINT NOISE	1-3 cm
RANGING MACHINE ACCURACY (SYSTEMATIC ERRORS)	2-4 cm
RANGE MODELING ERRORS (REFRACTION)	0.5 cm
UT1	.1 ms
Δ LATITUDE	5 cm
GEODETIC PRECESSION (MOON)	2 %
NORDTVEDT EFFECT	<5 cm

Current Limiting Factors and Activities Underway: Satellite Laser Ranging.
    1. *Limited Number of Satellites.* For measurements of crustal motion and Earth rotation, the Williamstown Report recognized the need for a complex of three retroreflector satellites in high orbit to provide sufficient geometry and opportunity for observations. As of the end of 1988, there is a complex of satellites at lower orbital altitudes but only one satellite, LAGEOS, at a higher altitude. As a result, it can take several months to acquire sufficient data to accurately determine station positions and changes in position.
    A number of satellites to be launched over the next five years will radically change the situation. These include:

High Satellites		*Low Satellites*	
METEOSAT P2 (ESA)	1988	STELLA (France)	1990-92
ETALON (USSR)	1989	ERS-1 (ESA)	1990
LAGEOS II(Italy/US)	1991	TOPEX (France/USA)	1991
LAGEOS III (Italy/US)	1992 (?)		
ACRE (US)	1993 (?)		

The complex of higher satellites will provide both the orbital geometry and nearly continuous opportunity for observation which should allow SLR site occupations to be reduced to a week or less, provided that tracking support is available at required accuracy for a sufficient complex of these satellites.
    2. *Refraction.* The commonly used Marini and Murray model for refraction delay correction has an estimated accuracy of 0.5 cm at 45° elevation (Marini and Murray, 1973). The error is much worse at lower elevations, but the symmetry of the passes provides considerable benefit through data averaging.
    Efforts are underway now to adapt some of the improvements developed in the CfA 2.2 refraction model (Davis et al., 1985) for VLBI to the optical region. Since the optical region is less influenced by water vapor, the improvements may not be as dramatic, but may still offer considerable benefit.
    The most promising technique for refraction corrections in the optical region compatible with millimeter accuracy is direct along the path dispersion measurements using multiple frequency ranging (Abshire and Gardner, 1983). The technique uses the difference in round trip arrival time

of frequencies selected to accentuate the dispersive effects of the atmosphere, and thereby give a direct measure of the effect of the intervening atmosphere. Work is currently underway at CNES with 1.06 micron/5300Å and at GSFC with 5300Å/3533Å. Other groups are pursuing similar systems, and satellite ranging is anticipated within the year.

Because of the nature of the measurement, dispersion measurements must be made to sensitivities of 10-20 times greater than the required accuracy (Abshire and Gardner, 1985). Millimeter accuracies require dispersion measurements of .05-.1 mm and probably ranging at three frequencies is likely to be necessary to reduce the sensitivity required and to allow more than one dual-frequency estimate to be made for comparison and control of systematic errors at the .2 mm level. Fortunately, commercially available streak cameras are already providing subpicosecond resolution (.1-.2 mm), and the technology is evolving very rapidly to address the systematic error sources at this level.

At the moment, it is unlikely that multiple frequency ranging machines would be required at every site. The first order of business is to develop the machines and use them for model evaluation and development. Once the limit of refraction corrections based on ground-based data can be evaluated at a variety of sites and conditions, then the economics and the value of going to multiple frequency systems could be evaluated. Model improvements alone may yield accuracies of a few mm for properly instrumented single frequency ranging systems.

3. *Spacecraft Center-of-Mass Correction.* Spacecraft center-of-mass correction is measured in the laboratory before launch. For a geometrically simple spacecraft like LAGEOS, the correction is taken as independent of aspect, and constant over lifetime. For more complex spacecraft, such as TOPEX, the correction depends upon aspect and even inertial fuel consumption. The most accurate correction determined to date is that of LAGEOS which is estimated at about ± 3 mm, (Fitzmaurice et al., 1977) and this limitation was imposed by the laboratory equipment used in 1976 prior to launch. More elaborate equipment, including multiple frequency capability, is now being made available for tests on LAGEOS-II, which should give us improved corrections for both LAGEOS-I and LAGEOS-II.

Until these measurements are conducted, we will not know the limitation of the LAGEOS array, but it appears that with averaging corrections the 1 mm or better should be attainable. At some point, however, limitations in accuracy will be imposed by polarization effects on the non-silvered retroreflectors.

4. *Geodetic and Orbital Modeling.* In spite of the fact that the ranging machines have demonstrated centimeter accuracies, it takes a considerable amount of data to reach 1 cm accuracy baselines and station coordinates. This is partially due to the geometry limitation and partially due to the quality of the geodetic and orbital models.

Historically, this modeling has been driven by the data quality and quantity. As the 1 cm data base grows and additional satellites are introduced, we anticipate the evolution of model quality to continue. In particular, the 1 cm data has been available from a fairly wide network for only a couple of years, and as a result, model development which relies on the historical data base is dependent upon a considerable amount of lower quality and quantity data. This situation, however, is changing rapidly, and line and station coordinates will not be model-limited above 1 cm for very long. Millimeter accuracy baselines may only be available over baselines short enough to allow geometric strengthening which reduces the impact of orbital uncertainties. For LAGEOS and for ETALON these could extend to nearly global baselines (8-12 thousand kilometers).

5. *Operational Limitations.* Limitations are imposed on the performance of the global SLR network as a result of operational and technical differences among stations. For the most part, SLR stations are locally built using the same general philosophies, but differ in configuration, components and subsystems, calibration, operating procedures, and software. In addition, these independently built and operated systems are operated with very different budget and manpower constraints which lead to differences in temporal coverage, system reliability, and implementation of upgrades for improved performance. As a result, there are differences in: data quality, data quantity, daylight ranging capability, and data turnaround.

Considerable effort is being made within the international community through organizations such as the IAG/COSPAR Commission on the International Coordination of Space Techniques for

Geodesy and Geodynamics (CSTG) to introduce more standardization, particularly in: calibration, data handling procedures, and specialized components (such as the micro-channel plate tube). There is a strong tendency for new stations to be built by a few manufacturing groups (such as BFEC, TPD, EOS, and Intercosmos) that are now following much more common paths. In addition, the older, more unique and problematic SLR systems are tending to fall into disuse or are being totally replaced by new systems.

An example of activities presently underway to foster standardization is the development of UNIX-based software for common use on site for:

. Data Evaluation
. Machine Performance Evaluation
. Orbit/Prediction Update
. Normal Point Generation
. Multisatellite Tasking/Interleaving
. Communications

6. *Geographic Distribution.* There are some areas of the world, such as Africa, the Indian Ocean, South America, and Indochina, where there is little or no SLR coverage for spacecraft tracking, global crustal motions, or local measurements. This situation will be relieved somewhat over the next five years through fixed systems now under development in/for: Japan, China, USSR, Saudi Arabia, Italy, and India. In addition, mobile lasers now under development in Italy, France, and the USSR should provide greatly enhanced coverage.

7. *Limitation in Manpower.* With the additional satellites and the demand for more comprehensive spatial and temporal coverage, stations will be required to extend hours of operation and reduce downtime. Since manpower is already the greatest expense at most stations, operations are limited to 1-2 shifts (40-80 hrs.) per week. The tendency now is to implement more automation, first for preparing and validating data to speed throughput and monitor system performance, and second for reducing manpower during data acquisition. Some systems now under development envision hands-off operations for at least part of the daily data acquisition and maintenance activities.

8. *Cost.* The single most important factor limiting SLR-determined crustal motion is cost, which limits the number of deployments and hence the spatial and temporal resolution of the measurements. GLRS (see next section) may significantly reduce the costs of laser-based measurements of crustal deformation.

Current Limiting Factors and Activities Underway: Lunar Laser Ranging.
Many of the issues and activities discussed under SLR, such as refraction, modeling, and operational considerations, are pertinent to LLR. In addition:

1. *Limited Number of Stations and Geographic Coverage.* At the moment, there are only three LLR stations in routine operation and no Southern Hemisphere coverage. This places very severe limitations on both global kinematics and lunar science.

The addition of operating stations at the Crimea, Orroral Valley, Wettzell, and Tokyo will provide comprehensive longitudinal coverage and some data from the Southern Hemisphere. This will constitute a very significant improvement in geographic coverage. There would still be a very strong desire for a second Southern Hemisphere site in either South America or Africa.

2. *Limited Data Yield.* Lunar ranging is difficult, requiring a highly optimized system for routine success. This, coupled with data limitations due to phase of the moon, daylight conditions, cloud cover, and seeing condition (atmospheric propagation), means that only groups with adequate technical and financial support will be active and successful participants. The threshold for success has been very high, and in the past LLR groups have been required to do considerable sophisticated technical development on their own.

The availability of commercial companies now willing to furnish part or all of the LLR system may relax the development requirement on individual groups. Some economies based on sharing software and hardware design/fabrication will be forthcoming. On the other hand, the low data yield is pressing for more and more sophistication to expand data yield and data quality.

Future Improvements in Ranging System Performance. The available technology has evolved beyond the stage of current SLR and LLR equipment in the field. To reach 1 mm ranging accuracy will require an order of magnitude improvement in hardware performance, but even with modest projections based on current technology, it does not appear that the ranging machine itself will be inhibited from reaching this capability.

The main areas where improvements will be implemented are: (1) laser pulse width, (2) time-of-flight electronics, (3) detectors, (4) calibration, (5) epoch synchronization, (6) laser pulse repetition rate, and (7) system reliability. Lasers in the field are currently operating with 200 ps pulse widths at 5-10 pps. Reduction in pulse width reduces range rms, but more importantly, it allows more detailed observation of systematic signatures for understanding and remedy. Field-durable lasers are currently available with 50-100 ps pulse widths, and an addition factor of 2-3 seems quite reasonable based on laboratory systems. The limitation here will be the trade-off between pulse width and output signal strength, but the current feeling is that the practical operating pulse width might be in the range of 25-40 psec.

Field time-of-flight electronics, including discriminators and either time interval counters or epoch timers, are currently of 30 ps (4 mm) quality. Hardware with 5 ps precision has been developed in prototype. This and upgrades planned by vendors indicate that considerable improvement is underway in this area, with the ultimate limitation being imposed by the discriminator.

The best detectors in the field now for SLR and LLR are the Micro-Channel Plate (MCP) tubes. The impact of the MCP is to reduce the internal system delay dependence on return signal strength. Tubes now in the field with 12 micron channel width constrain the range dependence on amplitude to about 3 mm with a quantum efficiency of 10%. Tubes are now being built with 6 micron channels and quantum efficiencies of 20%, and additional improvements may be possible. Other classes of photodetectors, such as the Avalanche Photodiode, also offer promise of improved performance but mostly in LLR where further improvements in quantum efficiency (as high as 50%) would help the current weakness in data yield.

The most accurate ranging systems in the field today are using either internal calibration or the close "Nelson Pier" external calibration target. At millimeter accuracies, measurements will be corrupted by temporal, spatial, and environmental variations, requiring full environmental monitoring including real-time or shot-by-shot monitoring of internal delay, plus system eccentricity within a local, stable reference frame. No new technology will be required to perform this monitoring, but the system configuration may have to be modified to accommodate the additional measurements necessary.

Epoch sychronization through GPS is currently at 100 nsec with improvement to 50 nsec or better within the current capability. This is sufficient to meet the 1 mm accuracy requirement.

With the large complex of satellites to be available within the next five years, interleaving of passes will become necessary during data acquisition. It may be advantageous to increase the data rate to as much as 25-50 pps and reduce the time span for each normal point. A LAGEOS normal point might be reduced from two minutes to, perhaps, 15-30 seconds. The impact of this on the system would be the requirement for higher pulse repetition rate lasers and faster data recording systems.

Lunar ranging would also benefit greatly from higher repetition rates for increased yield. In addition, the next five to ten years should see increases in laser output power and higher detector quantum efficiency in LLR systems.

Projections of System Performance. Projections of ranging system capability depend upon our estimates of how the hardware and modeling upgrades and development will evolve. After examining each of the issues and activities underway, we developed a projection of capability for the next decade. Table 5 shows some of the key parameters with overall ranging accuracy reaching 1-2 mm by 1998. The largest issues remain the atmospheric refraction and orbital modeling.

Table 5 Projection of Laser Ranging Capability for the Next Decade

	Present	*1993*	*1998*
SINGLE SHOT NOISE (mm)	7-30	2-3	1
NORMAL POINT NOISE (mm)	2-4	1	<1
RANGING MACHINE ERRORS (mm) (SYSTEMATIC ERRORS)	5-10	2-5	
ATMOSPHERIC REFRACTION (mm):			
MODEL	5	4	3
MULTICOLOR		2	1
S/C CENTER-OF-MASS CORRECTION (mm)	3	1	
EPOCH	100 ns (65 mm)	50 ns (63 mm)	10 ns (.1mm)

3.2 SPACEBORNE LASER RANGING

Potential applications of spaceborne laser ranging can be seen within the following three domains of solid Earth physics: (i) recent crustal movements, (ii) Earth's gravity field and (iii) time transfer.
Active study of the design and potential application of a spaceborne laser measurement system began in the early 1970s (Vonbun, 1972) and led to a conceptual system design in 1975 followed by proposals for a Shuttle-based experiment (e.g., Mueller et al., 1975). In these studies, a high accuracy laser distance measuring system, carried onboard a space vehicle, measures the distance between the spacecraft and laser reflectors on the ground at selected geodetic reference points. Placing the expensive laser tracking system onboard the spacecraft and the relatively inexpensive retroreflector targets on the ground, permits the establishment and location of far more ground points than with the normal SLR tracking method. The targets can be distributed in dense arrays over local deformation areas or can be separated from one another over distances from a few hundred to greater than 1000 km for regional crustal motion studies. In seismically active fault zones such measurements are important in earthquake studies and can be used to study the cycle of strain accumulation and release, the interaction of discrete geological blocks, temporal variations in rates of strain, strain migration, creep events, pre- and postseismic strain transients, and other fault zone features. On a regional scale measurements can be used to study intraplate deformation and plate boundary deformation. There are many other tectonic and non-tectonic crustal movements that we can study including volcanic inflation, subsidence due to fluid withdrawal, and so forth.

Table 6 GLRS Technical Specifications

Laser	Nd:Yag; Diode pumped; Frequency doubled & tripled Q switched, Mode locked, Cavity dumped
Pointing Capability	± 60 Degrees along-track ± 50 Degrees across-track
Maximum Pulse Rate	40 pps
Beam Divergence	0.15 Millirad (532 nm & 355 nm) 0.10 Millirad (1064 nm; 80 m Footprint) 0.20 Millirad (1064 nm; 160 m Footprint)
Telescope Diameter	18 cm (Ranging Function) 50 cm (Altimetry Function)
Weight:	380 kg (Est.)
Power	650 w (Peak)
Laser Lifetime	> 1 Billion Shots (5 Years)
Pulse width	100 Picoseconds (FWHM)
Energy	120 Millijoules (1064 nm) 60 Millijoules (532 nm) 20 Millijoules (355 nm)
Ranging Mirror Tracking Precision	0.01 Milliradians
Altimeter Pointing Knowledge	0.025 Milliradians
Timing Resolution	10 Picoseconds (Coarse Range Receiver) 2 Picoseconds (Streak Camera Receiver) 300 Picoseconds (Surface Altimeter Receiver) 500 Nanoseconds (Cloud Altimeter Receiver)

Depending on the spacecraft height, platform stability and orbital configuration such ranging systems can also be used for satellite-to-satellite tracking, gravity field modeling, non-conservative force field modeling, and satellite self-tracking to aid other sensors. Retroreflectors can be placed on ice sheets to monitor ice flow. A precision onboard clock can be used to transfer time from location to location and to conduct experiments in spatial and general relativity. Spaceborne laser rangers designed with nadir looking or pointable altimetric capabilities can also be used to measure land and ice topography, determine cloud top heights, measure atmospheric pressure, determine geoid heights, and measure wave heights.

Although engineering studies on a spaceborne geodynamics laser system started already in 1975 (Fitzmaurice et al., 1975) and on a satellite-to-satellite laser experiment in 1978 (Balmino et al., 1978), not one of these concepts has been realized in a mission up to now.

Current Status. Presently a Geodynamics Laser Ranging System (GLRS) is being developed by NASA as part of the Earth Observing System (Degnan and Cohen, 1988). The performance specifications for GLRS are shown in Table 6 and a conceptual block diagram of the system is shown in Fig. 3. GLRS uses a Nd: YAG laser which, under current development plans, will be pumped by a laser diode array. Its infrared output pulse is frequency doubled and tripled to produce 100 picosecond pulses at 1064, 532 and 355 nm. The green (532 nm) and ultraviolet (355 nm) pulses are used for laser ranging and the infrared pulse is transmitted at nadir for altimetry. The ranging signal is collected by an 18-cm diameter telescope. A portion of the return green pulse is used in a coarse receiver consisting of a microchannel plate photo-detector, constant fraction discriminator, and time interval unit. This subsystem has a timing resolution of 10 picoseconds. A 2 picosecond resolution streak camera is used to detect the difference in the arrival time between the green and ultraviolet pulses and thereby make the atmospheric delay correction. A variable optical delay is used on the transmitted pulse to assure that both of these pulses return within a time of a single sweep of the streak camera.

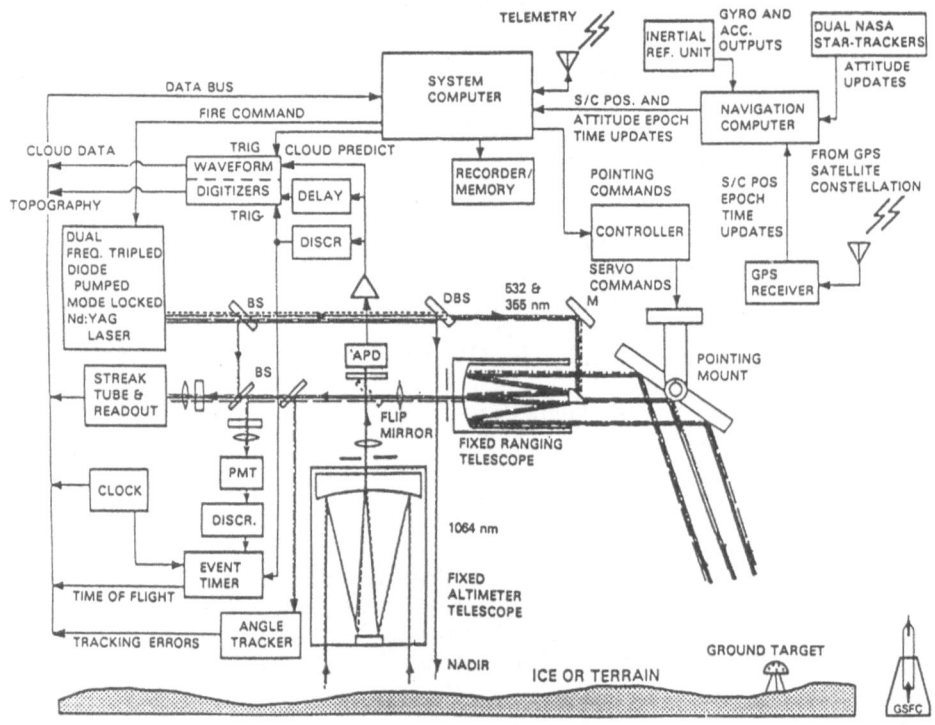

Fig. 3 GLRS Block Design

The altimetric signal (1064 nm) is collected by a 50-cm diameter telescope and detected by an avalanche photodiode, timing circuitry, and waveform digitizers. For surface altimetry the timing resolution is 300 picoseconds while for cloud tops it is 500 nanoseconds.

The system is pulsed at 40 pulses/sec. It has a ranging system pointing capability of 60 degrees along track and 50 degrees across track. Approximately 120, 60, and 20 millijoules are transmitted at the 1064, 532, and 355 nm wavelengths, respectively The beam divergence for the 532 and 355 nm ranging pulses is about 0.15 millirad, as dictated by eye safety considerations while for the altimeter pulses it is selectable at 0.1 or 0.2 millirad. These latter two divergences provide surface spots of approximately 80 and 160 meters in diameter from the Eos altitude of approximately 824 km. At a 40 pulses/second laser firing rate, the larger divergence provides nearly contiguous spot along track profiling.

Positioning and attitude information for GLRS are provided by an onboard GPS receiver, dual star-trackers, and 3-axis gyros. The detailed design of GLRS is being developed at the time this report is written; thus it is likely that some of the system specifications and design details will be changed as further progress is made in technology and design.

Key Technology Elements. The major requirements on the laser are high reliability and long life and high electrical and thermal efficiency. In addition, the requirement for subcentimeter accuracy from a system operating without ancillary meteorological data requires the use of a two or more color system. Several candidate laser schemes appear to offer the potential to achieve these requirements. Lasers using a Nd:YAG rod pumped by an array of AlGaAs laser diodes have demonstrated lifetimes in excess of 3×10^{10} shots and lifetimes in excess of 10^{10} shots are likely. The implications of these lifetimes are quite significant.

Alternative systems which appear to be quite promising include injection locking of a Nd:YAG regenerative amplifier with a seed pulse from a InALP laser and the use of a Nd:YAP laser with passive optical dye modelocking.

Since most current designs for spaceborne laser ranging systems involve sequential acquisition of individual targets rather than multibeam tracking, the pointing system must be capable of providing: 1) tracking of a given target with few arcsecond accuracy, 2) rapid slewing between targets, and 3) rapid settling into the tracking mode of newly acquired targets. Prototype pointing systems having this slew and settling capability were developed in the early 1980's. A typical prototype system can slew at a rate of 200 deg/sec or over a 100 degree pointing angle change in 0.5 seconds leaving more than 1 second for target observations. Longer dwell time can be achieved or more sites surveyed by combining data from successive passes in which only a portion of the target grid is viewed in one pass.

At least two receiver modes of operation are possible for both spaceborne and ground-based two color ranging systems. One large mode measures the round trip time of flight of the two pulses. This mode is capable of absolute range accuracies of about 5 mm. In the other mode the time of flight of both pulses is measured. This mode provides a range accuracy of a few mm. Most of the residual atmospheric propagation error is due to the different dispersion properties of the water vapor and dry components of the atmosphere. Correction for this term would require the use of multicolor or, possibly, combined laser and microwave instruments.

For determination of the differential time of flight in either mode, a timing resolution of a few picoseconds is required. Available streak cameras and recently developed timing units can achieve this resolution. It should be noted that the application of the differential correction may require capturing and processing all or most of the return signal pulse and processing of the waveform data either in real-time or a subsequent processing stage.

GLRS System Performance. Covariance analyses of the recovery of baseline lengths and relative heights indicate that with the use of short arc techniques the effects of gravity field uncertainty and laser noise dominate other error sources such as uncertainty in drag and solar radiation pressure. For distances up to several hundred kilometers, the baseline length uncertainty is less than 10 mm. At longer distances, the effects of gravity field uncertainty dominate the error budget. Vertical accuracies are poorer than the horizontal ones but remain less than 10 mm out to distances of about 300 km. Long baselines that are parallel to the spacecraft orbital path are recovered to a greater accuracy than those orthogonal to it. The effects of unmodeled spacecraft motion due to changes in center of mass and the residual effects of atmospheric propagation delay that are not removed by two color ranging have not been considered in this analysis.

Future Applications and System Concepts. Subpicosecond time transfer would be possible with a future laser ranging system. Such a capability would be quite useful for communications particularly in view of the fact that the time measurements could be completely integrated with position. Navigation and anticollision systems are possible.

A Continuous Wave Laser Interferometer system could be used for satellite-to-satellite relativity experiments. Range rates of 0.01 micrometers/sec are possible.

A Spaceborne Laser Ranging System Constellation could, for example, be based on GLRS hardware and be upwardly compatible. It could provide total global monitoring at the few mm level.

Some of the technology developments which appear to be quite promising for developing subpicosecond ranging systems include: streak camera detectors, integrated optical processing subsystems, solid state detectors, GaAs and other advanced materials and multiwavelength systems.

4. MICROWAVE TECHNIQUES

**Thomas A. Clark, William G. Melbourne, Christoph Reigber,
Larry E. Young, and Thomas P. Yunck**

In this section three general categories of geodetic measurement systems are considered, all involving the use of radio signals in the microwave spectrum (e.g., 1-20 GHz):

    a. Systems involving natural radio sources (VLBI).
    b. One-way satellite-based systems (GPS, GLONASS and DORIS).
    c. Two-way satellite-based systems (PRARE).

While these systems differ in the source of the radio signals, they share common characteristics: they operate by determining the signal's "carrier" phase,its time derivative (i.e., Doppler shift) and/or its derivative with frequency (i.e., group delay and range). All require high stability, low noise microwave hardware to extract these raw observables. Given the performance levels of modern electronic hardware and these measurement systems, the principal error source limiting the microwave system accuracy is the ability to calibrate the effects of the atmosphere.

4.1 VERY LONG BASELINE INTERFEROMETRY (VLBI)

The VLBI technique grew from developments in Radio Astronomy in the late 1960's. VLBI makes observations of extragalactic radio sources (quasars) at the far fringes of the universe; these sources provide a superior tie to the inertial celestial reference frame. However, these sources are intrinsically weak and require large, high sensitivity radio telescopes at each VLBI facility.

VLBI Systems. The "Williamstown Report" indicated the possibility of VLBI performing direct determination of contemporary plate motions with accuracies better than 1 cm/year, and systematic studies of irregularities in the Earth's rotation. In the ensuing two decades, instrumental precision and repeatability have improved by two orders of magnitude from about 1 meter to <1 cm in the determination of baseline length and the horizontal components of baseline orientation (Rogers et al., 1983, Clark et al., 1985, Davis et al., 1985, Clark et al., 1987).

A number of steps were necessary to achieve these performance improvements. First at the data acquisition end of the VLBI system, some of the evolutionary changes that have occurred are:

1. Adoption of "bandwidth synthesis" to allow the use of the group delay observable; modern VLBI observations span upwards of 400 MHz bandwidth to yield measurement precision of 10-30 psec (3-10 mm) for a single 1-10 minute observation.
2. The use of dual-frequency techniques to calibrate biases due to the ionosphere; most geodetic VLBI observations are now made simultaneously at S-band (2.2 GHz) and X-band (8.4 GHz).
3. Development of high stability instrumental calibration techniques. In order to make use of group delay observables, the fundamental time epoch at each station is defined by a train of fast pulses (~10 psec rise time) generated in the receiver. Variations in the epoch of this pulse with respect to the hydrogen maser frequency standard must be calibrated.
4. Development of more reliable, stable hydrogen maser frequency standards.
5. Development of reliable cryogenic low noise receivers using GaAs Field Effect Transistors (FETs) and High Electron Mobility Transistors (HEMTs).
6. Development of wide-bandwidth, dual-frequency feeds for dish antennas.
7. Development of VLBI data acquisition hardware designed to meet geodetic requirements (the Mark-III VLBI system).
8. Outfitting new fixed VLBI observatories dedicated to geodetic VLBI.
9. Development of small, mobile VLBI systems to make in situ measurements at locations of geophysical interest.
10. Implementation of national and international coordination of networks of VLBI stations to make temporally dense, synoptic measurements.
11. The development of automatic meteorological instruments and high-stability microwave "water-vapor" radiometers which remotely sense the troposphere in order to calibrate propagation biases.

Over the past two decades, improvements at the VLBI observing stations have been matched with a corresponding series of improvements in experiment planning, data analysis and data interpretation. These include:

12. Improvements in sensitivity have allowed the use of a large number of weaker radio sources. This in turn has allowed much improved observing geometry during each measurement session. Better observing geometry has enhanced the precision of daily measurement sessions and has permitted improved calibration of atmospheric biases.
13. Dedicated global networks operating on regular schedules (POLARIS, IRIS and CDP) have led to vast improvements in the determination of Earth orientation variations and the terrestrial reference frames. This in turn has permitted smaller networks (including those using mobile stations) to measure vector variations in regional-scale baselines.
14. The availability of the long-term VLBI data set has led to improvements in the basic physical models of nutation, precession, relativity, tides, plate tectonics, etc. The continuing improvement of data analysis techniques and models has allowed the reanalysis of older VLBI data so that data acquired circa 1981 can now be fit about twice as well as the initial analysis.
15. Improved atmospheric models, combined with better observing scenarios and the use of stochastic parameter estimation techniques (such as the Kalman filter estimator) and water vapor radiometers, have led to significant improvements in the repeatability (and presumably accuracy) of the VLBI technique.

With all these improvements over the past two decades, VLBI has demonstrated its ability to perform measurements on trans- and intercontinental scale baselines at levels of repeatability of

1-2 cm. Fig. 4 shows the repeatability in baseline length for 147 baselines measured since 1984 following a scaling law:

$$\text{Repeatability} = (5.3 + (\text{Length}) \times (2.4 \times 10^{-9})) \text{ mm}$$

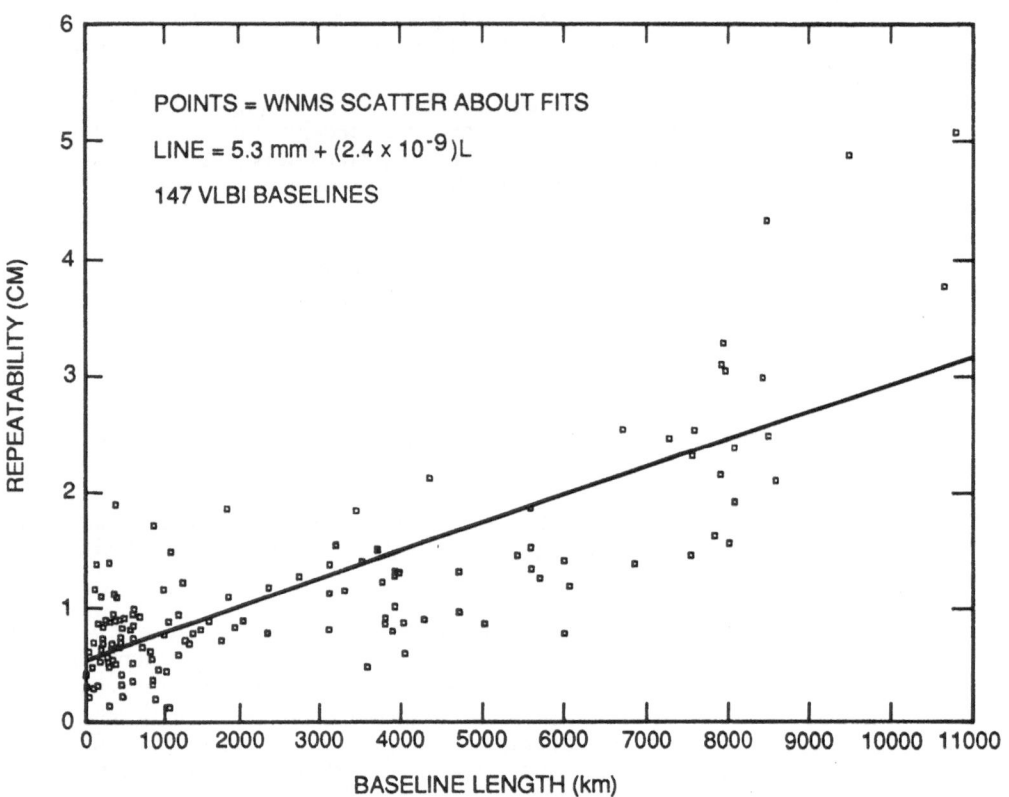

Fig. 4 VLBI Baseline Length Repeatibility

   This analysis is based on the large volume of data in the NASA Crustal Dynamics Project's (CDP) archives. Each baseline determination reflects about one day of data; no editing based on weather conditions was done. The data for a given baseline was fit with a linear rate of change to allow for steady tectonic motions, providing that the archives contained at least five determinations of the baseline. The shortest baselines are dominated by the mobile VLBI systems which have limited sensitivity due to their small size. For baselines longer than 6000-7000 km, the length repeatability is limited by inaccuracies in calibrating atmospheric biases. A similar analysis of the recovered station heights shows repeatability of about 3 cm.

The quality of these results is shown by their repeatability. The understanding of the intrinsic accuracy requires comparisons with other techniques. Under the aegis of the NASA CDP, a series of comparisons between VLBI and Satellite Laser Ranging (SLR) have been undertaken on baselines ranging from 1000 to 7000 km. In all cases the agreement between the two techniques is at levels of a few cm or better on data taken through 1986. In 1987 and 1988 the CDP sponsored a major VLBI/SLR intercomparison program to continue these investigations.

The NASA CDP has embarked on a program to enhance the intrinsic precision of the VLBI technique. Hardware is being developed to improve the precision of individual 100-1000 second measurements by a factor of 2-3 by increasing the observing bandwidth and improving the accuracy of the instrumental calibration systems. Plans also call for measurements to determine the geometric stability of the fundamental antenna reference point.

VLBI Networks. The early geodetic VLBI results were based on the use of a number of radio telescopes (operated by the radio astronomy community) and deep-space-tracking stations on a part-time basis. The past decade has seen the development of a number of dedicated geodetic VLBI stations around the world. Development of dedicated stations began in the USA under the auspices of the NASA Crustal Dynamics Project and the NGS Project Polaris. By 1984 four dedicated stations were operating in the USA, and were soon joined by other dedicated stations in Europe and Asia:

Westford, MA	(18 m)	Kashima, Japan	(26 m)
Fairbanks, AK	(26 m)	Ft. Davis, TX	(26 m)
Mojave, CA	(12 m)	Vandenberg AFB, CA	(9 m)
Richmond, FL	(18 m)	Maryland Point, MD	(26 m)
Wettzell, FRG	(20 m)	Shanghai, China	(25 m)

In addition, NGS and NASA have operated two mobile VLBI systems (3 m and 5 m in diameter) in measurement programs in California, the Western USA, Alaska and Canada. A number of radio telescopes and tracking stations around the world are part-time participants in geodetic measurements:

Haystack, MA	(37 m)	Goldstone, CA	(26 & 34 m)
Medicina, Italy	(32 m)	Onsala, Sweden	(20 m)
Owens Valley,CA	(40 m)	Hat Creek,CA	(26 m)
Hartebeesthoeck, S.Africa	(26 m)	Kauai, Hawaii	(9 m)
Tidbinbilla, Australia	(34 m)	Roi Namor, Kwajalein	(26 m)
Madrid, Spain	(34 m)	Pie Town, NM	(25 m)

New VLBI facilities which should be operating in 1989 and which will contribute to geodetic programs include two Italian telescopes: Matera (20 m) and Noto (32 m). The U.S. Naval observatory plans to reactivate a 26 m telescope at Greenbank, WV. The Pie Town, NM antenna is the first of a series of ten antennas being constructed by the U.S. National Radio Astronomy Observatory as a dedicated Very Long Baseline Array (VLBA). When completed in the early 1990's, this facility should provide much useful geodetic data from the antennas located in New Mexico (2 antennas, including Pie Town), Arizona, California, Iowa, Texas, Washington, New Hampshire, Hawaii, and the Virgin Islands.

The Canadians have plans to reactivate the Algonquin Park (46 m) telescope. The Japanese areplanning several new facilities on islands near Japan and in the Antarctic. The Chinese have plans for two additional antennas. Groups in Brazil, Saudi Arabia and India all hope to have operational facilities in the 1990's.

A Prognosis for the future. The two radio geodetic techniques -- VLBI and GPS -- should not be viewed as competitors. Each has its own unique strengths and weaknesses. Future measurement programs which require high accuracy will make use of the combination of the two.

The VLBI systems offer a superb tie to an inertial celestial reference frame based on extragalactic radio sources (quasars) at the farthest reaches of the universe. Therefore, VLBI will continue to be used to maintain the terrestrial reference frame. VLBI operations will be concentrated at facilities which are members of dedicated networks; these networks will produce data for earth orientation and global-scale plate tectonics.

In the two decades since the birth of geodetic VLBI, accuracies have improved by a factor of ten per decade. Presently VLBI can be used to determine the horizontal position of a station with a repeatability of a few mm during a one day measurement session. Measurements in the vertical are poorer by a factor of 4-5 due to inaccuracies imposed by the calibration of atmospheric path delays. Several activities are underway to make further improvements in the next few years.

The use of extragalactic radio sources enables VLBI to have superb inertial ties; these radio sources are very weak. VLBI observatories need to use large (typically 10-30 m) diameter radio telescopes equipped with state-of-the-art low-noise microwave receivers. The cost of building and operating such facilities limit the number that can be available to support geodetic programs. Such considerations suggest that VLBI networks of the future will continue to be sparse with stations spaced a few thousand km apart.

In the USA, two mobile VLBI stations have been developed using antennas of 3 m and 5 m diameter. These antennas have been employed for studies of the deformation along the Pacific-North American plate boundary (in California and Alaska), measurements of the stability of the North American plate in the western USA, and to establish additional fiducial sites. The size and complexity of these systems limit their ability to perform more than 20-30 measurements per year.

VLBI system accuracies and operational costs can be improved with the development of a next generation VLBI data system using recent advances in solid state electronics. The spanned band width of the current Mark III system could be doubled thereby enabling dual band phase delays to be obtained over a broad range of baselines lengths. This would improve the precision of the delay measurements made by VLBI by roughly an order of magnitude and should lead to intrumental accuracies that should approach 1 mm in the horizontal baseline components.

4.2 ATMOSPHERIC PROPAGATION

Errors. The principal limiting error source shared by all the radio geodetic measurement systems is the inability to calibrate biases caused by the propagation of the radio signals through the troposphere. Two separate tropospheric biases can be considered:

1. The "dry" tropospheric contribution arising from neutral molecular constituents of the atmosphere (primarily nitrogen and oxygen). At the zenith the added "dry" path delay is about 7 nsec (equivalent to about 2 m).
2. The "wet" tropospheric contribution arising from water vapor in the lower troposphere. The "wet" delay is highly variable with site, season and weather conditions, but typically ranges from 0.1 - 1.5 nsec (3-50 cm).

These propagation biases primarily affect the determination of the vertical components of the baseline vector at each station. To first order, the zenith "dry" component can be determined by measuring the atmospheric pressure with a high quality barometer to "weigh" the total atmospheric mass overhead; one mbar of pressure corresponds to about 2 mm of zenith path delay. The zenith path correction is then applied at other parts of the sky by using a scaling law which models the actual distribution of the neutral atmosphere with height, and by assuming the atmosphere has azimuthal symmetry. Fig. 5 (courtesy G. Elgered) shows the three year history of the dry tropospheric corrections deduced from meteorological data at three sites: Fairbanks, Alaska; Landvetter (Gothenburg), Sweden and West Palm Beach, Florida.

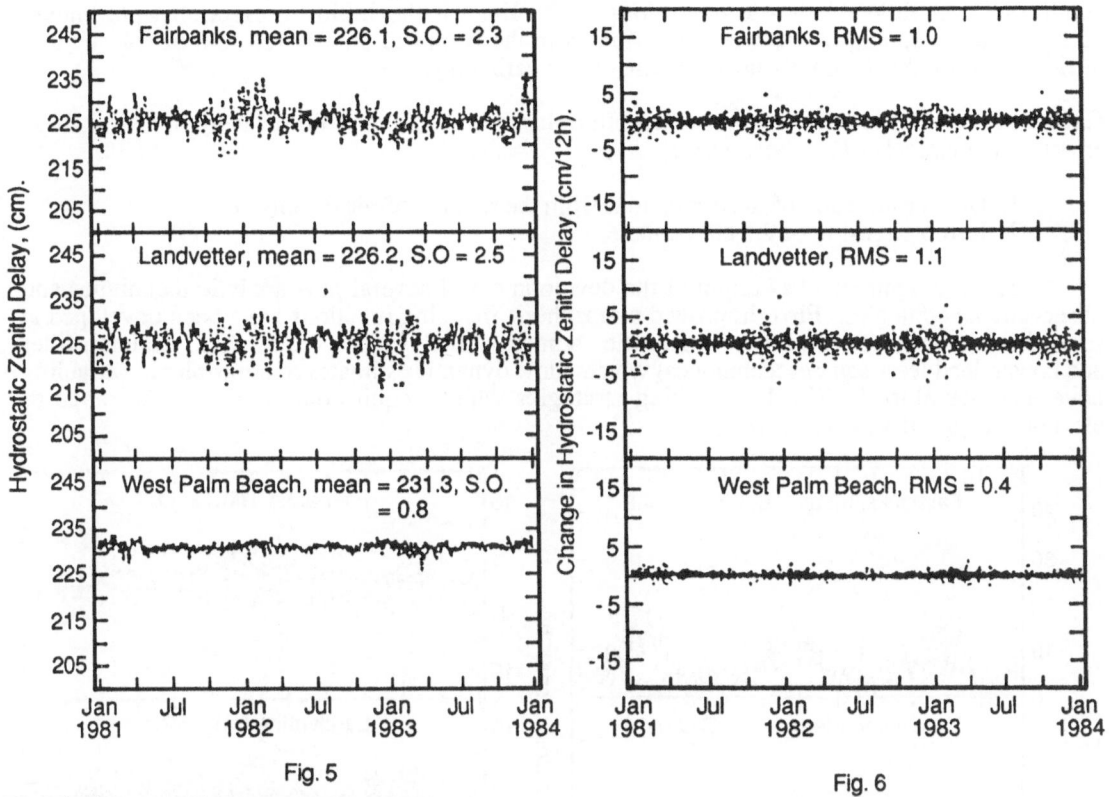

Fig. 5

Fig. 6

Fig. 5 Three year history of tropospheric corrections (from meteorological data)

Fig. 6 12 h variability

There are several problems with this simple approach to calibrating the "dry" propagation effects. First, the presence of winds and moving weather patterns demonstrates that the assumption of symmetry is inadequate to fully characterize the dry tropospheric corrections. In the future it may be necessary to instrument arrays of pressure sensors around principal observing stations to map the spatial variability of the dry component. Some indication of the magnitude of this variability can be deduced from meteorological data. Fig. 6 shows the data of Fig. 5 plotted in terms of variability between successive 12 hour measurements.

A second problem arises from the mapping of the zenith path delay to other elevations. This mapping requires a detailed knowledge of the mass distribution of the atmosphere with height. This mapping function varies between sites, with season, and with the detailed temperature and wind profiles aloft.

The "wet" contribution to the observed delay presents additional problems. The wet contribution arises from the vibrational resonance of water molecules at 22.2 GHz. Atmospheric water in the form of droplets has little effect on path delay since the droplets become dielectric (Mie) scatterers. Fig. 7 presents synoptic meteorological data obtained from routine weather balloon launches at the same three sites, while Fig. 8 presents the 12 hour variability. It should be noted that while the magnitude of the wet zenith delay is much smaller than the dry component, its variability on all time scales is much larger.

While fairly accurate dry corrections can be inferred from surface meteorology, the wet delay correction cannot. The vertical distribution of water vapor is highly variable and surface conditions bear little relation to conditions aloft. Even if the zenith correction were known perfectly, the mapping of the wet correction from the zenith to the desired line of site is more difficult than for the dry component because of this variability.

Calibration and Modeling Approaches. In order to correct the radio measurements for these errors, two approaches have been used:

1. Direct estimation of the corrections from the radio geodetic data itself.
2. Remote sensing of the atmosphere.

The first approach has required the development of several new analytic techniques and observational strategies. First, improved atmospheric mapping functions have been developed at several institutions. Second, new stochastic estimation analysis procedures (e.g., Kalman filter and constrained least-squares techniques) which allow dynamic estimates of atmospheric variability have been developed. Third, observing strategies which acquire data over a wide range of elevation angles have been developed.

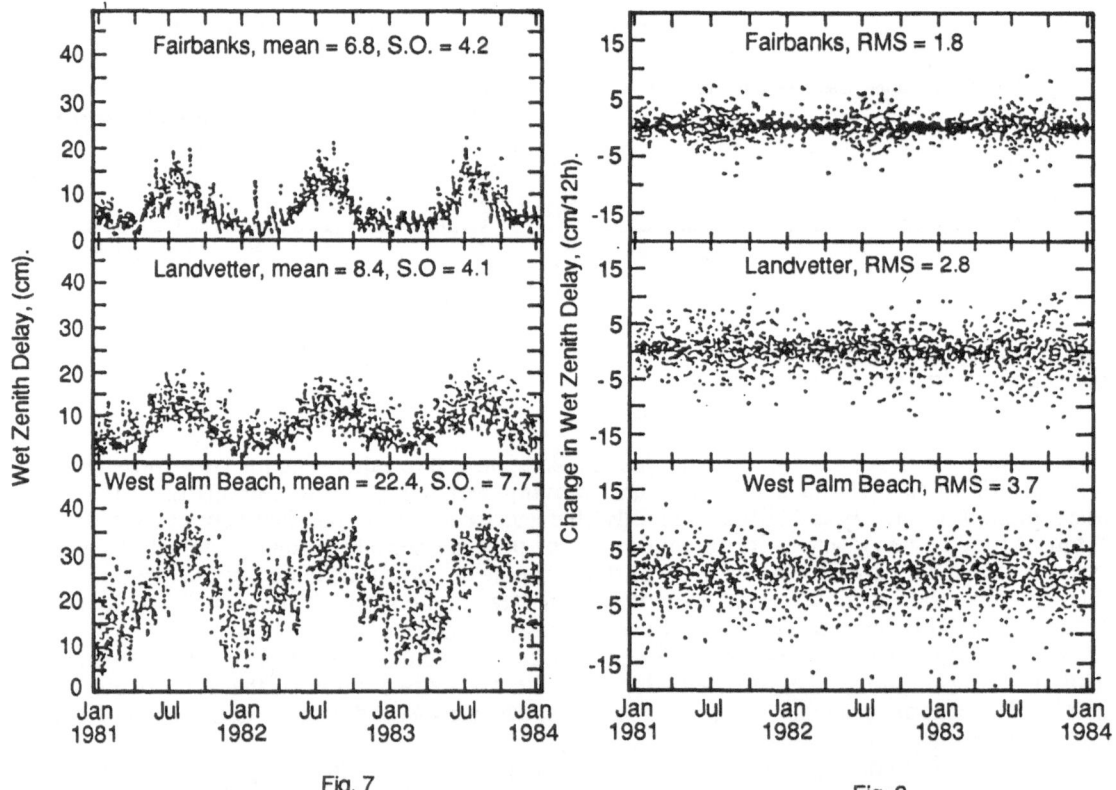

Fig. 7

Fig. 8

Fig. 7   Three years history of wet tropospheric corrections (from balloon weather data)

Fig. 8   12 h variability

The second approach has resulted in augmenting the meteorological sensors at the stations with passive microwave radiometers operating at about 21 GHz (near the 22.2 GHz water vapor absorption line) (Jannsen, 1985, Elgered, 1989). These water vapor radiometers (WVRs) measure the component of sky brightness due to resonant absorption. The WVRs include a second channel away from the absorption line (typically 31.4 GHz) in order to separate the sky brightness contributions of water vapor from droplets in clouds. A combination of the brightness temperatures in the two channels is used to estimate the integrated water vapor along the line of sight, which in turn can be used to estimate a path length correction.

The inversion of brightness temperature to provide an estimate of the integrated water vapor content requires a knowledge of both the distribution of water vapor with height and the atmospheric temperature profile. The inversion coefficients are dependent on the meteorology at that site; the coefficients applicable in desert environments are not applicable in the arctic or on tropical islands. The derivation of these coefficients has typically involved statistical studies of synoptic upper atmosphere meteorological data obtained from routine balloon launches by national weather bureaus. Once the integrated water vapor is inferred from the WVR data, the path delay is computed using microwave refractivity data obtained in the laboratory.

In order to measure path delay corrections to a one cm accuracy, the WVRs must be able to measure sky brightness to an accuracy of about 1 K in each channel. Absolute brightness temperature measurements to this level place stringent demands on the stability and calibration of the WVR instruments. In an attempt to verify instrument performance, an intercomparison between the Swedish ASTRID WVR and the NASA JO-3 WVR was conducted during the summer of 1988 at Onsala, Sweden. Each instrument was operated with its own independently derived calibration and inversion parameters. The two WVRs were found to agree with a bias of < 5 mm referred to the zenith. Similar comparisons between other WVRs have been conducted, but the data analysis is not complete.

When the atmospheric calibrations obtained directly from the radio geodetic data are compared to those obtained by WVRs, differences of about 1 cm (referred to the zenith) are typically seen. There may be several sources for these differences:

1. The stochastic estimation process may be inadequate because the mapping functions used to relate observations at one elevation angle to those at another may have errors. These errors may reflect atmospheric variability from day to day.
2. There may be other error sources which have signatures in the data similar to the atmosphere. An example might be structural deformations of the antennas used in VLBI or multipath induced errors in GPS observations.
3. The presence of horizontal gradients in either the wet or dry atmosphere may cause errors.
4. The inversion and calibration coefficients used to reduce WVR data may be in error.
5. The stability and calibration of the WVR instruments may be inadequate to allow determination of path delay at levels better than about 1 cm.

An error in calibrating the zenith path delay allowed to propagate into the determination of geodetic site or baseline parameters has an effect on the vertical coordinates about three times as large. The inability to calibrate tropospheric delay corrections to better than 1-2 cm (in the zenith) is the largest single error source for any of the radio geodetic systems. The Committee on Microwave Techniques places continued research and development activities as the highest priority task to improve system accuracy.

4.3 GPS-BASED GEODESY

Goals. As the goals of tectonic physics evolve from confirmation of plate motions to study of the driving forces, a shift is occurring in the requirements for geodetic instrumentation. In particular, to test tectonic models, the goals for GPS instrumentation will include the ability to make baseline measurements that are dense in space and time over interplate deformation zones spanning up to several hundred kilometers. The magnitude of such campaigns requires the development of instrumentation with, a) improved accuracy, b) expanded functionality and autonomy, and c) greatly reduced cost. The low strain rates in these deformation zones, in many instances a few millimeters per year, call for improvements in current system accuracy to reduce the span of time required to determine these velocities. Baseline measurement accuracies on regional scales of a few millimeters appear achievable within the next five years with relatively modest improvements in GPS instrumentation and in network design and data analysis strategies. Receivers with expanded functionality and autonomy would operate unattended, recover from signal loss and/or cycle drop-outs, transmit data, and perform chores such as observation scheduling and self-diagnosis. The staggering potential volume of data from a dense network of such measurement systems requires the development of "smart receivers" which perform on-line editing, data compression, calibrations, and some pre-processing of data. The capital equipment costs of these receivers must be greatly reduced to make densification economically viable. To make the majority of the Earth's tectonically active regions accessible to measurement, hybrid GPS/underwater acoustic systems for sea floor geodesy must be developed. The ability to accurately track the position and attitude of kinematic platforms with GPS receivers will also be required to support other types of instruments, e.g., satellite-based altimetry, gravity mapping from ships or from space, synthetic aperture radar mapping of ocean surface currents, and ship-based oceanic acoustic tomography.

Present Status: *System Accuracy.* In the past year the accuracy of GPS-based measurement of Earth baselines has reached the level of 1 cm in the horizontal and length components over distances up to about 1000 km, and about 1 part in 10^8 over longer distances. The accuracy of the vertical component is typically 1-3 cm. This has been assessed by direct comparison of GPS-based measurements with a time history of VLBI measurements of the same baselines (Beutler et al., 1987; Blewitt, 1989; Lichten and Bertiger, 1988; Dong and Bock, 1989). Comparable repeatabilities have been achieved in regions where VLBI measurements are not available for ephemeris control or for comparison (Tralli and Dixon, 1988). Fig. 9 shows the level of agreement between GPS and VLBI measurements as a function of baseline length for a set of baselines measured in the continental U.S. during experiments in November 1985 and June 1986. The accuracy of survey ties between GPS benchmarks and intersections of VLBI antenna axes is not confidently less than 1 cm at this time. Note that there is no significant trend in the GPS-VLBI length discrepancy as a function of distance, indicating consistent scales for the two measurement systems. Receivers themselves are only one component of the elaborate data collection, calibration, modeling, and analysis systems needed for today's centimeter baseline measurement accuracies. Data must first be gathered with a well-planned network of receivers extending over a sizeable geographic area, and a portion of the collected data must be used to improve the ephemerides of the GPS satellites. A variety of different approaches to system design and data analysis have proved successful. In one approach, three or more receivers placed at well known reference ("fiducial") sites, with at least one stable clock, are held fixed during a grand simultaneous solution for all other receiver locations and GPS ephemerides. In addition, parameters for stochastic effects such as zenith tropospheric delay, solar radiation pressure, and other unmodeled GPS satellite accelerations are adjusted. In a variation on this approach, fiducial data are processed first to produce accurate GPS ephemerides which are then held fixed during a second solution for the baselines of geodetic interest. Experiments have shown the latter (somewhat simpler) approach to be virtually as accurate as the full simultaneous solution, so long as the geodetic baselines are within or close to the fiducial set. In both of these approaches, the fiducial network defines the reference frame in which the GPS ephemerides and baselines are determined. An alternative approach, in its most extreme form, frees up all receiver coordinates except (necessarily) the longitude of one reference receiver, allowing the considerable strength of the GPS data and a fixed value for the GM of Earth to determine all orbits and receiver positions in a self-consistent, properly scaled reference frame. This eliminates dependence upon predetermined fiducials, at the cost of introducing potential differences with the reference frames established by VLBI and SLR techniques. Variations and hybrids of these techniques, in which different numbers of fiducial parameters are freed and constraints of varying levels of severity are applied, are also being investigated. Identifying a quasi-optimal strategy for establishing a globally consistent geodetic reference for all investigators at all times will be an important pursuit in the next decade.

Several special characteristics of system design and techniques of data analysis have proven particularly advantageous in high performance GPS geodesy. These include, a) carrier cycle ambiguity resolution or "bias fixing" (and the algorithms and careful geometric arrangement of the network of receiving sites that make it possible), b) stochastic modeling of zenith tropospheric delay and the non-gravitational forces acting on the GPS satellites, c) judicious use of water vapor radiometers to calibrate the wet tropospheric delay at humid sites having high spatial and temporal variability, d) multi-day GPS orbit solutions for measurements of long baselines, and e) the combined use of dual frequency carrier phase and pseudorange data.

Carrier phase ambiguity resolution has now been achieved over distances of up to 2000 km. This has been accomplished using algorithms that formally reflect the integer character of the cycle ambiguity problem, combined carrier and pseudorange data, and an arrangement of receiver sites permitting short and intermediate baseline resolution as preliminary steps to long baseline resolution (Blewitt, 1989; Dong and Bock, 1989). For baselines of a few hundred kilometers, solutions after ambiguity resolution are typically 3-5 times more accurate than solutions treating the carrier phase bias as a continuous variable. Carefully tuned stochastic troposphere models have proved most effective in reducing tropospheric contamination, improving baseline repeatability by as much as a factor of two over constant delay models and surface meteorology calibrations (Tralli et al., 1988).

The use of high precision pseudorange data in combination with carrier phase confers a number of benefits. These include, a) greatly simplifying data editing and recovery from cycle breaks, b) improving the speed and reliability of ambiguity resolution, c) improving solutions for clock offsets between receivers, d) improving the carrier phase bias solution when the ambiguity cannot be resolved, e) enabling precise kinematic tracking of moving platforms, and f) generally reducing the system sensitivity to data outages and other operational problems. Several studies have shown that pseudorange of subdecimeter accuracy achieved by the new high performance receivers and antenna systems, when used together with ambiguous carrier phase, can yield baseline accuracies approaching those currently achievable only with resolved carrier phase.

Equipment Characteristics. These levels of baseline measurement accuracy have been achieved with a variety of dual-band receiver designs, both code-tracking and codeless, developed in the early to mid-1980's. The accuracy of these receivers in measuring carrier phase, while clearly adequate to yield high quality baseline results, is far below what will be required for millimeter accuracies or for autonomous functions such as phase reconnection or self-diagnosis. Aside from media delay calibration errors, the current instrumental factors that limit accuracy for the best receivers, in both the carrier phase and pseudorange measurements, are multipath and antenna "phase center" variation. For current commercial receivers such as the TI-4100, multipath is near the centimeter level for carrier phase and usually at least several decimeters and often higher in pseudorange. Phase center variations with azimuth and elevation can amount to several centimeters but these variations tend to be nearly identical for two antenna/backplane systems of identical design and installation. Also, most of this effect can be calibrated. Multipath and thermal data noise, typically at the meter level on pseudorange measurements in past experiments, has been reduced by an order of magnitude in field tests of recently developed receivers and antenna/backplanes (Young et al., 1988; Thomas, 1988).

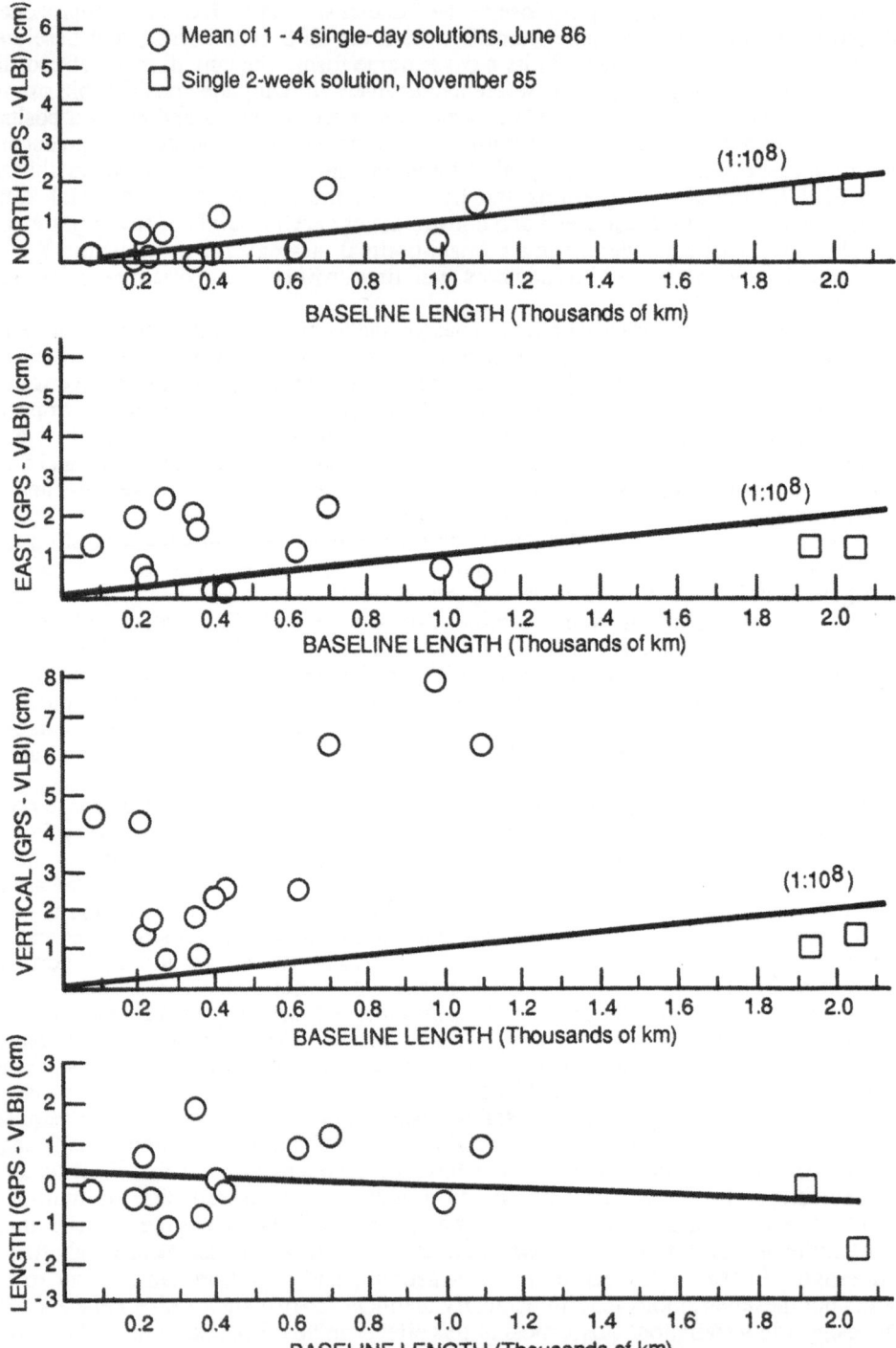

Fig. 9 Agreement of GPS and VLBI results as a function of baseline length

Current GPS receivers cover a wide spectrum of accuracy, operability, and cost. Receivers that obtain dual frequency carrier phase and dual frequency P-code pseudorange data are necessary for geodetic measurements of the highest accuracy and the widest range of applications. The ability to track signals without cycle dropouts through accelerations up to several g's due to platform dynamics or ionospheric scintillations, which is necessary for reliable data reduction, is available in certain recently developed receivers. Some receivers can be operated by remote control. None of these receivers is capable of achieving all of the autonomous and adaptive operations goals cited above.

Cost Drivers. The typical cost range of current dual-band receivers is $ 70,000-100,000. The costs of GPS measurement campaigns are distributed over equipment, operations, and data processing. For the occasional experiments used to measure average plate motion rates, the costs of data analysis outweigh operations and pro-rated equipment costs. Attended operation of a spatially and temporally dense network would cause the operations costs to dominate. Current operations costs (assuming nominal burdened labor rates, a typical mix of occupation sites in terms of accessibility and travel costs, and including all aspects of management, preparation, deployment, logistics, data acquisition, and equipment maintenance) run between $400 and $600 per station-day depending on the complexity of the campaign. The practical operation of dense GPS networks requires the development of smart receivers for unattended operations, and streamlined processing techniques, as well as low cost receivers.

Future Prospects: *Costs.* The progress of both design and manufacturing technology in the field of very large-scale circuit integration (VLSI) continues to be very rapid. This has allowed the development of GPS receivers that use digital baseband signal processing techniques, removing the phase and delay variation errors that are present in analog circuits. It has also led to major reductions in cost, weight and power. Next generation receivers will use VLSI advances to digitize the incoming GPS signals directly at L-Band frequencies, as well as to further reduce size and cost (Thomas, 1988). An order of magnitude reduction in equipment costs appears likely within the next five years. Because of the availability of more powerful onboard microprocessors, as well as the shifting of some receiver calculations to special purpose hardware, these next generation receivers will have much greater capability for autonomous operation. The availability of these "smart receivers" will allow some data analysis to be performed in real-time. When this advanced receiver processing is combined with the use of artificial intelligence in post-processing, the total cost per GPS baseline determination will decrease dramatically. Another promising technique for reducing operations costs, if cycle ambiguities can be reliably removed with combined pseudorange and carrier phase data, is the use of carrier range to provide near instantaneous geometric differential positioning. This capability would significantly improve productivity costs by reducing the dwell time at each measurement site.

Accuracy. A wealth of approaches to refining the design and composition of antennas and backplanes promise major future improvements in phase center and multipath errors. These advances coupled with the high instrumental accuracies provided by all-digital systems, are expected yield advanced GPS receivers and antenna systems capable of carrier phase accuracies near the millimeter level and pseudorange accuracies near the centimeter level. A digital front end located at the antenna, with a long fiber optic link to the baseband processor, will allow the antenna site to be optimized for low multipath. The design architecture of next generation receivers, using combined carrier phase and pseudorange measurements under wide dynamical conditions, will be adaptable to specialized applications such as positioning ocean platforms in sea floor geodesy, or satellite-based applications. Most of these capabilities should be commercially available within the next 3 years.

Based on recent very short baseline tests with high performance laboratory developed GPS receivers, and with recent experience in cycle ambiguity resolution, the potential baseline accuracy on regional scales due to instrumental errors can be improved to the millimeter level. However, tropospheric delay and GPS ephemeris errors will limit the realization of this accuracy depending on the region and its scale. In relatively benign environments such as the Western U.S.(with zenith wet delays of 0-20 cm and a few centimeters of spatial and temporal variability), the current baseline errors due to the troposphere are in the range of 0.5-1 cm for the horizontal components and 1-3 cm for the vertical. These figures tend to be somewhat worse for regions with wetter climates and/or higher variability. For benign environments stochastic modeling without WVR calibration has proven as effective as WVR calibration alone - at the current level of GPS system accuracy. For tropical regions a combination of WVR calibration and stochastic modeling of residual WVR errors has proven the most effective (Tralli et al., 1988). Substantial progress can be made in more sophisticated stochastic tropospheric modeling including azimuthal variability, in water vapor radiometer instrumentation, in region-dependent recovery algorithms, and in the use of synoptic data. Understanding how to use sophisticated modeling of and direct solution for tropospheric delay in conjunction with in situ WVR measurements, possibly aided by synoptic observations, to reach the millimeter horizontal accuracy level, particularly in the more humid and meteorologically unstable regions, is another challenge requiring thorough investigation over the next few years. Additional improvement in performance could be realized with receivers that simultaneously track GPS and Global Navigation System (GLONASS) satellites. This will help protect against satellite outages, improve overall data strength, and, by effectively doubling the sky sampling density, improve our ability to recover spatially variable tropospheric delay with the satellite observations themselves. Ionospheric scintillations at high latitudes during high intensity periods may also limit the tracking accuracy of the receiver unless it is designed and configured to track high signal accelerations.

For baselines longer than roughly 1000 km, GPS ephemeris errors tend to dominate. The control of GPS ephemeris errors (including the variability of satellite phase center to center of mass offset) through concurrent tracking from fiducial stations with strong geometry and with careful dynamical and observable modeling, may yield subcentimeter accuracies for transcontinental baselines.

Future Applications: *Continuously Active Remotely Monitored Arrays.* The use of smart receivers with streamlined post-processing algorithms will be tested by the implementation of a Continuously Active Remotely Monitored Array. This will consist of a dense network of receivers capable of determining baselines with 1 mm precision at 4 hour intervals, given appropriate coverage by GPS/Glonass satellites. These receivers will be connected to a central processing site by a low rate data link. The data processing will be highly automated, and capable of keeping pace with the rate of data collection. Outputs will be in a form that is appropriate for the scientific user. These networks may be considered as extensions of existing seismic arrays to the low frequency regime.

Sea Floor Geodesy. A combined GPS/underwater acoustic ranging system will be developed to allow geodetic measurements to be made between points on the sea floor. The GPS receiver(s) will be used to provide the location and orientation of a surface platform in the GPS reference frame. The acoustic system, supplemented by pressure gauges, will measure the offset of a point on the surface platform from the reference mark on the sea floor.

Satellite Precision Orbit Determination. By extending the geodetic fiducial network globally and placing a flight rated GPS receiver onboard an orbiting satellite, GPS can be used to perform precise orbit determination (POD) as described elsewhere in this report. The first major goal will be to achieve sub-decimeter orbit determination on TOPEX/Poseidon, a NASA/CNES oceanographic satellite to be launched in 1992. This goal will be tightened to 1-3 cm on future satellites such as the platforms of the Earth Observing System (Eos). The primary driver for such high orbit accuracy is precise ocean altimetry in which centimeter accuracy in the sensor orbital altitude must be maintained to observe centimeter-level features in the ocean topography. To achieve such performance it will be necessary to develop space qualified versions of the highest performance GPS ground receivers now under development.

Synthetic Aperture Radar Platform Positioning. There are at least two applications of GPS to synthetic aperture radar (SAR) that should be developed. The first is the use of dual SAR systems operating in a differenced mode to make direct global maps of ocean currents. GPS data are required to provide the orientation and velocity of the satellite as a calibration to the differenced SAR data.

A second application of high accuracy GPS to SAR would be the combination of images from a single SAR element. Here, these images taken during successive orbits are combined after the use of GPS tracking to determine the relative SAR antenna trajectories to much better than a wavelength. For targets for which coherency is preserved this would allow synthesis of apertures of the size of the separation of the orbits.

Ocean Tide Gauges. Another future application will involve the development of free floating high performance GPS receivers, deployed in the oceans, to serve as floating tide gauges. Measurement from arrays of floating receivers can be tied with centimeter accuracies or better to land-based fiducial points over great distances, enabling precise monitoring of ocean tides far from the complicating influence of land masses. It would also enable highly dense monitoring near land masses.

Earth Orientation Monitoring. When the full Block II constellation is deployed and a global network of GPS reference stations is operating, the tracking data from this system will provide a valuable addition to the monitoring of Earth orientation variability, currently conducted using VLBI and SLR through the International Earth Rotation Service. GPS tracking will provide complementary and accurate short period information and longer period information of comparable accuracy. Although the GPS (and the SLR) requires VLBI for maintaining the inertial reference system, its introduction into Earth orientation monitoring will fundamentally change the mix in the use of these three techniques for this application.

Super GPS. A future system of satellites specifically designed to support geodesy could be implemented which would allow the use of much simpler receivers. Choice of more favorable signal frequencies would allow carrier cycle ambiguities to be determined with only 1 second of data, and multiple tones would allow the more accurate calibration of ionospheric delays. In addition, the design of a dedicated satellite for geodesy would provide the opportunity to reduce satellite associated errors, such as mismodeled accelerations due to drag, radiation and outgassing, and the multipath originating from the satellite antenna.

4.4 OTHER SATELLITE-BASED MICROWAVE SYSTEMS

DORIS. The French DORIS measurement system involves one-way Doppler measurements designed for precise orbit determination for altimeter missions with a goal of 5-10 cm radial accuracy. The design stresses an accuracy goal of 0.3 mm/sec in the determination of range rate. The system involves one-way transmissions from ground-based transmitters, with the receiver on the satellite. Both the transmitter and receiver have high-stability crystal oscillators (5×10^{-13} over 10 minutes). The transmitter radiates about two watts at UHF (400 MHz) and S-band (2.1 GHz) to support dual-frequency ionospheric calibration. Uplink signals involve brief transmissions from each participating terrestrial beacon as scheduled by a timer in the transmitter. The uplink signals also transmit digitally encoded meteorological and calibration data.

The first DORIS package will be launched on the SPOT-2 satellite in mid-1989. The goal is for the early deployment of some 45 ground-based transmitter packages to work with the SPOT-2 DORIS receiver; about 15 of these will be located at VLBI and SLR stations and at worldwide tide gauge sites. It is also planned to carry DORIS hardware on the TOPEX/POSEIDON and SPOT-3 missions.

PRARE. The West-German PRARE (Hartl et al., 1986) is a spaceborne two-way, two-frequency (S/X-band) range and range-rate measuring system, which will be flown for the first time on ERS-1 in late 1990.

PRARE is a self-contained system, being able to up- and downlink all relevant data from and to the ground stations on its tracking loop. Included in its space segment is sufficient data memory to store the onboard regenerated tracking data and transmitted corrective data from the global station network for transmission to the master ground station during overflight times. Through the master station commands and broadcast data are transmitted into the on-board memory and are disseminated from there to the ground stations.

The ground stations operate as regenerative coherent transponders. They are of low weight, highly mobile and can be operated unattended. The equipment consists of an antenna unit (0.6 m diameter steerable parabolic dish), an electronic box and a weather unit.

The PRARE system will operate

- at X-Band (in the 7.19 - 7.235 GHz region) with 10 MHz bandwidth in the up-link
- at X-Band (in the 8.45 - 8.50 GHz region) with 10 MHz bandwidth in the down- and up-link respectively and
- at S-Band (in the 2.20 - 2.29 GHz region) with 1 MHz bandwidth in the down-link.

The procedure to perform range and range-rate measurements is as follows: Two signals are sent to the Earth from the satellite, one of which is in the S-band (2.2 GHz), the other in the X-band (8.5 GHz). Both signals are modulated with a PN-code (pseudorandom noise) for the distance measurement containing data signals ("broadcast information") for the ground station operation. The time delay in the reception of the two simultaneously emitted signals is measured in the ground station and retransmitted to the onboard memory for the later ionospheric correction of data. In the ground station the received X-band signal is transposed to 7.2 GHz, coherently modulated with the regenerated PN-code (or with one of three orthogonal copies for code multiplexing) and retransmitted to the space segment where the PN code is fed into a correlator to determine on board the two-signal delay, which is a measure for the two-way slant range between the satellite and ground station. In addition, the received carrier frequency is evaluated in a Doppler counter to derive the relative velocity of the spacecraft to the ground station. Four independent correlators and four Doppler counters allow the simultaneous measurements with four ground stations in code multiplex. Geometric position and orbit determination approaches are applicable in this case.

The PRARE measurement accuracy is estimated to 0.1 mm/s for X-band Doppler (integration interval of 30 seconds) and 3 to 7 cm for X-band ranging (one measurement per second); the main error source is the tropospheric refraction (2 to 5 cm).

Future Prospects. The principle error source limiting the microwave system's accuracy is the ability to calibrate the atmospheric refraction effects. Therefore it is planned to extend the PRARE-system with the help of two additional optical uplinks. With this upgraded equipment it will be possible to determine the tropospheric correction to better than 5 mm, so that the resulting ranging accuracy will be better than 1 cm. This PRAREE (= PRARE Extended Version) is planned for post-ERS-1 missions.

5. TOPOGRAPHIC MAPPING TECHNIQUES

Timothy H. Dixon, Philipp Hartl, Duncan Wingham

Topographic Mapping is defined here as the measurement of the 3-D coordinates of points on the Earth's surface. Mapping techniques differ for land, ocean, and ice surfaces. Over land, the topography can be very complex, and horizontal resolutions of order 100 m or better, and vertical accuracies of 1 m or better may be required.

The ocean surface topography is relatively smooth, and the required horizontal resolution is modest, several kilometers or more. The accuracy requirements for the vertical components are two orders of magnitude higher than over land, of order 1 cm.

The resolution requirement for ice surfaces lies generally between the extremes of land and ocean surfaces. In the following, only methods for topographic mapping of land and ice are discussed in detail, as ocean altimetry has been addressed extensively elsewhere (e.g., Stewart, 1986).

5.1 PRESENT STATE OF LAND AND SEA SURFACE MAPPING FROM SPACE

Land Mapping. Digital 3D-topographic data are mainly acquired by digitizing existing contour maps and by aerial photogrammetric data. Large digital elevation data bases are available for the USA, Canada, Europe and Australia (Kubik, 1988). Manpower-related costs for manual and automatic digitizing are similar at present. According to Kubik (1988) for manual digitizing they include

- 50 - 70 hours for manual digitizing,
- 0.5 hours of scanning,
- 8   hours of vectorization on VAX, and
- 8 - 20 hours editing on an interactive work station

assuming a map sheet of 50 x 70 cm^2 with 100 m sampling.

This effort will certainly be reduced in the next years due to the development of improved automatic and semiautomatic processes and increasing computer power, but it will remain large. The above mentioned techniques will not allow development of a global 3D-digital map within reasonable time and cost limits. In addition, present methods will not lead to a data base with the required internal uniformity and consistency (accuracy).

In contrast to spaceborne 2D-mapping (imaging), where large sets of data have been collected and analyzed in the last decade, 3D-mapping of the Earth's land surface has not been performed. However, feasibility demonstrations and experimental tests have been conducted.

Sea Surface. Pulse-limited radar altimetry has been used very successfully over the ocean and, to some extent, over ice and land. The measurement of the marine geoid and mapping sea surface dynamic topography, and by inference currents and the tides, can be performed with this type of instrument. SEASAT radar altimetry data, collected during its 100 days of operation in 1978, was an important milestone for these types of investigations. The U.S. military GEOSAT satellite is presently in operation and is being used by a number of scientists and organizations for scientific and operational applications. TOPEX/POSEIDON (U.S. and France), and ERS-1 (ESA) will offer more extensive pulse-limited radar altimetry with very high accuracy. It is estimated that the height errors of measurements over the ocean are as small as 7 cm for the ERS-1, and 3 cm for the TOPEX/POSEIDON radar altimeter, provided that accurate orbital ephemeris data and the propagation errors through the atmosphere can be compensated. The ionospheric error compensation requires either two-frequency radar altimeters (TOPEX) or some two-frequency ranging system (like PRARE in ERS-1).

Current radar altimeters are essentially profilers that determine the surface height along the subsatellite track by measuring the two-way propagation delay (Satellite -> Surface -> Satellite). Profiling from a single platform demands an unsatisfactory compromise between the distance of adjacent tracks and repetition period, leading to tradeoffs between spatial and temporal sampling, and aliasing of certain features, e.g., meso-scale eddies. It is difficult for existing systems to achieve both a sufficient repeat cycle and satisfactory horizontal track density from a single spacecraft.

5.2 INSTRUMENTS FOR LAND APPLICATIONS

1. A variety of optical imaging techniques can be used to generate stereo pairs, from which height information can be extracted. These include:

a. **Stereophotogrammetry.** Photographic methods have the advantage of extremely high geometric resolution. The disadvantages are the requirement for film material, which has to be recovered from space, and the "analog" nature of film. It requires much effort for geometric corrections, detection of reference points and conversion of brightness data into digital information.

USSR missions perform photogrammetric mapping on an operational basis.and the products are commercially available. Experimental systems have also flown in Shuttle missions using the West-German MC camera and U.S. Large Format Camera. Planimetric accuracies range from 3 to 26 m and height accuracy is about 5 to 32 m. No further camera flights are expected in the near future on the Shuttle.

b. **Electromechanical Line Scanner.** The electromechanical line scanner is used in the Landsat-sensors MSS and TM. The 2-dimensional images are acquired line by line by scanning a mirror mechanically across track. Where the swaths of two adjacent subsatellite tracks overlap, stereo pairs are obtained. Overlapping areas are small (15 %) near the equator and large in the polar regions. If the scanning mirror could be tilted off-nadir (not possible for the Landsat/sensors), then the overlapping areas and the baseline to height ratio would be improved.

c. **Electronic Line Scanner (Pushbroom).** In this approach an array of Charged Coupled Devices (CCD) is used instead of a single detector and scanning mirror. This has several advantages relative to the electromechanical scanner, including the absence of moving elements, longer integration times per pixel, and better geometric performance. The disadvantage comes from the fact that the number of detectors in the array is still modest and the number of pixels per line limited to a few thousand, limiting the resolution of the system. Photographic cameras are superior in this respect.

The pushbroom principle is used by various types of sensors, the most prominent one being the HRV of the French SPOT satellite series. In this particular sensor a mirror can be tilted off-nadir and used for the acquisition of stereo images from adjacent swaths.

Various test results have been published. The accuracy actually achieved depends on the accuracy of the geometric image correction, the baseline to height ratio, and the attitude and orbit accuracy of the satellite. With a horizontal resolution (pixel size) of 10 m the position accuracy for SPOT in theory can be as good as 0.5 to 1 pixel, i.e., 5 to 10 m rms, with a corresponding rms height error of about 5 m to 6 m, for a baseline to height ratio of about 1. This can be achieved by use of many ground control points and for limited areas (Priebbenow and Clerici, 1988, Day and Müller, 1988). More typical results range between about 24 and 17 m vertical accuracy (Arai et al., 1988).

d. **Stereo Line Scanner.** Three (3) lines of semiconductors are arranged in the focal plane of the pushbroom scanner perpendicular to the direction of flight. The 3 lines scan simultaneously, in forward-, down-, and aft-direction. Three images of the same scene are acquired from the same (not from adjacent) swath(s). The stereo images are taken simultaneously. The baseline to height ratio can be chosen by proper adjustment of the fore and aft look angle and is independent of latitude. Digital correlation for control points and stereo measurements are more easily performed than in the single line pushbroom case. First tests of an airborne stereo line scanner have been performed successfully, and a spaceborne instrument is feasible (Hofmann, 1988).

The major disadvantages of all these stereo-optical systems, whether analogue or digital, are limitations due to cloud cover (both images must be largely cloud free to extract good elevation data) and the amount of data processing which is very large. These problems prohibit full global coverage. In addition, ground control points are generally required, limiting accuracy (if unavailable) or greatly increasing costs if properly implemented.

2. Scanning Laser Altimetry. Laser altimetry is the only way to obtain elevation data with centimeter-level vertical precision and high horizontal resolution. Several aircraft laser altimeter systems have been used for precise terrain mapping of relatively small areas. Low-altitude systems can produce topographic maps using a helical scan approach, and high altitude systems with 10 m footprints and 50 cm vertical precision have operated in a profiling mode for traverses of up to 100 km in length.

Although limited by cloud cover and instrument lifetime, aircraft laser altimetry produces very high resolution topographic information and complements coverage from other techniques. However, spaceborne lasers are not an immediate prospect for global topographic data acquisition. Cloud cover precludes global mapping with lasers in any short-duration mission. On long-duration missions, current generation lasers would also present reliability and lifetime problems. However, a laser-ranging system is a candidate instrument for the Earth Observing System (late 1990's time frame) for geodetic applications and offers the potential of high-resolution topography for selected local areas by operating in a profiling mode.

3. Synthetic Aperture Radar Mapping. Synthetic aperture radar systems generate microwave images which can be acquired independently of weather conditions. By using the images from adjacent tracks one can again produce stereo maps. However, observation of the two overlapping images must occur from the same side. Data reduction costs are also high, and vertical accuracy is poor.

4. Radar Interferometer. The radar interferometer applies two radar receiver units and a baseline across track. The phase differences as a function of location are used for direct measurement of the elevation of the backscattering surface. Demonstration tests with airborne (Zebker and Goldstein, 1986) and spaceborne instruments (Li and Goldstein, 1988) have shown the feasibility of this concept. Good horizontal resolution (~ 30 m or better) makes this concept attractive, but cost and extremely high data rates are problematic.

5. Scanning Radar Mapper. All spaceborne radar altimeters flown to date have been wide-beam systems limited in accuracy by their pulse duration (pulse-limited altimeters). Such altimeters are useful for smooth surfaces, but are ineffective over relatively high relief continental terrain. The analogous problem of wide-beam echo sounding of the ocean floor has been largely solved in the last decade with the introduction of narrow-beam, phased-array sonic transducer systems. This approach is possible from space using narrow beam radar altimeters and is an attractive technology for acquisition of global, high resolution topographic data.

A fundamental obstacle to narrow-beam spaceborne radar altimetry is the physical constraint of antenna size. Large antennas are required for small radar footprints, making deployment in space difficult. However this problem can be alleviated through use of higher operating frequencies (beam width is proportional to $1/f \cdot d$, where f is frequency and d is antenna diameter) and synthetic aperture techniques. To achieve the horizontal resolution described earlier, the radar beamwidth must be very narrow, $\leq 0.05°$. Beam scanning across track would be required to ensure full global coverage from most orbital platforms.

High frequency, high-reliability radars corresponding to an atmospheric transmission window at the Ka-band (≈ 37 GHz) have been flown on aircraft. To obtain high horizontal resolution at this frequency from orbital altitudes, a real aperture, mechanically stable $\approx$ 10 m x 10 m antenna would be required, not possible with current technology. However, an antenna with dimensions, 10 m by 0.4 m is feasible, and can yield high horizontal resolution parallel to the long axis (cross-track) with the real aperture beam and high resolution parallel to the short axis (along-track) with synthetic aperture techniques. Depending on altitude, such a system is capable of achieving horizontal and vertical resolution of 100-200 meters and 1-2 meters, respectively. Millimeter-wave (~ 90 GHz) radar systems may also prove feasible, potentially leading to even better resolutions.

6. POTENTIAL FIELD MEASUREMENT SYSTEMS

Jose Achache, C.W. Francis Everitt, Ho Jung Paik, Christoph Reigber, David Sonnabend

This section deals with the measurements from space platforms of global fields of interest to geophysics, geodesy and crustal dynamics. The fields include gravitational, magnetic and electrostatic. In most cases, a global field determination requires an ensemble of instruments; and the discussion below is divided into these measurement ensembles. For instance, the topic of gravity gradiometry includes essential support measurements of attitude and position to high accuracy.

The section is divided into five parts. The first two, Satellite to Satellite Tracking and Gradiometry, are concerned with proposed methods for obtaining a high resolution gravity field. A section on drag free systems follows, as this unique technology is central to both gravity measurement methods. Following this is a part on magnetometry, summarizing the status of instruments and proposed new missions in that field. Finally, there is a more speculative part on future technologies, and their application to field measurement systems.

6.1 SATELLITE-TO-SATELLITE TRACKING SYSTEM

Current Status. The first ideas about the application of satellite-to-satellite tracking (SST) as a substitute for satellite tracking from ground stations and as a means for measuring the Earth's gravity field were published in the 1960's (Wolff, 1969). Both the high-low and low-low SST concepts were introduced at this time and experimentally proven for the Earth gravity recovery in 1975.

SST requires at least two satellites. The low-low concept is based on two satellites following each other along the same orbit, a few hundred kilometers apart. One- and two-way microwave intersatellite tracking systems have been used so far to measure their relative velocities. Continuous tracking is possible for such a configuration. The irregular variations of this velocity convey gravitational information, and the lower the satellites' orbit the more pronounced and detailed this information becomes. Alternatively, a main spacecraft (such as an orbiting platform) may track two or more small orbiters in the same orbit. Studies were conducted for such a configuration building an optical interferometer from three spacecraft, with the active laser system installed on the center spacecraft (Balmino et al., 1978).

The high-low concept describes the situation where a high orbiting satellite carrying an intersatellite measurement device tracks a low Earth orbiting spacecraft. With one high orbiting satellite tracking data coverage is limited, only about 1/5 of the coverage from a low-low configuration. Data coverage could be continuous around the globe if the low satellite is equipped with a receiver for a high multi-satellite tracking system such as GPS. A variant of the high-low SST concept was applied early in the APOLLO program to obtain information about the lunar gravity field (Muller and Sjogren, 1968).

The first application of high-low SST between two satellites in orbit around the Earth took place in April 1975, with a tracking experiment between the geostationary ATS-6 satellite and GEOS-3 at an altitude of 840 km (Schmid et al., 1975). The key instruments for this SST experiment were a 9 m diameter steerable antenna on ATS-6 with its accurately programmed scan and monopulse pointing capability and a transponder on GEOS-3 to receive signals from ATS-6 and to generate a return signal. This historic first was followed by the ATS-6 tracking of the NIMBUS-6 meteorological satellite, which was launched in June 1975.

The SST data have been used for the orbit computation of the lower-altitude satellites, but the ATS-6/GEOS-3 SST data have also been used extensively to obtain information about the Earth's gravity field. During this SST experiment both one-way and two-way range and range-rate data were obtained over the link ground station/ATS-6/GEOS-3. The specially processed range-rate measurements applied for the gravity field determination had a precision of about 0.3 mm/s.

A second gravity experiment with SST, now in both the high-low and low-low modes, took place during the APOLLO-SOYUZ mission at an altitude of about 223 km in July, 1975. During this mission, high-low SST range and range-rate data were collected for 108 orbital revolutions, where ATS-6 again served as a tracking and relay satellite. The Doppler tracking data, with a precision of about 0.5 mm/s, have been used to recover gravity field information (Kahn et al., 1982).

The pioneering experiments involving ATS-6 have demonstrated that the high-low SST mode is very attractive for the orbit determination of the user satellite and capable of producing good-quality information about the gravity field. The gravity signal recovery results from the low-low SST APOLLO-SOYUZ experiment were not conclusive, because the range-rate tracking signals were found to be corrupted by an unexpectedly high noise level of about 3 mm/s, which prevented the unambiguous identification of gravity anomaly signatures. Table 7 summarizes these past experiments.

Limitations and Error Sources. Limitations for the past experiments are primarily the weak data coverage, the limited precision of the range-rate measurement and with regard to an analysis of the details of the gravity field the medium altitude of the GEOS-3. Main error sources were signal propagation errors and mismodeling of surface forces of the tracked satellites.

Future Missions and Technologies. As shown in Fig. 10 gravity field recovery with a resolution and precision compatible with the science requirements is only possible if the satellites tracked are at very low altitudes (200 km), if tracking system performances of 1-10 μm/sec are achievable and if the non-gravitational acceleration is measured or eliminated with extreme precision (cf. Section 6.3). Various mission concepts of this performance level have been studied in the past ten years by NASA and ESA. The most developed ones are NASA's Geopotential Research Mission (GRM) (Keating et al., 1986) and the POPSAT high-low mission of ESA (Schürenberg et al., 1985). For various reasons both concepts will not be pursued, but will be replaced by gradiometry missions as described in the next section.

Nevertheless the GRM concept is still a viable technical option to model the Earth gravity field's precisely . In this concept, the observable would be provided by a low-low SST Doppler link between two coorbiting GRM satellites in a 160 km altitude circular polar orbit. The separation between the two satellites is adjustable from 100 to 600 km. The variation of the relative velocity of both satellites is measured through a two-way SST-link, operating at frequencies of 42 and 91 GHz. These high frequencies effectively eliminate any residual ionospheric propagation errors in the measurements. The tracking system is based on the concept that a continuous wave signal is transmitted by one of the satellites to the other satellite, which receives it and compares it to an onboard generated signal. At the same time this satellite radiates an incrementally frequency-shifted signal to the other satellite, where it is compared to its onboard generated signal. The resultant continuous comparison of the signal serves to measure velocity changes with a precision of 1 μm/s.

Table 7 Results Achieved by Various Research Groups from Past SST Experiments

Experiment	Mode	Experimenter Reference	Range Rate	Result
Apollo: Soyuz	1 1	SAO (Weiffenbach et al., 1976)	±48 mm/sec	no gravity parameter recovery possible
Earth: lunar orbiters IB, IIB, IIIB, IVC, V-C	h 1	JPL (Muller and Sjogren, 1968)	±0.3 mm/sec	lunar mascons (structure of the moon's gravity field)
ATS-6: Apollo-Soyuz	h 1	GSFC (Vonbun et al., 1977)	±0.3 mm/sec	recovery of 5° gravity anomalies with ± 7 mgal
ATS-6: GEOS-3	h 1	JPL Sjogren et al., 1978)	±0.5 mm/sec	qualitative analysis above the American continent
ATS-6: GEOS-3	h 1	OSU (Hajela, 1977)		recovery of 5° gravity anomalies with ± 12 mgal
ATS-6: GEOS3: Apollo	h 1	GSFC (Kahn et al., 1982)		selected set of 5°x5° mean gravity anomalies with ± 5 mgal

Because of its low altitude, relatively large atmospheric forces would act on the GRM spacecraft. To effectively eliminate the effects of these and other surface forces on the satellite orbit, both satellites would be equipped with the DISturbance COmpensation System (DISCOS).

One of the most likely developments in SST is to make use of the US NAVSTAR GPS System or the USSR GLONASS System. One low spacecraft could carry a GPS/GLONASS receiver to simultaneously track several of the very high (20000 km) spacecraft. Data coverage could be continuous, because several of these high satellites would be constantly in view. In this way one would get the equivalent of instantaneous fixes on the position vector of the low spacecraft, using the P-code signal, or on the velocity vector, using carrier Doppler. The orbits of the GPS satellites would be determined with tracking data from ground stations, possibly with VLBI-determined coordinates and hydrogen maser clocks, to provide very precise ephemerides and timing. With the spacecraft in a near polar orbit a dense and uniform coverage with range or range-rate data would be achievable, and such a data set could be analyzed to recover the finer details of the gravity field which would depend on the altitude of the low satellite and the instrument precision. One possible mission, with this type of satellite in low polar orbit, could be the Gravity Probe-B, whose main purpose is to test general relativity by measuring the effect of Earth rotation on the local inertial frame of the satellite. As the orbit altitude would be near 600 km, this will not be a subsitute for a dedicated GRM type mission, but it can serve as a possible interim attempt to expand our knowledge of the gravity field before more ambitious projects come about.

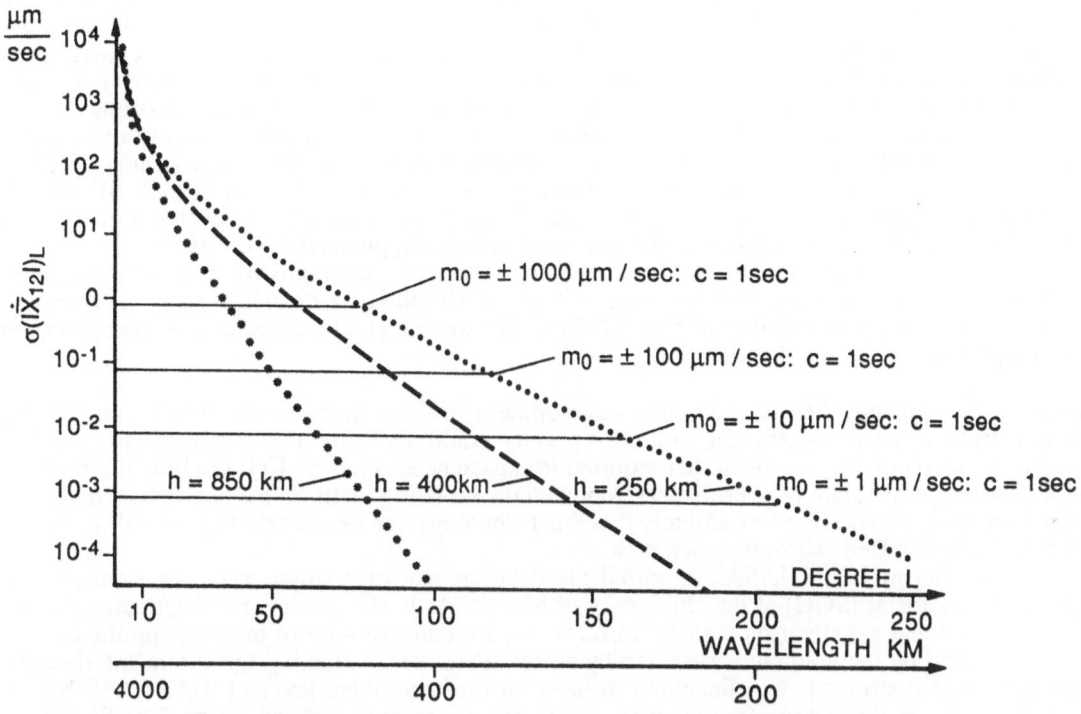

Fig. 10 Velocity difference spectra at different altitudes vs. noise spectra; high-low mode

Conclusions. The ATS 6/GEOS-3 and ATS 6/Apollo experiments have proven that gravity information can be extracted from SST data. Basically the analysis of SST observables for gravity parameter recovery is understood. Recognizing the important scientific returns extractable from intersatellite tracking measurements on Earth orbiters and space probes parked in orbits around bodies in the solar systems, it is considered important to

- develop space-qualified microwave trackers (multi-channel, carrier phase tracking GPS, two-way PRARE, etc.) for the gradual gravity field improvement of the Earth and planets from intersatellite range or range rate data,
- develop extremely precise and stable accelerometers,
- investigate CO_2 PN-coded laser intersatellite ranging techniques for SST gravity missions,
- study space-qualified laser interferometry systems,
- make satellite-based data relay systems available for fast and reliable communication of data from SST and gradiometer missions.

6.2 SATELLITE GRAVITY GRADIOMETRY

Background. A pair of separate accelerometers can measure the space rate of change of gravity, which is proportional to the second gradient of the gravity potential. The measurement is a combination of gravitational and rotational effects; so to get information on gravity, it is necessary to eliminate the effects of rotation. A gradiometer tends to be more sensitive to short wavelength components of the Earth gravity, compared to SST. Unlike SST, gravity gradiometry requires only a single spacecraft, and could therefore be extended more easily to planetary missions.

In the early 1970's, various studies were conducted to review the feasibility of an Earth orbiting gravity gradiometer mission. Although there were three DoD funded programs (Hughes, Draper Lab, Textron-Bell) to develop gradiometers of 1-10 $E/Hz^{1/2}$ sensitivity, a .01 $E/Hz^{1/2}$ instrument, required for satellite gradiometry, was deemed to be too challenging at that time (1 $E = 10^{-9}$ s^{-2}); and SST was chosen for the gravity mission (GRM), planned for the 1980's.

Various requirements in solid Earth geophysics and oceanography of the 1990's, and early 21st century, correspond to a gradiometer sensitivity of 10^{-2} to 10^{-4} $E/Hz^{1/2}$ the range representing the desired horizontal resolution of 50 - 100 km. The instrumental requirement also depends on spacecraft altitude and mission duration.

Review of Current Status. Intense development efforts, made in the 1970's, led to the demonstration of a successful moving base gravity gradiometer by Textron-Bell, albeit with a sensitivity of 10 $E/Hz^{1/2}$ far from that required for space geodesy. The Bell gradiometer consists of four room temperature accelerometers, mounted on a rotating platform for heterodyne detection of the gravity gradient. It seems unlikely that this technology can be extended by over three orders of magnitude, but Bell claims they can do it.

Starting in 1980, NASA supported the development of a single axis super-conducting gravity gradiometer (SGG) at the University of Maryland. In this device, persistent currents are used to difference acceleration signals from two super-conducting proof masses, and the current signal induced by the gradient is detected by a SQUID (Superconducting QUantum Interference Device). This instrument was designed to have an intrinsic noise level of .03 $E/Hz^{1/2}$ but its performance was limited to 10 times that, due to uncompensated rotation errors. The SGG now appears to have reached the marginal sensitivity level needed to yield 100 km horizontal resolution; but much engineering needs to be done to qualify the cryogenic instrument for space.

Current Development Program. There have been numerous attempts to develop gravity gradiometers for terrestrial and space applications. At present, there are two main projects being sponsored for space geodesy. One is an improved superconducting gradiometer, at the University of Maryland, funded by NASA and the US Air Force. The other is a room temperature gradiometer called GRADIO, at ONERA in France, supported by ESA and the French government.

GRADIO (Balmino et al., 1985) is based on an improved version of the French CACTUS accelerometers, previously flown to measure radiation pressure and air drag on spacecraft. Eight electrostatically levitated three-axis accelerometers are located on the corners of a cube, to make redundant measurements of all five independent components of the gravity gradient tensor. The projected sensitivity of GRADIO is .01 $E/Hz^{1/2}$. A two-dimensional version of this concept is the candidate for ESA's ARISTOTELES mission, planned to fly about 1994.

The University of Maryland device (Paik and Richard, 1986) incorporates a superconducting "negative spring," which effectively turns the SGG into a free mass system, thus improving the sensitivity of the instrument further. So modified, the projected sensitivity of the device is 10^{-4} $E/Hz^{1/2}$. A six axis superconducting accelerometer, capable of measuring six degrees of platform motion will simultaneously provide linear and angular acceleration signals of sufficient sensitivity for drag free operation and attitude control of the instrument. The flight hardware will consist of three diagonal component SGGs, and the six axis accelerometer. This combined instrument is the candidate for NASA's Superconducting Gravity Gradiometer Mission

(SGGM) (Morgan and Paik, 1988) which is envisioned to fly in about 2000. A precursor space test of this instrument is now being planned.

In both GRADIO and the SGG, instrument performance could be limited by angular motion sensitivity and self-gravity effects from the spacecraft. Providing an adequate dynamic environment for these sensitive gradiometers will impose severe requirements on spacecraft design and attitude determination and control.

Future Technology. Certain disciplines of geophysics, including oceanography, require 25 km horizontal resolution of gravity at the 1-2 mgal level. This implies a gradiometer with sensitivity better than 10^{-6} E/Hz$^{1/2}$ at 160 km altitude, or flying even lower, perhaps on a tethered satellite.

It is possible to envision improvement of the SGG beyond 10^{-6} E/Hz$^{1/2}$ by achieving a better quality factor (Q > 10^7) for the proof masses, and employing lower noise SQUIDs, which already exist as laboratory models. However, the more stringent attitude corrections, imposed by such an instrument, may demand better estimation methods (Sonnabend and McEneaney, 1988) or unconventional attitude sensors, such as the superconducting gyroscope or the cryogenic quartz telescope developed for the GP-B mission.

A tethered satellite approach requires active six axis stabilization of the platform, with large dynamic range. However, the gradiometer's inertial instrument ensemble should be sufficient to sense the platform motions at the required level.

Planetary missions will require much less stringent gradiometer sensitivity, perhaps 0.1 - .01 E/Hz$^{1/2}$. Several of the room temperature gradiometers under development should be capable of this. Much more sensitive gravity missions, with superconducting gradiometers, may be envisioned, if mechanically quiet closed cycle refrigerators, operating below 10 K, become available for space applications; or high temperature superconductors (above 90 K) for sensitive instruments.

A challenge in designing the gradiometer system will be slosh and the tidal motion of liquid helium. Helium is the hardest liquid to control, because of its extraordinarily low viscous damping (fortunately there is some damping of large motions in bulk superfluid), and extraordinarily low surface tension (two orders of magnitude below water). To see the problem, the addition of 1 cc (0.13 gm) of helium liquid, 50 cm from the gradiometer, produces a gradient equal to the 10^{-4} E resolution of the instrument.

Conclusions. Gravity gradiometers and spacecraft control technology have reached a stage that the highly sensitive gradiometer mission can now be planned. Advanced gravity mapping missions, such as ARISTOTELES and SGGM, will contribute greatly to our understanding of the Earth, and prepare the way for studies of the planets. It is therefore recommended that the highest priority be given to the development of the necessary instruments and ancillary technologies. These and the design of the mission, the spacecraft, and the data analysis techniques would best be accomplished by pooling the resources and talents of ESA, NASA, and other space agencies and scientific institutions. Although less sensitive, ARISTOTELES would serve as a valuable, perhaps necessary, precursor for the more demanding SGGM, needed to meet the gravity measurement requirements of the 21st century. Therefore, it is recommended that both missions be pursued with the highest priority.

6.3 DRAG FREE SYSTEMS

A "drag free" spacecraft is one that flies a purely gravitational trajectory, in spite of uncertain disturbances, such as air drag and radiation pressure. These systems generally involve a small, dense, free floating proof mass, contained in a larger structure, which protects it from disturbances. When a collision is imminent, thrusters on the spacecraft are fired to keep it away. Thus, the proof mass follows an almost perfect gravitational trajectory, and the spacecraft does so on average. As both the SST and gravity gradiometry satellites depend critically on drag free systems, they are discussed separately here.

Review of Current Status. The first drag free satellite was TRIAD, a member of the US Navy's Transit Improvement Program series, flown in 1973. The proof mass was a 22 mm diameter, solid, platinum-gold sphere, contained in a 40 mm spherical cavity. Six plates inside the cavity comprised a three axis RF capacitance bridge to measure relative displacement. Propulsion was by impulsive cold gas (Freon). A performance of better than 10^{-10} m/s^2 average nongravitational acceleration was established by external tracking. The Navy's intent was to extend orbit predictions of navigational satellites, thus easing the tracking and operational requirements. Since then, the Navy has developed a single axis drag free system, which was tested on TIP-II, and used operationally on the NOVA (Transit) satellites.

Gravity Probe B (GP-B) is to use essentially this scheme, to reduce the centering forces of its electrostatically supported gyros to a very low level (Young, 1986). The principal differences are a much smaller gap, and the use of boil-off helium in continuous proportional thrusters for centering.

The GRM SST (Keating et al., 1986) originally employed TRIAD-like schemes in each satellite. The main differences were that hydrazine propulsion was to be used, and that a much larger hollow proof mass was intended, to increase the capacitances and thus improve the location accuracy of the proof mass, relative to the cavity. It was later recognized that a substantial improvement in system accuracy could be achieved by a two stage drag free system. In this arrangement, an intermediate body, containing the tracking system antennas, is closely centered on the proof mass by magnetically pushing on the outer satellite. Collision between the intermediate and outer bodies is again prevented by an impulsive hydrazine system. In the later GRM studies, the two stage arrangement was clearly preferred.

Space gravity gradiometers pose a somewhat different requirement. Here, "scale factor errors," arising from mismatch of accelerometer gains, or misalignment of their input axes, can be controlled by operating at very low levels of non-gravitational acceleration. Since only the proof mass is really drag free in a TRIAD type system, either the whole ensemble of instruments becomes the proof mass, or a two stage drag-free system is needed, with the instrument package on the intermediate body.

Two proposed gradiometer missions employ forms of the two stage drag-free system. The now dormant ESA/NASA collaboration would put the GRADIO and associated instruments on the intermediate body of the system studied for GRM, while the NASA/University of Maryland SGG would float the whole cryogenic dewar as the intermediate body, and employ its six-axis accelerometer proof mass as its inner body. Control forces and torques on the dewar can be provided either by the helium boil-off, as on GP-B (Parmley, 1986), or by eddy current repulsion against the outer body.

Error Sources and System Performance. The criterion for drag-free performance is how low the non-gravitational accelerations of the proof mass can be made. The two worst sources of error are the build up of electric charge on the proof mass, and the gravitational attraction between the proof mass and the rest of the spacecraft. Other error sources, such as magnetic effects and proof mass asphericity are well controlled in current designs.

Proof mass charge can result from high energy radiation causing electrons to transfer between the proof mass and its surrounding cavity. A potential arises if this two-way process is unsymmetrical. Even then, there is no disturbing force unless the proof mass is off center. No error from this source was discerned on TRIAD; however, the intervals between tracking passes were much greater than the times between freon firings, so the effect, if present, could have averaged out in the data. Charge effects have been observed in lab tests of electrostatically supported accelerometers; and small disturbances attributed to charge were detected on the French CACTUS flight test. This experience has led ONERA to consider grounding the proof mass by a very thin wire in the GRADIO accelerometers. While the charge problem needs careful attention in future programs, it will probably not limit the performance of any future mission, and it is certainly well in hand for GP-B.

Unlike charge effects, self-gravity is essentially out of the hands of the instrument designer, and must be dealt with at the spacecraft level. For size, 1 kg, 1 m from the proof mass causes an acceleration bias of 6.67×10^{-11} m/s^2, worsening with the inverse square of the distance. This is distinct from the effect on a gradiometer, where the same mass gives a gradient bias of .0667 E (tensor coefficient), worsening with the inverse cube of the distance. The primary method of dealing with these biases is an accurate mass survey of the spacecraft, which, for TRIAD, was a very involved undertaking. Moving masses are the worst problem, particularly propellants in partially empty tanks. On GRM, the tanks were to contain a flexible diaphragm, and were to be placed as far outboard as possible.

There is another way to get relief from self-gravity - frequency separation. If the purpose of the mission is the measurement of gravity variations, then the desired signal is contained in a frequency band, bounded below by a few times orbital frequency, and above by the time needed to fly a ground distance equal to the altitude, perhaps .05 Hz at the lowest practical altitudes. In this case, a high bandwidth, low displacement drag compensation system will move vehicle displacements above the measurement band, and discourage high frequency excitation of propellant motions. Tank baffles and diaphragms can also be selected with the same frequency separation idea in mind.

A possible extension of existing technology may be propulsion by thrusters where solid teflon is vaporized. The specific impulse is about a factor of ten higher for teflon than for bipropellants. Teflon thrusters have been demonstrated in orbit. Their advantage for geodesy missions would be either in extending the lifetime at a given altitude, or in operating at lower altitudes. Against this is the need for more power, and hence larger solar panels, which might appreciably increase the drag.

6.4 SOLID STATE MAGNETOMETRY

The Earth's magnetic field is made up of three contributions: the main field generated by a self-sustaining dynamo in the liquid core, the surface anomaly field resulting from the magnetization of crustal materials, and the external field, due to ionospheric, magnetospheric, and field aligned currents. This report addresses only the solid Earth problems, i.e., the first two components.

Any magnetic survey (ground, air, or spaceborne) should separately measure the three vector components of the magnetic field **B**, with an accuracy better than 1 nT (i.e., 1 part in 60,000). Under special circumstances, and for the study of the anomaly field only, one may proceed with measuring only the magnitude of **B**.

Instruments. Many spacecraft magnetometers have been built and flown, both for Earth orbit and deep space. The main types include flux gate, atomic vapor, and search coil magnetometers. Flux gates (Acuna, 1974) are one axis devices, usually combined into vector instruments. Atomic vapor types have been based on helium (Keyser et al., 1961; Smith et al., 1975), rubidium (Bloom, 1962; Farthing and Folz, 1967), and cesium (Mobley et al., 1980; Langel et al., 1982) and both scalar and vector instruments exist. Search coil instruments are for the measurement of AC fields, and are not of much interest here.

Current Status. The first substantial measurements of the Earth's field were in the late 1960's on the Orbiting Geophysical Observatory (OGO), employing a flux gate magnetometer. As the inclination was low, and the orbit highly elliptical, the mission fell well short of global coverage. The coverage improved with the three POGO satellites, flown in about 1969. While polar, they were also highly elliptical, and didn't give global coverage. A scalar magnetometer was used.

A big step forward was MAGSAT (Mobley et al., 1980; Langel et al., 1982), a sun synchronous, low altitude satellite, flown in 1979. It carried a 2 nT cesium vapor scalar magnetometer, and a 6 nT fluxgate vector magnetometer. The instruments were on a boom to avoid the spacecraft field, and star cameras and an attitude transfer system were used to obtain accurate attitude for the vector instrument. The result of the mission was an International Magnetic Reference Field. No similar mission has flown since, and no space geomagnetism program is currently being funded.

Current Development Program. Several studies and programs exist at different levels of preparation and study.

MAGNOLIA/MFE (Phase B completed) is a joint French/US program for long-term (more than 5 years) monitoring of the main field and its temporal (secular) variations. The intent is a global description of secular variations for accurate mapping of the motions of the core. Vector measurements are intended, with an accuracy of 3 nT, and scalar measurements with 1 nT accuracy, both including all sources of error. The mission objectives include a) Main field model updating; b) Secular variation study; c) Core motion determination, and correlation with the length of day variation and Chandler wobble excitation; and d) Electrical conductivity of the mantle.

EMF is a NASA project, to study both the main and external fields, and plasma phenomena. Vector and scalar measurements with accuracies of 5 and 2 nT, respectively are planned. The mission objectives include a) Main field model updating; b) Ionospheric, field aligned, and magnetospheric current systems studies; and c) Auroral and plasma studies.

ARISTOTELES (Phase A completed) is an ESA project to measure the gravity and magnetic anomaly fields simultaneously, at 100 km resolution, on a global scale. Only the magnitude of **B** will be measured, using two magnetometers in a gradiometric configuration, to remove the field generated by the spacecraft. The nature of the magnetometer is not yet decided. Mission objectives: a) Thermal structure of the continents; b) Rheology of the continental lithosphere.

GOS (Geomagnetic Observing System) is a NASA proposal for the EOS US polar platform. The desired system performance is 2 and 1 nT, for the vector and scalar instruments, respectively. The mission is intended to be a continuation of the objectives of both MAGNOLIA/MFE and EMF.

GOS-L is a NASA proposal, identical to GOS, for the ESA polar platform. It will allow real time studies of the local time dependence of the external field, and related phenomena (the NASA and ESA platforms are at different local times).

TEM (Tethered Equatorial Magnetometer) is a NASA proposal to fly a scalar magnetometer at the end of a 100 km tether below the Space Station. It would provide high accuracy mapping of the surface anomaly field over one-third of the Earth's surface. It would also permit a close look at the equatorial ionospheric currents, known as the electrojet and counterelectrojet. The instrument is a helium vapor scalar magnetometer.

GASP and POGS: The US Navy has two projects within a series of short lifetime free flyers, launched either by the shuttle (GASP), or by any vehicle (POGS). They will carry low precision vector magnetometers, mainly intended for updating navigational charts. Scientific applications are unlikely.

6.5 FUTURE TECHNOLOGIES

Seven products may be looked for from future technologies, as applied to field measuring instrumentation: 1) higher measurement precision; 2) easier, cheaper measuring techniques; 3) more compact instruments; 4) measurement of new quan-tities; 5) lower altitude missions; 6) longer duration missions; and 7) improved methods of handling large quantities of data. Item 7) is covered in the next section.

In what follows, the discussion is focused mainly on 1), 5), and 6). More profound is the invention of instruments to measure quantities not previously investigated, e.g., the cloud chamber in its impact on nuclear and high energy particle physics. These are also the hardest technologies to predict. As to cheaper, easier, and more compact; very rapid progress may be expected when the micro-circuitry revolution finally reaches space. Not to be ignored is the less well known revolution in precision machining techniques - diamond turning, numerical control, new metrology devices - which should enhance the performance, simplicity of design, and ease of manufacture of many instruments.

Three technologies seem likely to provide bases for many advances over the next 10 - 15 years: cryogenic instrumentation, drag free control, and "tether."

Cryogenic Instrumentation. Operation at temperatures attainable with a liquid He^4 cryostat (1 - 4.2 K), should contribute greatly to magnetometry, gravity gradiometry, and infrared surveys of the Earth. In broad terms, cryogenics offers the following potential advantages: a) reduced thermal noise; b) greatly improved temperature stability, due to better thermal isolation, increased thermal conductivities in certain metals, and the extraordinary thermal behavior of superfluid helium; c) greatly improved mechanical stability, in combination with the much reduced low temperature thermal expansion and creep coefficients; d) great reductions in radiation pressure on suspended bodies, and in background radiation in detectors; e) operating pressures as low as 10^{-15} torr; f) SQUID magnetometers, either directly for magnetic field measurements, or indirectly for ultra precise displacement measurements; g) superconducting magnetic shields to create regions of extremely stable magnetic fields (10^{-12} of Earth's field has been demonstrated) and, if necessary, extremely low fields (below .01 nT has been demonstrated); and h) exploitation of superconductivity in many devices, e.g., magnetic bearings, precision subtraction circuits, ultra precise gyros, specialized position actuators, etc.

The development of SQUID magnetometers is one of the triumphs in this field. It has been underway for over two decades, and for most applications SQUIDS can be regarded as fully developed instruments. The resolution far exceeds that of any other existing device, and appears to exceed any foreseeable requirement for studying the Earth's magnetic field. The detection of a 10^{-15} T field change in a 100 Hz bandwidth has been routinely demonstrated.

Resolution at high frequency is not everything. Also important are dynamic range, null stability, absolute calibration, and 1/f noise limitations on long-term integration of signals, all under active investigation. Intrinsically, the SQUID is a device of very wide dynamic range, wider by far than conventional instruments even in principle, by making use of "flux counting" techniques. Although not yet done, there is little doubt that a SQUID instrument could be developed for measuring the Earth's field with full dynamic range, absolute calibration, and a resolution of .01 nT or better - a factor of 100 beyond current limits. SQUIDS also provide the bases for the measuring systems in the projected gravitational missions: the Maryland gradiometer, the Stanford relativity gyroscope, and the Stanford equivalence principle accelerometer. Other applications abound.

In assessing the state of cryogenic technology in space we should note that many fundamental problems remain. First, long hold time helium dewars for space exist. The Infrared Astronomy Satellite (IRAS), launched in January 1983, with a superfluid helium dewar, operating at 1.7 K, had a demonstrated hold time of more than 10 months. The Cosmic Background Experiment (COBE) dewar, scheduled for 1989, and the GP-B dewar, scheduled for 1993, have projected hold times of 14 and 28 months, respectively. Dewars not containing a large aperture telescope could certainly be made with significantly longer hold times.

On the other hand, no present cryogenic program does all this at once. IRAS and COBE only depend on a) and d). A magnetic mapping mission similarly would depend only on f). Missions such as GP-B and the cryogenic gradiometer are much harder. They depend on successful integration of most of the listed technologies. The first such demonstration will be in the engineering test of the GP-B instrument, manifested for February 1993.

Tethered Satellites. "Tether" is a funded US-Italian program to develop coupled satellites in which a low altitude vehicle is drawn along by one at higher altitude, by a cable 10 to 200 km long, depending on the application. For geodynamics, the advantage would lie in providing a means for operating at a much lower altitude. The dynamics of tethers are complex, but have been extensively studied. Unfortunately, an instrument placed on a tethered platform would not be drag free at the tether altitude, and would require an active six degree of freedom platform with a large dynamic range. Nevertheless, the concept has sufficient potential to deserve careful study. Magnetometers, fortunately, do not need to be drag free.

New Concepts. Predicting new concepts is hazardous, but the following longer term possibilities deserve notice:

High temperature superconductors. Much publicity has been given to the discovery of ceramic materials that are superconducting at temperatures near, say, liquid nitrogen (77 K), and claims have been made that room temperature superconductors impend. If so, they would have many advantages in space instrumentation for magnetometers, gravity gradiometers, inertial navigation instruments, etc. They would not, however, embody all of the advantages of cryogenic operation listed above, such as low thermal noise and extreme mechanical stability. The future of high temperature superconductors is murky. The extraordinarily rapid, and highly publicized early progress has not been maintained. Many difficulties lie ahead in reducing the new discoveries to practice. The time scale is probably ten, or even twenty years.

Operation at temperatures below 1 K. An advance the other way would be to operate at much lower temperatures, say a few mK, to take advantage of still lower thermal noise and increased mechanical stability. The obvious means is developing a helium dilution refrigerator to operate in space, and such refrigerators are under study. The time scale here is also long, probably ten years or more. No obvious application to geodesy yet exists.

Long-term cryogenic systems. Planetary missions requiring a superconducting gravity gradiometer would need a lifetime much longer than five years; and a very long life Earth orbiting mission may also be needed. For Earth orbit, refilling the helium dewar might be feasible, and various schemes for doing so are under study. For a planetary mission, it would be necessary to carry an onboard closed cycle refrigerator. Although it would be desirable for the refrigerator to be mechanically quiet, that would not be an absolute necessity, if transfer to a floated dewar proves feasible. Much development will be needed.

7. DATA MANAGEMENT SYSTEMS

Peter Wilson

The application of space techniques for geoscientific purposes confronts the user community at large with data management problems of varying magnitudes. These problems are generally associated with data collection, preprocessing and distribution and not with archiving. The entire spectrum of data collecting rates is encountered when listing the various techniques to which the user can be exposed during particular missions. This is illustrated in Table 8, which lists typical examples of the data collection and transmission rates encountered in the exploitation of single satellite mission, here ERS-1.

For all applications there is an understandable delay in transferring the data from the recording facility at the measuring instrument to the user. Data management problems are encountered in different degrees, dependent on the type of observation performed (which determines the quantity of data to be recorded) and the facilities created and used for the tasks of transmission and preprocessing. Though it is not surprising that more money and formal effort is put into setting up lines of communication and establishing processing facilities for onboard satellite instrumentation, too little support is provided for speeding up the transfer of relatively low density data from the instrument to the end user.

It is the intent of this section to identify the sources of the delays most frequently encountered and to propose realistic means for alleviating the problems identified. The conclusions drawn in this section reflect the requirements of operational groups responding to ongoing demand for specific results rather than the scientific investigator not tied to repetitive deadlines, for whom a more relaxed time line for disseminating the data is visualized.

7.1 THE CATEGORIES OF PROBLEMS ENCOUNTERED

The delays in disseminating data to the user community fall into a number of categories, corresponding to the nature of the problems responsible for creating the delay. The difficulties encountered most frequently are associated with these problem areas:

1. the communication facilities in use between field station, data collecting center and user;

2. the operational design and complexity of the processing algorithms;

3. the data storage media put to use at the different phases of the transfer process;

4. the use of nonstandardized equipment and processing software;

5. the proliferation of nonstandardized data recording formats;

6. the incompatibility of communications networks.

Whereas some of these categories are correlated, for the most part they are technique dependent. Problem area 1 is particularly critical for transmitting control information and data between remote field sites and the operational control center; the next four categories are technique dependent and the effects of the last category are felt most in the collecting center to end-user link.

The delays which result, inhibit the work of the operations and analysis centers, add significantly to the cost of operational programs and even bring confusion to the analysts. Ultimately, they determine the viability of applying specific techniques for routine operations.

7.2 DISCUSSION

The quantities of observed data requiring centralized collection, preprocessing and distribution are constantly increasing and the sources from which they originate are becoming more dispersed around the globe. In the near future observing systems will be deployed on the ground and in space for the continuous monitoring of changes taking place at the solid surface of the Earth and in the oceans. The demand for near real-time accessibility of the data collected will increase.

Several approaches to speeding up the overall distribution of information can be identified and it is apparent that most problem areas could be relieved or even disposed of completely by the introduction of a few relatively inexpensive but far-reaching modifications to the current hardware and procedures. However, the success of most of the solutions hinges on dialogue between the manufacturer of the equipment, the organization producing the information and the end user, as well as on the level of international cooperation achievable at each of the foregoing levels.

The most significant change could result from the consequent application of state-of-the-art communications facilities. Whereas the time spent on design and manufacture of any instrument makes it inevitable that the technology employed lags behind state-of-the-art development of individual components, the gap between the availability and application of advanced communications systems for collecting and disseminating large and small quantities of data is too large.

Furthermore, efforts have to be made to introduce higher levels of standardization for output formats, preprocessing software and for interfacing between different national and international communications networks. These requirements presuppose an intense dialogue with industry and the international user community as well as the consequent adoption and use of the standards agreed upon.

In addition, much can be achieved for easing the communication of data acquired by certain techniques by introducing more sophisticated computer facilities and software at the instrument with the objective of condensing the data sets on site to more manageable proportions. This approach will be extremely important e.g., for the deployment of GPS instruments at remote sites for the continuous monitoring of tectonic features.

7.3 CONCLUSIONS

It is concluded that the international data user community should strive to achieve the following objectives in the development of data management systems to be operated until the end of this century and beyond:

1. Commitments should be made to support the use and further development of vendor independent computer networks and electronic network communications facilities with standardized interfaces for operational use by the community for the two-way transfer of information between the field, the collecting and archiving centers and the analysts, irrespective of location.

2. A dialogue should be established between the manufacturers and users of both ground and spaceborne instrumentation to achieve a higher level of standardization of output data formats.

3. More effort should be put into extending on-site/onboard computational facilities to reduce the overall size of the data set and, as required, convert it to standardized output formats.

Table 8 Typical Collection Rates Encountered in Geodetic Space Activities.

(Information based on values provided for the ERS-1 mission for which the typical down-link rates will be 1.2 MB/s, with a dump-rate of 15 MB/s at Kiruna.)

Application	*Bit-Rate*	*Storage and Transmission*	*Distribution Delay Incurred*
SAR	110 MB/s	Dedicated RF link; no onboard storage; specialized ground facilities	mag.tape via mail
Scatterometer	0.5 MB/s	data multiplexed and collected at specialized ground facilities	mag. tape via mail
Radar altimeter	25 KB/s	onboard storage; data dump at selected overflight positions with specialized ground facilities	mag. tape via mail
GPS	1.2 KB/s	onboard storage; data dump on command from specialized ground facilities	mag. tape via mail
PRARE	2 KB/s	onboard storage; data dump on command from specialized ground facilities alternate data recording mode at ground receiver collection via mail	mag. tape via mail
SLR	1-2 KB/s	mag. tape or floppy disk at ground station; data collected via mail	mag. tape via mail

8. CONCLUSIONS AND RECOMMENDATIONS

William G. Melbourne and Christoph Reigber

8.1 TECHNOLOGY DEVELOPMENT

Recommendation: Technology development in microwave and laser geodetic systems should continue with the goal of achieving, as appropriate for each system, millimeter-level accuracies, lower capital and operating costs than current systems, and an expanded range of applications, including high temporal and spatial resolutions.

Rationale: Current space geodetic system accuracies, 1-3 cm, are limited by instrumental and environmental errors that must be eliminated or reduced, if the long and short-term solid Earth measurement requirements are to be achieved. Moreover, the requirements for temporal and spatial resolutions down to hours and kilometers currently can not be economically or technically achieved.

Innovations such as coherent cw lasers and carrier phase-delay VLBI systems are technically feasible and would enable millimeter-level instrumental accuracies. A concomitant development and modeling of environmental errors should enable millimeter-level system accuracies.

Dense arrays of continuously operating and remotely monitored microwave receivers (e.g., GPS) that are highly automated, very low cost and millimeter-level accuracy, are economically and technically feasible.

GLRS and GPS on EOS and other platforms would provide the required accuracies and resolutions in the late 1990's. A mix of laser, VLBI and GPS/GLONASS tracking systems would provide Earth rotation variability, at the 0.1 mas accuracy level and at hourly resolutions. GPS positioning would support ocean acoustic ranging systems for sea floor geodesy at the required accuracies and would provide kinematic positioning required for airborne and/or spaceborne gravimetric and mapping systems.

8.2 SATELLITE-BASED MICROWAVE TRACKING SYSTEM

Recommendation: Future geodetic and ocean altimetry missions should carry flight quality receivers for tracking navigation satellites such as GPS and a corresponding global network of ground tracking stations should be maintained to enable acquisition of continuous differential observations in support of the high accuracy orbital position and gravity recovery requirements of such missions.

Rationale: Orbit determination continues to be the major error source in determining sea surface topography from altimetric satellites. Moreover, errors in the ocean geoid model at mesoscale and larger wavelengths limit the accuracy of ocean circulation recovery from the sea surface topography. Future accuracy requirements of oceanographic missions will be at the centimeter level for radial orbital position and for marine geoid heights. A GPS-based tracking system should yield radial orbit accuracies in the range of 1-5 cm. For drag-free missions such as Gravity Probe-B, and also for a satellite with low non-gravitational perturbations such as TOPEX, this tracking system would yield significant improvements in the geopotential coefficients down to wavelengths roughly equal to the satellite altitude. This would provide valuable interim gravity information prior to the flight of a high performance gradiometry mission in the late 1990's. This would in turn improve orbit determination accuracies of other missions.

8.3 GRAVITY GRADIOMETRY

Recommendation: The development of gravity gradiometer instruments and ancillary technologies should be supported and detailed studies of gravity gradiometry missions should be completed at the earliest possible date, with the goal of flying one or more such missions before the end of the century.

Rationale: The lack of a high resolution (50-100 km) global gravity map at milligal accuracies is impeding the advance of many branches of solid Earth physics and physical oceanography. Low altitude (160-200 km) gradiometer missions of the ARISTOTELES and the SGGM type are the logical steps to meet these requirements.

8.4 TOPOGRAPHIC MAPPING

Recommendation: New dedicated microwave and optical mapping systems for topographic studies of land and ice should be developed and flown to provide a global database with a minimum 100 m horizontal resolution and 1 m vertical precision within the next decade.

Rationale: Global, uniform and consistent land and ice elevation data are required for various geoscience applications, including studies of possible secular warming trends. Such a database cannot be obtained within reasonable processing time and cost limits with existing instrumentation. The following systems have a potential to acquire such a database: narrow multiple and/or scanning radar or laser altimeters, and stereo radar or optical imaging systems. Such instruments would also provide data of importance to oceanography, and provide useful constraints to sea floor bathymetry.

8.5 TROPOSPHERIC CALIBRATION

Recommendation: Calibration techniques for tropospheric delay, including instrumentation, modeling, and observational approaches, must be improved by an order of magnitude over existing capability. Both the wet and dry components must be addressed to reach millimeter-level position accuracies with space geodetic systems.

Rationale: For baselines beyond the wet tropospheric correlation length, current wet tropospheric calibration techniques limit microwave geodetic systems to 2-5 cm accuracies in the vertical direction depending on the system, local and meteorological conditions. Horizontal accuracies are limited to 0.5-1 cm. Current limitations in calibration of the delay due to the dry component would limit vertical accuracies of microwave and optical systems by several millimeters in certain conditions. These calibration techniques must include further development of remote sensing instrumentation such as improved multi-channel water vapor radiometers, Raman scattering lidars, multi-color laser ranging and coherently combined laser/microwave systems. This new instrumentation must be used in conjunction with improved modeling and mapping algorithms including the use of optimal observation strategies and ancillary measurement systems, possibly including synoptic observations.

8.6 SEA FLOOR GEODESY

Recommendation: A combined GPS/ocean-acoustic-ranging prototype, and the required environmental calibration equipment (e.g., sound velocity meters), should be developed and tested. The development of short baseline GPS interferometers to determine the attitude of the surface platform is required to tie the GPS antenna position to the acoustic transmitter/receiver located on the hull of the platform. After prototype validation is complete, operational systems should be developed and deployed.

Rationale: The ocean covers more than two thirds of the Earth's surface, and a variety of interesting geophysical phenomena to which geodetic measurements are applicable, occur in oceanic regions. These include the vast majority of the world's spreading centers, subduction zones and transform faults, i.e., most tectonic plate boundaries. Baseline measurements with few-cm accuracy are required between reference marks on the ocean floor, separated by distances of 10 to 1,000 km.

REFERENCES

Abshire, J. and Gardner, C., 1985, IEEE Trans. Geoscience and Remote Sensing, **23.**, No. 5.

Acuna, M., 1974, *IEEE Trans. Magn., 10,* No. 3.

Arai, K., Fujimoto N., Sato, H. and Koizumi, S., 1988, Proc. Intern. Society of Photogrammetry and Remote Sensing, Tokyo.

Axelsson, S., 1988, Proc. Intern. Society of Photogrammetry and Remote Sensing, Tokyo.

Balmino, G., Barthel, G., Castel, L., Halldorson, T., Hufnagel, H., Leisbold, G., Meissner, D., Reigber, C., Rummel, R. and Schmidt, C., 1978, SLALOM Report, ESA Contract No 3483/78/F/DK(SC), Munich.

Balmino, G., Barlier, F., Bernard, A., Bouzat C., Rummel, R. and Touboul, P., 1985, Bureau Gravimetrique International, Toulouse.

Beutler, G., Bauersima, I., Gurtner, W., Rothacher, M. and Schildknecht, T., 1987, *J. Geophy. Res., 92*.

Blewitt, G., 1989, Carrier Phase Ambiguity Resolution for the Global Positioning System Applied to Geodetic Baselines up to 2000 km, *J. Geophys. Res. (in press)*

Bloom, A., 1962, *Applied Optics, 1*, No.1.

Buchroithner, M., 1988, Proc. Intern. Society of Photogrammetry and Remote Sensing, Tokyo.

Christodoulidis, D., Smith, D., Kolenkiewicz, R., Klosko, S., Torrence, M. and Dunn, P., 1985, *J. Geophys. Res., 70*, B11.

Clark, T. et al., 1985, Precision Geodesy Using the Mark-III Very-Long-Baseline Interferometer System, *IEEE Transactions on Geoscience and Remote Sensing GE-23*, pp. 438-449,

Clark, T. et al., 1987, Determination of Relative Site Motions in the Western United States Using Mark-III Very Long Baseline Interferometry, *J. Geophys. Res., 92*, pp. 12741-12750.

Davidson, J and Trask, D., Utilization of Mobile VLBI for Geodetic Measurements, *IEEE Transactions on Geoscience and Remote Sensing GE-23*, pp. 426-437.

Davis, J., Herring, T., Shapiro, I., Rogers, A. and Elgered, G., 1985, *Radio Science, 20*.

Day, T. and Müller, J., 1988, Proc. Intern. Society of Photogrammetry and Remote Sensing, Tokyo.

Degnan, J., 1985, *IEEE Trans. Geoscience and Remote Sensing, 23.*

Degnan, J. and Cohen, S., 1988, *SPIE, 889.*

Dickey, J., Williams, J., Newhall, X. and Yoder, C., 1984, Proc. Int. Assoc. Geodesy Symp., IUGG XVIIIth General Assembly Hamburg; Dept. of Geodetic Science and Surveying, Report Ohio State University, Columbus.

Dong, D. and Bock, Y., 1989, GPS Network Analysis with Phase Ambiguity Resolution Applied to Crustal Deformation Studies in California, *J. Geophys. Res.,* (in press).

Elgered, G., 1989, Tropospheric Radio Path Delay from Ground-Based Microwave Radiometry, in M. Janssen (ed.), *Atmospheric Remote Sensing by Microwave Radiometry,* Wiley & Sons.

Farthing, W. and Folz, W., 1967, *Rev. Sci. Instr., 38.*

Fitzmaurice, M., Minott, P. and Kahn, W., 1975, NASA TM X-723-75-307, Greenbelt.

Fitzmaurice, M., Minott, P., Abshire, J. and Rowe, H., 1977, NASA TP-1062, GSFC, Greenbelt.

Hajela, D.P., 1977, Recovery of 5° Mean Gravity Anomalies in Local Areas from ATS-6/GEOS-3 Satellite-to-Satellite Range-Rate Observations, Dept. Geodetic Science, 259, The Ohio State University.

Hartl, Ph., Schäfer, W., Reigber, Ch. and Wilmes, H., 1986, ESA SP-1080.

Hofmann, O., 1988, Proc. Intern. Society of Photogrammetry and Remote Sensing, Tokyo.

Janssen, M., 1985, A New Instrument for the Determination of Radio Path Delay Due to Atmospheric Water Vapor, *IEEE Transactions on Geoscience and Remote Sensing GE-23,* pp. 485-490.

Kahn, W., Klosko, S. and Wells, W., 1982, *J. Geophys. Res., 87.*

Keating, T., Taylor, P., Kahn, W. and Lerch, F., 1986, NASA TP-86240, Greenbelt.

Keyser, A., Rice, J. and Shearer, L., 1961, *J. Geophys. Res., 66.*

Kubik, K., 1988, Proc. Intern. Society of Photogrammetry and Remote Sensing, Tokyo.

Langel, R., Ousley, G. and Berbert, J., 1982, *Geophys. Res. Let., 9,* No. 4.

Li, F. and Goldstein, R., 1989, *IEEE Geosci. Rem. Sens.,* (in press).

Lichten, S. and Bertinger, W., 1988, Demonstration of Sub-Meter GPS Orbit Determination and 1.5 Parts in 10^8 Three-Dimensional Baseline Accuracy, submitted to Bull. *Geodesique.*

Marini, J. and Murray, C., 1973, GSFC Techn. Report X-591-73-351, Greenbelt

Mobley, F., Eckard, L., Fountain, G. and Ousley, G., 1980, *IEEE Trans. Magn., 16.*

Morgan, S. and Paik, H. (ed.), 1988, NASA TM 4091.

Mueller, I., van Gelder, B. and Kumar, M., 1975, *Dept. of Geod. Sci. Rep., 230,* Ohio State Univ., Columbus.

Muller, P. and Sjogren, W., 1968, *Science, 6.*

Murad-Al-Shaikh, M., 1988, Proc. Intern. Society of Photogrammetry and Remote Sensing, Tokyo.

Paik, H. and Richard, J., 1986, NASA Contractor Report 4011, University of Maryland.

Parmley, R., 1986, *Proc. Photoopt. Instr. Engin., 619.*

Priebbenow, R. and Clerici, E., 1988, Proc. Intern. Society of Photogrammetry and Remote Sensing, Tokyo.

Rogers, A. et al. Very-Long-Baseline Radio Interferometry: The Mark-III systems for Geodesy, Astrometry and Aperture Synthesis, *Science, 219,* pp. 51-54, 1983.

Schmid, P., Trudell, B. and Vonbun, F., 1975, *IEEE Trans. Aerosp. Electr. Sys., 11,* No.6.

Schürenberg, B., Lopken, L., Schüssler, H., Ockert, R., Reigber, Ch., Pavlis, E., Raimondo, J., Müller, H., Wakker, K., Ambrosius, B., Ranieri, R., Esposti, M., Hartl, Ph., Meyer, H. and Oerder, M., 1985, ESA Contract No. 5715/83/NL/MS.

Sjorgen, W.L., Laing, P.A., Liu, A.S., Wimberly, R.N., GEOS-3 Experiment Final Report: Analysis of SST Doppler Data for the Determination of the Earth Gravity Field Variations, pres.: GEOS-3 principal investigators meeting, Wallops Island, 1978.

Smith, E., Connor, B. and Foster, J., 1975, *IEEE Trans. Magn., 11.*

Sonnabend, D. and McEneaney, W , 1988, JPL EM 314-441, Jet Propulsion Laboratory, Pasadena.

Tapley, B., Schutz, B. and Eanes, R., 1985, *J. Geophys. Res., 90*, B11.

Thomas, J. 1988, *JPL Publ. 88 -15*, Jet Propulsion Laboratory, Pasadena.

Tralli, D., Dixon, T. and Stephens, S., 1988, *J. Geophys. Res., 93*.

Tralli, D. and Dixon, T., 1988, *Geophys. Res. Letters, 15*, pp. 353-356.

Vonbun, F.O., Kahn, W.D., Wells, W.F., Conrad, T.D.: Gravity Anomalies Determined From Tracking the Apollo-Soyuz, NASA Technical Memorandum 78031, GSFC, 1977.

Vonbun, F.O.,1972, *Astro. Res.*, Reidel Publ., Dordrecht.

Weiffenbach, G.C., Grossi, M.D., Shores, P.W.:Apollo-Soyuz Doppler Tracking Experiment MA-089, SAO, 1976.

Wolff, M., 1969, *J. Geophys. Res., 74*.

Young, S., 1986, *Proc. Photoopt. Instr. Engin., 619*.

Young, L., Meehan, T. and Spitsmesser, D., 1988, *Proc. of ANTEM 88*, Winnepeg.

Yunck, T, Melbourne, W. and Thornton, C., GPS-Based Satellite Tracking System for Precise Positioning, *IEEE Transactions on Geoscience and Remote Sensing GE-23*, pp. 450-457, 1985.

Zebker, H. and Goldstein, R., 1986, *J. Geophys. Res., 4*.

Chapter 6

DATA ANALYSIS

Francois Barlier, Ron Beard, Alberto Cenci, Oscar Columbo, Richard Eanes,
Clyde Goad, Steve Klosko, Maria Marsella, Haim Papo, Cheng Pengfei,
Richard Rapp, George Rosborough, C.K. Shum, S.H. Ye, Taizoh Toshino,
Tom Yunck

1. INTRODUCTION

The Data Analysis Panel (originally convened as the Engineering Panel) was tasked with discussion and documentation of techniques for determination of geophysically and geodetically useful parameters from space geodetic data, i.e. using space geodetic data given in engineering units to derive geodetic descriptors which are useful for additional scientific insight to the geophysicist, oceanographer, etc.. The panel set up three main topic areas for study— measurement/environmental monitoring, gravity, and orbital analyses. These are the three major subject areas of the Data Analysis Panel in this report. This breakdown was natural considering the areas in which space geodetic data are used in geophysical modeling.

While all three are closely tied together, these do seem to be logical dividing classes for documentation purposes. For instance, the Measurement/Environmental monitoring is common to all space tracking systems which broadly includes correlation techniques, such as very long baseline interferometry and connected element interferometry, and ground based satellite tracking systems (radiometric and laser). For these and Earth based measurements such as gravity, tilt, strain, etc., proper modeling of ground motions (and gravity) are required. Gravity is a topic well covered in this document. However, the optimum way to obtain gravity information from space based techniques must be documented in order to design future missions if knowledge of the Earth's gravity is a fundamental goal of such missions. It is now clear that our knowledge of the Earth's gravity field (minimum wave number depending on satellite altitude) must be known better in order to extend current information obtainable from satellite tracking. However, to obtain such information, better satellite tracking systems must be deployed with more global coverage.

2. MEASUREMENT/ENVIRONMENTAL MODELING

2.1 STATEMENT

Space geodetic technologies and the understanding of the site positioning provided by these methods have a critical dependency on the monitoring of local meteorological phenomena and environmentally induced station motions which have a nonpredictable behavior over measurement averaging intervals and/or short period time scales. The environmental noise can be divided into two categories: (1) that affecting the transmission and propagation of signals;

and (2) that contributing to nontectonic motions of the reference markers. In order to isolate site displacements arising from tectonic activity at the 1 mm level, to accurately monitor periodic and long-term gravitational changes, and to achieve the proper referencing of geodetic data, various transient environmental signals must be eliminated. This poses a major engineering requirement for improving environmental modeling algorithms given that contemporary capabilities yield unacceptably large uncertainties for these effects.

2.2 OBJECTIVES

The determination of strategically located points on the Earth's surface and monitoring their motion at the 1 mm level with high temporal resolution is a major goal of future space geodesy measurement systems. This level of site positioning will sense a wide range of signals. The local displacement of a station by Earth tides, tidal loading, atmospheric loading, thermal changes, groundwater retention and wind stress must be understood in order to filter out four-dimensional nontectonic signals and thus isolate the tectonic station motions of geophysical interest.

Station positioning at the 1 mm level will require additional instrumentation to provide observations to monitor environmentally induced effects; this will place further financial and strategic requirements on regional space geodetic networks and measurement campaigns. Additionally, the instantaneous location of the observing site in an inertial reference frame, where the geodetic measurements are evaluated, will require much more detailed resolution of the Earth's pole position and orientation. This is especially critical in measurement technologies (e.g. SLR, GLRS, GPS) which rely on satellite dynamics to interconnect sequential observing intervals spaced in time.

All existing space geodetic positioning technologies require significant improvements in the modeling of signal propagation effects to approach the 1 mm level of site positioning. The modeling of the propagation delay of microwave, optical and microwave signals passing through the atmospheric medium and obtaining the supporting meteorological data for the computation of corrections for these effects is a major obstacle for present measurement systems. Within available geodetic observation technologies, uncertainty in the area of atmospheric refraction modeling is one of the major, if not, the major limiting factor on resulting measurement accuracies.

Instrument calibrations, network time transfer and data quality control are significant engineering concerns for geodetic system performance at the millimeter level of data accuracy and validation. Intersite collocation experiments will be needed for the verification of performances within a system and among systems at this accuracy level.

2.3 BACKGROUND

The proper utilization of geodetic data from space technologies requires a series of measurement corrections for an unambiguous referencing of the observations. The space technologies of the future will be required to yield significantly improved station location accuracies with higher temporal resolution to satisfy the more stringent demands of the geophysical modelists. Satellite laser ranging, VLBI, and GPS data analysis strategies all currently rely on favorable averaging of "environmental" errors to produce improved accuracies for site positioning. These errors are attributable to imperfect location of the observation (in either time or position), anomalous tracking system performance, uncertainty in the corrections for signal propagation effects, errors in theoretical models for station displacements due to tides, ocean loading and unmonitored station motions due to the local response of the Earth to various other sources of loading. It is the magnitude, character and variability of these errors which ultimately determines the temporal resolution and accuracy obtainable from these measurement systems.

Station Motion Modeling. The spatial distribution and time dependence of deformations across and within zones of tectonic activity requires a rich spectrum of measurements, for events of interest which occur on all length scales and on time scales ranging from a few minutes (e.g., earthquakes) to millions of years (plate tectonics). Measurement models and environmental corrections must be advanced to support observations of short duration relying on increasingly more precise space geodetic measurements. Unfortunately, many of the errors arising from imperfect measurement models and unmodeled station displacements caused by environmental factors can appear to be quite systematic with insufficient sampling or averaging. Minimally contaminated site positioning thereby requires extensive improvement in these environmental corrections and less reliance on averaging over variable error behaviors to support higher temporal resolution in geodetic positioning. Furthermore, characterizing the site motions measured by these systems and attributing them to tectonic sources will depend on monitoring local meteorological loading effects, ground water retention, and the high frequency variations in the Earth's rotation and orientation over measurement averaging intervals and/or short-period time scales.

Knowledge of the site displacements resulting from Earth tides and ocean loading can be obtained from improved models of the ocean tides and regional tidal loading models and more complicated and detailed theoretical models of the tidal response of a laterally inhomogeneous and inelastic Earth. Other signals, such as those resulting from atmospheric loading, wind stress, thermal sources and ground water retention, will require in situ measurement campaigns using supporting instrumentation deployed over regional networks, and models of the regional loading responses which utilizes these auxiliary measurements. Environmental monitoring will be necessary in order to eliminate effectively these transient signals when studying station deformations in tectonically active zones. Regional networks containing heretofore unavailable instrumentation (including, where applicable, tidal gravimeters, tide gauges, meteorological stations, tiltmeters and other motion-sensitive instrumentation) will be required for integrated measurement campaigns. Modeling of the deformations due to changes in local water tables, monitoring of monument integrity, and correction for station displacements caused by loading effects are all necessary when assessing 1 mm geodetic positions for tectonic signals. Comparable attention needs to be given to the signal content and correction algorithms of the local gravity and tiltmeter measurements taken in support of these regional networks.

Atmospheric Refraction Modeling. As a signal traverses the atmosphere, it experiences a propagation delay and bending due to the variable characteristics of the medium through which it is passing. The Earth's ionosphere introduces significant systematic perturbations on all microwave tracking data. At lower frequencies (150 MHz) daytime ionospheric biases can easily reach several kilometers in range, several meters per second in range rate, and up to two or three milliradians in position. Since most of the effects decrease as the inverse square of frequency, modern radiometric systems track at dual and well separated higher frequencies. After removing the first-order ionospheric error using dual frequency algorithms, smaller residual errors remain due to higher order terms in the refraction index, bending of the optical path and magnetic field effects. However, at the frequencies now used in modern radiometric satellite tracking systems and VLBI, these higher order effects are seldom a problem. Unfortunately much remains unknown about the actual electron density profile at the point where a signal passes through the ionosphere which makes correction for higher order effects quite difficult. However, as shown in later discussions of the individual tracking technologies, the modeling of the higher order ionospheric refraction effects is not viewed as a major concern operating at frequencies > 1 GHz.

As a signal traverses the troposphere, it sees a varying refractive index resulting primarily from the spatial variations in atmospheric pressure, temperature and humidity. For radiometric technologies, variations in the water vapor content is of chief concern. Optical signals have a much weaker dependence on water vapor. The varying refractive index influences the propagating signal in several ways. For optical signals, the most important effect is the varying group velocity where the pulse speeds up as it travels from the ground station to low pressure regions at the higher altitudes. This change is a consequence of Snell's law of refraction which predicts the bending and speed of a light ray as it moves through atmospheric layers which have differing refractive indices.

Similar propagation properties affect radio frequencies. However, the higher sensitivity of radio signals to the atmospheric water vapor gives rise to large wet in addition to the dry tropospheric refraction components which are impossible to completely model because the mixing ratio of water vapor and dry air is very variable. Most analysis strategies for radiometric signals apply some form of empirical modeling to accommodate errors arising from the wet term in the tropospheric refraction effects, while for laser data, this is seldom the case. Means for accommodating atmospheric modeling errors are discussed by Davis et al. (1985) for VLBI, Abshire and Gardner (1985) for SLR, and Lichten and Border (1988) and Janssen (1985) for GPS.

System Calibration, Data Validation and Quality Control. Extensive monitoring of instrument behavior and the quality testing of resulting data is required for space systems located in nonlaboratory environments. These systems are subjected to thermal shock and systematic changes in local weather. They acquire data during different parts of the day yet they often poorly sample daily environmental effects. Occasionally, these systems are moved over large distances. Often they have numerous calibration procedures requiring great care to ensure proper system performance. Similarly, while it is ideal to look upon a system's performance as having a certain level of observation noise, systematic corruption of the data and the acquisition of spurious values commonly degrades the experimental result beyond that expected from noise effects. Therefore, the calibration of these systems, their intercomparison through collocation of systems, and the validation and quality control of the observations play a very important role in improving geodetic solutions and better understanding the true error in the estimation procedures. This is especially important if differing mobile systems are to occupy a given site periodically over the course of an experiment; in such cases, system-specific tracking anomalies translated into site locations can alias the tectonic signal sought from these systems.

2.4 PROSPECTS

Station Motion Modeling. The object of station motion models is to compute the position and velocity of the stations (and baselines) in the inertial reference frame at the time of the observations. The models can be divided into four classes:
(1) Displacements of the individual stations,
(2) Rotation of the entire Earth,
(3) Modeling the motion of the satellite or individual source positions, and
(4) Modeling the position of the Earth with respect to the solar system barycenter and the motion of other planets within the solar system.

The (3) and (4) classes of motion modeling are discussed in detail elsewhere in the orbit modeling and reference frame portions of this document. They will only be mentioned briefly herein.

Periodic and Environmentally Induced Local Site Displacements. A terrestrial (crust fixed) reference system must be realized where the station markers are described and in which ideally, through accurate modeling and elimination of short period tidal and loading effects, the station coordinates are invariantly defined except for motions due to tectonic processes. Precise station positioning would then yield an unambiguous description of the site's tectonic motion and deformation to the precision and temporal resolution provided by the relevant space technologies. At the millimeter level, the local and regional station loading displacements are largely unmonitored and inadequately modeled, and the routine acquisition of in situ data needed to support this level of modeling is generally unavailable even over the most extensively instrumented regions. Present models of Earth tides and the ocean loading are also inadequate for supporting geodetic technologies at this level of site recoverability. The status and prospect of improving the Earth and ocean loading tidal models are discussed below. It should also be noted that the "noise" sources are also geophysical signals through which we can learn about the Earth.

The upper layers of the Earth crust on which space geodesy stations are monumented should be regarded as a dynamical continuum. Due to a variety of factors, some identified but inadequately modeled and some unidentified, there are residual relative motions in three-dimensional space between parts of the upper crust with amplitudes and frequencies which are relevant for 1 mm geodesy. The small area occupied by the monument of a space geodesy instrument may not be representative of the behavior of the surrounding portions of the upper crust. Thus, the super precise space geodesy measurements between points/stations hundreds or thousands of kilometers apart may give a false picture of the actual crustal motions. It is imperative to develop concepts and methodologies of an "area-station" (footprint) which will replace the conventional and currently standard point-station concept. Extensive studies have to be undertaken to investigate the stochastic (nonpredictable) behavior of each space geodesy monument with respect to a dense network of markers within its "area-station". After these nonpredictable relative motions within the area-station are understood and modelable with supporting in situ measurements, such motions can be filtered out of the time history of site locations. Only then can one speak of motions between "points" in a network of such area-stations at the 1 mm level. Conventional (classical) geodetic measurement techniques as well as any relevant and motion-sensitive auxiliary measurements will be required as input for assessing the nonpredictable motions within an area-station. The frequency and accuracy required of these measurement systems will have to be determined experimentally on a site by site basis. These hybrid (heterogeneous) input data must be processed using linear dynamic modeling within the area-station geodetic networks. Many of these observations may be sensitive to an integrated series of environmental and tectonic effects. Data analysis strategies and application of evidence combination or like techniques will need to be developed to isolate the environmentally induced station signals using these hybrid measurement systems. The concept and theory of a dynamical network datum will have to be developed beyond the current standards. The ultimate result of all the aforementioned efforts will be the determination of small but highly significant (at the 1 mm level) corrections to the varying positions of space geodesy monuments so as to obtain a stable and area representative network of reference points.

New models or advancements in existing models need to be developed for filtering out stochastic, or in general these nonpredictable components of station motions due to atmospheric, thermal and other strictly local environmentally induced noise sources. This will further entail the exploration and standardization of the use of auxiliary data, dynamic modeling and autocovariance functions (autoregression models or Green's functions as used by Goad (1980) in his ocean tidal loading models) for their application in modeling station displacements arising from these environmental sources. These environmental signals, which depend on many variables (i.e., the wind loading and thermal distortion of a reference point of the observing instrument depend on the direction and strength of the wind, sunshine, and cloud cover etc.), give rise to station motions which contain a distributed and aperiodic component. Existing stochastic models are unavailable and even if developed, may not always provide satisfactory predictions. New and improved models beyond traditional linear prediction and adaptive filtering methods are expected to play an important role in this complex distributed time series analysis. Recent achievements with maximum likelihood analysis, evidence combination, parallel distributed processing, feature extraction and discovery, and nonlinear adaptive filtering, to name a few analysis techniques, may lead to significant improvement in the future analysis of local station motions and the elimination of these environmental noise signals.

Ultimately, constraints on the positioning of closely distributed stations using the "area-station" concept may be the only means for eliminating or averaging out the most localized of these environmental signals.

Precise monitoring of local environmentally induced station motions have been studied in depth over select regions when reducing absolute gravimetry, tidal gravimetry and tiltmeter observations. Local gravity and motion sensing technologies will have to be routinely incorporated into space geodetic networks to monitor site motions at the super precise levels and highly resolved time intervals sought from future space technologies. A brief discussion characterizing these environmental signals and the station motions which result, follows.

Solid Body Tides. By far the largest periodic displacement at the Earth's surface, globally speaking, is the rise and fall due to the tidal attractions of the sun and moon. This phenomenon is usually expressed as a scale factor times the tidal potential expressed by its degree expansion, U_n. The lower the degree, the larger is the effect. Earth models can be used to compute these degree parameters called Love numbers h_n, k_n, l_n (Munk and MacDonald, 1960). In addition to station displacements, altimeter measurements, and satellite orbits also respond. Checks can be made in that surface responses (stations and ocean surface) have a different frequency response from the satellite orbits. A complicating factor is that the response factors in position are different than from potential responses which affect satellite orbits. Proper Earth modeling can relate potential response to surface response. Although responses span the entire tidal spectrum, they do diminish significantly with increased degree. The most complicating factor is the frequency of the ocean loading which is the same as the solid contributions.

Ocean Tidal Loading. Similarly to the solid body response, a deformation will also occur due to the application of the tidal load. The response of the Earth to the load is described by the load deformation coefficients h_n', k_n', l_n' (Munk and MacDonald, 1960). More recent investigations have shown these Love numbers to be frequency dependent. Most of the uncertainty associated with ocean loading deformations arises from our uncertain knowledge of ocean tides in the open oceans.

Thermal Deformations. The solid Earth is subjected to thermal changes over a wide range of time scales. The signal which is found which is most periodic within a day looks very much like the S2 solar tide, and gives rise to a radial marker motion which is induced by daily temperature variations. The radiational heat flux through the Earth's surface causes temperature changes in the ground which are a source of time-dependent thermo-elastic deformations. The thermal disturbances can be highly localized and abrupt as local conditions change. On longer time scales, one finds significant seasonal changes in regional albedo and ground temperatures. Albedo itself can be constant for periods on the order of a few days to a few months, but it can also be highly variable over longer periods, especially during winter.

The localized heat flux through the Earth's surface causes a change in the temperature field by heat conduction resulting in thermoelastic stresses, deformations and displacements. These thermal changes have been found to displace markers by as much as 2 cm and also can have significant impact on the gravity instruments themselves causing linear drifts in the instruments with temperature. Correction for these effects can presently be made to the 1-2 µgals (0.5 cm or better) level as reflected in the thermal corrections now made in absolute gravity measurements.

Atmospheric Loading. The monitoring of the radial deformation of the solid Earth by the atmosphere have been done for mostly regional applications and then in connection with tiltmeter investigations. The mass of the atmosphere causes a perceptible deformation in the crust in response to this load. There is also a significant change in local gravity measurements due to the changing attraction of the atmosphere. Estimates of this effect indicate station displacements on the order of 15 to 20 mm assuming peak-to-peak pressure fluctuations of 40-50 mbar, with a capability to correct for these loading effects to the 2 mm level (see Van Dam and Wahr, 1987).

Station Motions Arising from Changes in Groundwater/Rainfall. The difference in the water table due to rainfall creates local motions which generally disappear within one week after the rainfall. Groundwater itself has two effects on local gravity measurements. (1) There is a variable gravitational attraction due to the change in water mass; and (2) the loading effect produces vertical soil displacements. This latter effect, which is of chief interest herein, depends strongly on the raising of the pore pressure within saturated layers of soil and rock, the time variations for these effects, and the bulk in-situ permeability of the soil and rock itself. There are also transient gradients found in the water level which induce tilt/strain anomalies which are somewhat correlated with rainfall, but are not highly predictable especially during times of thaw and after periods of low precipitation. Therefore, site selection should take into account factors like the depth of the water table, slopes in the water table topography, and time variations in the water tables should be accurately monitored. It is thought that present uncertainties in modeling the effects of groundwater and rainfall on absolute gravity measurements is causing errors on the order of 1-2 μgals and for geodetic positioning applications, errors in site motions of much less than 1 cm are now being achieved in well instrumented regions.

Polar Motion/Earth Orientation. Earth rotation and orientation information is needed to transform from the terrestrial coordinate system to a reference inertial coordinate system in which the geodetic observations are more easily calculated. These rotational transformations included long-term polar motion—the so-called polar wobble, diurnal polar motion, diurnal spin, nutations and precession. Accurate and highly temporally resolved Earth orientation parameters are required for locating sites continuously in a crust-fixed reference system when using space-based observations (see the section on Reference Frames elsewhere in this report).

Station Motion Modeling: Summary. In summary, it is not unrealistic to assume with the available super computers and well instrumented regional networks of the future, that detailed and highly accurate models of a station's motion can be achieved. This will entail calculating the time-dependent response of a point on a elliptical, rotating, laterally inhomogeneous, inelastic Earth with realistic ocean models subject to the tidal generating potentials of the moon and sun. This point will be subjected to loading from complicated ocean tidal behavior on near-by continental shelves, and will be enveloped by tidally and thermally perturbed atmosphere. The nonperiodic driving potentials of variable levels of groundwater, variable seasonal mass transport, wind stress and thermal flux will also require detailed monitoring to fully understand station motions due to environmental disturbances. However, 1 mm station positioning and environmental site motion modeling at this level will require considerable extension of existing technologies, models, and instrumentation.

Atmospheric Refraction Modeling. Calculations of both ionospheric and tropospheric contributions require a model of the atmosphere and use of the satellite (or quasar)/station geometry (mainly by elevation angle). Under certain circumstances where many "looks" over many azimuthal and elevation directions (for example in the case of GPS and VLBI), sufficient redundancy is available to estimate the contributions of the troposphere and possibly ionosphere. Constructing stochastic models for the permissible rate of change in these models conditions shows promise in estimating the behavior over time using a sequential filter. This has the obvious advantage that the model does not depend on just surface conditions. Otherwise, some additional information in the form of direct measures of the atmosphere or its effect on radio and laser measurements is required.

Tropospheric Refraction Corrections. As given in earlier, the most promising near term advances will come from the use of filtering, use of Water Vapor Radiometer for radio measures, or use of multiple wavelengths with laser measurements.

The modeling algorithms for tropospheric propagation delay most commonly in use today assume that the atmosphere is composed of spherically symmetric shells. Commonly, the models utilize pressure, temperature and humidity measurements taken locally at the site during the time of the observations. For technologies which track at radio frequencies, Water Vapor Radiometers being tested as a means of estimating the delay due to water vapor along the line of sight to the radio source. Empirical correction algorithms and filtering techniques have been developed for some radiometric data analysis strategies which assume that the atmospheric delay has a stochastic character with a specified power and decorrelation time. Such stochastic and/or filter models have shown significant promise for reducing errors in station positioning from refraction errors.

The refraction modeling algorithms and models themselves have generally kept pace with the observational precisions achieved within each of these space technologies, although measurement modeling clearly is already the limiting factor for VLBI, GPS and radar altimeter systems which operate at radio frequencies and lack routine consideration of regional meteorological gradients. However, even for SLR, uncertainty in the tropospheric refraction correction is becoming a leading error source as hardware advances are being achieved and millimeter level ranging capabilities are developing. A 1-2 cm uncertainty in the accuracy of the refraction correction is found within a typical SLR measurement taken at a 20 degree elevation angle, with a factor of two modeling improvement possible using future multicolor systems. Dual frequency lasers, (even after overcoming the problem of measuring the received time differential of the two pulses within a picosec), will probably not yield better than 3 to 5 mm refraction modeling errors unless additional information is obtained on the integrated water vapor content of the atmosphere along the ray path.

Therefore, of critical concern within future systems is the modeling of the propagation delay of radiometric, optical and microwave signals passing through the atmospheric medium and the supporting meteorological observations required for the computation of accurate refraction corrections which properly account for horizontal pressure gradients, nonisotropic water vapor distributions and turbulence in the atmosphere. Improved refraction models and improved routine methods for mapping regional behavior of the atmosphere are required to support an order of magnitude improvement in our ability to correct for these effects.

Ionospheric Refraction Corrections. The ionosphere is the term applied to the outer atmosphere where the atmospheric components are ionized by the action of sunlight. It is a plasma of free electrons in a constant geomagnetic field which changes the refractive index, n, of the medium from that of free space. The propagation of radio waves in this media is governed in the high frequency approximation by the Appleton-Hartree equation (Budden, 1985),

$$n^2 = 1 - X \left[(1 - \frac{Y_T^2}{2(1-x)} \pm \sqrt{\frac{Y_T^4}{4(2-x)^2} + Y_L^2} \right]^{-1}$$

where $X = (\frac{f_n}{f})^2$, $Y_T = Y \sin \theta$, and $Y_L = Y \cos \theta$ $(Y = (\frac{f_g}{f})$, f_n = plasma frequency,

f = frequency of propagation , f_g = gyrofrequency, and θ = the angle between the propagation direction and the magnetic field.

This equation shows the dependence on the geomagnetic field, but is independent of it to the first order. Higher order terms will depend upon magnetic field direction and strength as,

$$n \approx 1 - \frac{X}{2} + \frac{XY}{2} \cos \theta - \frac{XY^2}{4} \sin^2\theta + \ldots = 1 - \frac{f_n^2 + f_n^2 f_g}{2f^2} - \frac{}{2f^3} \cos \theta - \frac{f_n^2 f_g^2}{4f^4} \sin^2\theta + \ldots$$

Signals passing through the ionosphere will experience errors due to the changes in the refractive index, and for high frequencies the actual bending is small. The errors are due to the phase velocity differential from free space. The effective group delay, of the signal is then,

$$\Delta T = \frac{1}{c} \int_0^R (n-1) ds = \frac{1}{c} \int_0^R \Delta n ds ,$$

where, R is the total slant range to the satellite and c is the speed of light.

The deviation of the refractive index from unity can then be expressed to first order as,

$$\Delta n \approx \frac{f_N^2}{f^2} = \frac{\epsilon^2}{8\pi^2 \epsilon_0 m} \frac{N}{f^2} = \frac{40.3}{f^2} N ,$$

where, e is electron charge, m is electron mass, ε_0 is the permittivity of free space, and N is electron density, in electrons per cubic meter.

The group delay error is then expressed as,

$$\Delta T = \frac{1.34 \times 10^{-7}}{f^2} \int_0^R N\, ds = 1.34 \times 10^{-7} \frac{N_T}{f^2},$$

where,

$$N_T = \int_0^R N\, ds = Q\, N_{TEC}.$$

An integrated electron content of 10^{18} electron/m^2 corresponds to a GPS L$_1$ ionospheric group delay of ~ 16 meters. To generalize the expression for use in arbitrary positions anywhere in the world against a worldwide model of the ionosphere, the integrated electron content, NT, may be modified to the columnar total electron content, NTEC multiplied by an obliquity factor, Q. Modeling the ionosphere concerns primarily representing the distribution of the electrons in the various ionospheric layers and the concentration as a function of the geomagnetic latitude and longitude, solar input and other effects. The electron distribution, its variability, and fine structure (irregularities causing scintillation) is the challenge in developing an accurate global model.

The GPS control segment provides the coefficients of a model for the single frequency receiver in the data message. This model should compensate for about 50% of the ionospheric delay. Efforts in modeling the ionosphere has shown that for increasing accuracy, more complex models are required relying on more data. Active compensation has produced the best results since the environment is actually measured along the propagation path.

Many systems are providing the means for actively correcting the errors caused by the ionosphere. GPS transmits on two frequencies for the correction of these errors. The application of higher frequencies in RF systems is becoming easier to implement with the improvements in electronics. Work by the ionospheric physics community is continuing in understanding the basic mechanisms and variabilities of the ionosphere. The amount and quality of data to support these efforts have grown in proportion to the number of systems capable of measuring the effects. As the data and resultant knowledge increase, the accuracy of the developed ionospheric models combined with the investigations into the geomagnetic field will improve yielding higher order corrections.

2.5 TIME TRANSFER AND SYNCHRONIZATION

The origins of time scales are based in astronomical measurements from the unstable platform of the Earth. Today they have evolved to the point where time scales are determined independently of the motions of the Earth (Guinot and Seidelmann, 1988). Since 1955 atomic standards have been available and providing a more accurate basis for determining the second and hence the day. Until 1960 the second was defined as a specified fraction $\frac{1}{86400}$ of the mean solar day, and from 1960 to 1967 the second was defined as a fraction of the year. In 1967 the System International (SI) second was defined as the duration between two hyperfine levels of the ground state of the cesium 133 atom. The SI second is now the standard unit of time and the basis for International Atomic Time (TAI). TAI is maintained on an international basis by the Bureau International des Poids et Measures (BIPM) under the responsibility of the International Committee on Weights and Measures (CIPM) and of the General Conference on

Weights and Measures (CGPM). This was adopted by the International Astronomical Union (IAU) in 1985. The TAI time scale is a statistical combination of the atomic standards from most of the world's major timing centers and is the basis of Universal Coordinated Time (UTC), which relates time based on the rotation of the Earth (UT1) to TAI. UTC is now the basis for civil time and can be physically realized by local clocks properly synchronized to the timing centers. It is also stepped by leap seconds to keep it within 0.9 s of UT1. These time scales are likewise related to Ephemeris Time (ET) (H.M. Nautical Almanac Office, 1961), which has been the basis for all astronomical observations before 1955.

The maintenance of time scales and their international coordination is a function of the ability to intercompare clocks accurately on a global scale. The Navstar GPS, which is still a developmental system just beginning to enter its operational phase, provides the most readily available means of accurate intercomparison. It has already been used not only for geodetic measurements, but has become the preferred system for the dissemination of time. All the major observatories and timing centers now have GPS time transfer receivers and are routinely using GPS to intercompare the international time scales. Through GPS many centers can now provide accurate and timely information to BIPM which was not possible (or very difficult) previously. Many other users are using the system to maintain themselves in synchronization with UTC (USNO). GPS has an operational requirement to be within a microsecond of UTC (USNO), and the system is capable of being kept within 100 nanoseconds. Use of the system to transfer time between remote sites using a satellite in common view has been reported to be within 10 nanoseconds (Allan et al., 1984). Many of the satellite laser tracking stations are also equipped to time their observations by timing from GPS.

For future geodetic measurements the ability to use an accurate and common time base should enhance the performance and suitability of combining measurements from different sites. The IAU General Assembly in 1979, adopted new definitions of time scales relating the motions of the planetary bodies, correcting or discriminating between proper and coordinate times of physical clocks and maintaining TAI. These definitions are Terrestrial Dynamical Time (TDT) and Barycentric Dynamical Time (TDB). TDT becomes an ideal form of atomic time (or TAI) independent of the motion of the solar system bodies. TDB will define a dynamical time scale for ephemerides with respect to TAI, and the ready availability of UTC will provide continuity with the values and practice in the use of ET. Before 1955 the determinations of ET can be considered to refer to the new scale. These new scales will provide easy realization of time with actual clocks in the different environments and locations where they may be used. Correcting for the relativistic effects on these clocks should enable greater precision on the geodetic observables of interest.

2.6 USE OF SYNERGETIC/HYBRID MEASUREMENT TECHNOLOGY

There are three principal categories of space measurement systems used in geodetic applications, VLBI, SLR and satellite RF systems. The VLBI systems are capable of very high precision measurement over global distances using intergalactic radio sources and, in some cases, artificial satellites. SLR systems are of two types, fixed systems which have telescopes with large apertures to improve the overall optical gain, and mobile (or even highly transportable) systems which make use of compact electro-optics. The precision of the fixed SLR systems are rapidly approaching the few millimeter level but like VLBI and GPS, these measurements are significantly degraded by atmospheric propagation effects at these high precision levels. Satellite RF systems include Doppler systems (such as Transit and DORIS) and ranging systems (such as GPS and PRARE). The precision for the RF systems have been inherently less than the other two types, due principally to system design in the case of Transit, and technology in the case of GPS. In the last several years contemporary receiver technology has improved to the point that GPS receivers are approaching comparable precision to that of cm level SLR systems when used in the differential or interferometric mode.

In general, the measurements within a given class of systems are self-consistent but in many cases are relative measurements. For example, data taken between two GPS receivers in differential mode are relative to one another. What the absolute basis of the master or reference receiver is cannot be determined from the differential data alone. These measurements are still capable of very good precision and are of great value in determining geodetic observables.

However, full use should be made of the different techniques for network densification and system intercomparisons. Reducing network distortions due to differences between the reference systems is vital. Therefore, using different techniques and overlapping space systems is also important to eliminate intersystem biases that may result from using only one technique. For example, synergetic use of multiple technologies may be one of the most practical means for improving environmental corrections for geodetic data such as improving atmospheric refraction modeling.

The use of hybrid systems and experiments would tie the different techniques together within a unified system. An experiment such as that proposed for ACRE, for example, would be able to link the SLR and GPS networks together, since ACRE would provide a common source for the GPS signal and laser target. It has long been suggested that laser retro-reflections be added to all of the GPS satellites for the same purpose. The addition of a VLBI beacon on the ACRE satellite would then provide a common source for all three major systems.

The combination of different measurement systems on future satellites would be recommended to support the intercomparison of the different measurements and improve the uncoupling between signal propagation effects and true site motion signals in the space-based observations.

3.   GRAVITY

3.1 INTRODUCTION

An accurate and global knowledge of the gravitational field of the Earth, at spatial scales ranging from 10 to 40000 kilometers, and at temporal scales ranging from hours to centuries, is an important scientific and technical goal. The applications of such knowledge extend from the practical uses of leveling and of geological prospecting, to the primarily scientific study of the structure and properties of the solid Earth, the mapping of the global mean circulation in the oceans, and the investigation of time-dependent phenomena such as tides, polar motion, glacial rebound, variations in ice cover, etc.

Achieving the overall goal outlined in the opening statement involves the following:
   a) The global coverage of the Earth with gravity data of high resolution and accuracy, collected both at the Earth surface and from space.
   b) The development of measurement techniques (preferably on the basis of currently understood technology) for both terrestrial and space-based gravity field mapping.
   c) The improvement of data analysis methods for the processing of terrestrial and space data, particularly the large sets now becoming available, as well as those likely to result, from the dedicated gravity missions now contemplated by various space agencies.
   d) The further development of gravity field models combining space and terrestrial data for engineering and scientific applications, including the precise determination of satellite orbits.
   e) The accurate and detailed study of changes in the gravity field.

3.2 BACKGROUND

Terrestrial Gravity Information. Surface gravity data provide very valuable information on the short wavelength geopotential field. Overland gravity data can be used to validate crustal models and for the improvement of general purpose geopotential models. Marine gravity data is useful for separating short scale ocean from geoidal signals to precisely locate and study ocean currents. Unfortunately, surface gravity data comes from many sources and is often unaccompanied by adequate information as to how the data were taken, and is of extremely variable quality and absolute accuracy. Variations among local datums and survey networks is a persistent problem complicating the optimal utilization of this source of information.

Although surface gravity data are generally collected as in situ, or point, values, they are convenient to handle in the form of area mean values of gravity anomalies, distributed on a uniform geographic grid. Comparisons of mean values with satellite fields have been useful in locating poor data, especially in regions where the anomalies have been obtained using geophysical predictions (based on geological maps) rather than actual measurements.

Terrestrial gravity information can be used to gain knowledge that satellite data cannot provide to the same accuracy or resolution. Combining terrestrial and satellite data can result in better gravity field models than using either data type in isolation. Furthermore, terrestrial data can be used to check and calibrate the gravity field models obtained using primarily satellite data. This, in fact, is an area of work where physical and space geodesists have long been able to pool successfully their specialized knowledge and resources to advance science. The terrestrial data are usually averaged to form mean gravity anomalies in various cell sizes. The cell size needed for combination solutions is usually $1°x1°$ or $30'x30'$. A recent compilation of surface anomalies is shown in Fig. 1. These anomalies cover 61 percent of the land area and 79% of the ocean areas. However the accuracy in many regions is quite poor. For example, only 11% of the total area of the Earth is covered by anomalies whose accuracy is 5 mgal or better.

The anomaly estimates can be contaminated by a number of error sources that lead to correlated anomaly errors. Errors can be caused by incorrect base station connections, vertical datum inconsistencies, gravimeter drift, lack of point data in a cell, etc. In general there is good agreement between the terrestrial and satellite data when compared at the same wavelengths.

The land areas for which the coverage is weakest includes parts of South America, Greenland, Africa, Eastern Europe, Asia, and the polar regions. In the near future gravity data over Africa at the 30' resolution will become available. Data in Eastern Europe and Asia exists but such information is not released by many countries despite requests by individuals and by international scientific organizations. Gravity at sea is still collected by ships; the need for such data has been reduced, but in no way eliminated, by satellite altimeter data. In addition, the expensive ship cruises have been reduced, so that the acquisition of ship gravity data has slowed down. For this reason, it is important that gravity measurements be collected whenever scientific cruises are carried out.

For some applications the 30' mean values are most useful. However such data, on land, is now available only in North America, Europe, Australia, parts of Africa and a few other areas. The determination of additional values will be helped by the efforts of organizations such as the International Gravity Bureau.

Altimeter Derived Data. Geos-3, Seasat, and Geosat data have provided substantial information that has been used to derive gravity anomaly and deflection of the vertical values over much of the oceans. The accuracy of the anomalies derived from such data is (approximately) 12 mgal for point data; 4 mgal for 1 degree area means; and 7 mgal for 30' means. (These estimates can vary from one region to another, depending on the smoothness of the gravity field, the satellite track spacing, orbit error, data noise, and enviromental factors including oceanographic effects, which may be reduced by track averaging.) Of greatest concern is the track spacing, which can reach several hundred km at the equator. Such gaps leave many areas of the ocean uncharted; in them, local gravity signatures can only be estimated.

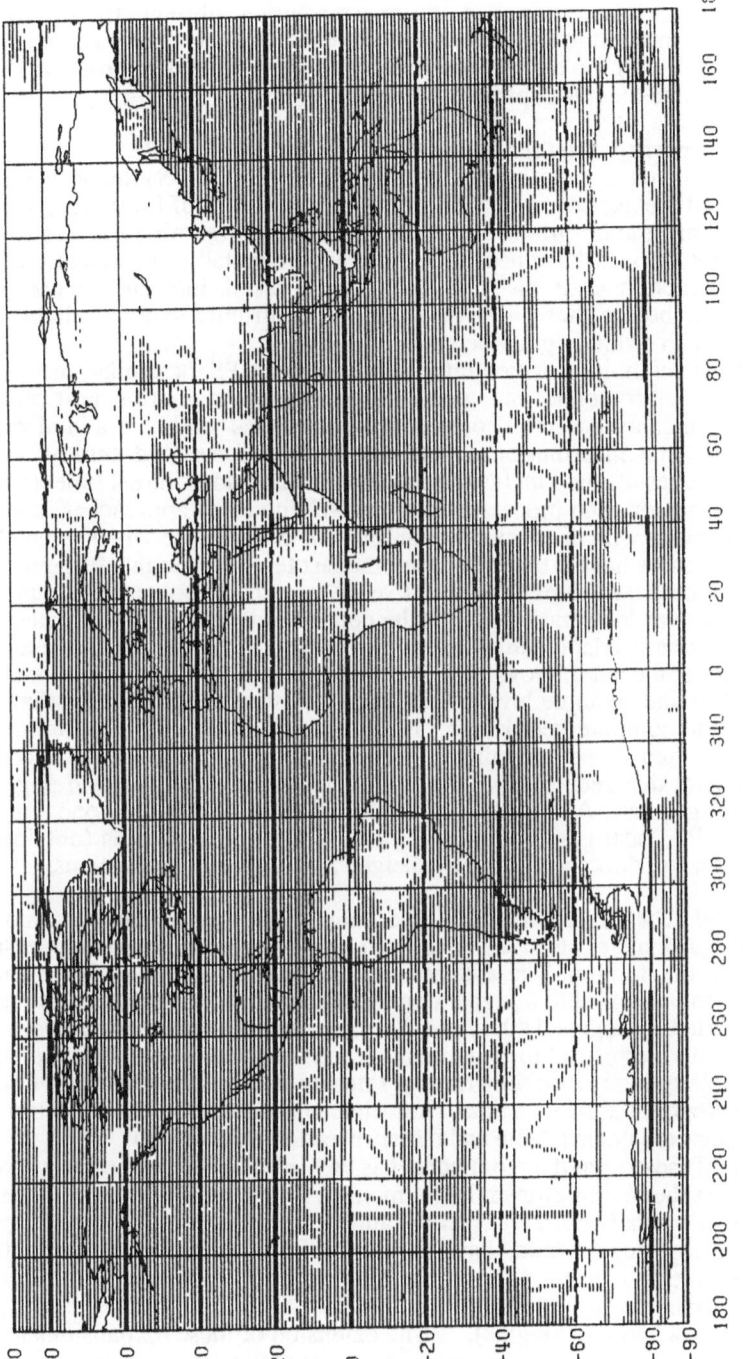

Fig. 1 Location of the 42624 1° x 1° terrestrial gravity anomalies in the
Ohio State Unversity June 1986 data set

Today (Geosat) and future (ERS-1, Topex/Poseidon) altimeter missions are not being planned with track coverages dense enough to improve resolution substantially. It is important to design a new altimeter mission (or modify one of those already scheduled) so that the final track spacing is consistent with the highest resolution allowed by the data noise. A desirable situation would be one where the track spacing is 10 km.

New Methods for Measuring Surface Gravity. Over the last decade, combinations of inertial positioning and space positioning technologies (e.g. GPS) have been under investigation, for fast and precise coverage with gravity observations of limited regions of the Earth's surface. Moreover, gravity gradiometry and gravimetry from aircraft, also involving inertial and space geodesy devices, have been developed experimentally and are currently being tested for a variety of applications. At this stage, known early results indicate that considerable work remains to be done before these new methods can contribute data as reliable as that collected on the ground with ordinary gravimeters.

The main idea is to combine position information from a GPS, or similar, receiver, with the measurements of accelerometers and gyros of an inertial navigation unit and (or) a gradiometer, the whole (including a microcomputer) being carried on board a land vehicle, a ship, or a plane. Additional information on position from precise land surveys could be incorporated to strengthen results, and the whole process could be done in real time, or the data logged in the computer memory and processed afterwards. With some approaches, gravity can be determined simultaneously with the position of ground stations traversed by the vehicle.

Considerable advances have taken place over the last few years in the design of precise sea-gravimeters (carried on board ships), with relative measurements now attaining precisions at the five milligal level over long traverses. For the control of terrestrial gravity nets, which contribute to define a precise geodetic datum, a new generation of absolute gradiometers, some portable, with accuracies at the ten microgal level, are being used now by some national survey agencies. Progress in fast and accurate leveling methods (relevant to defining a datum for the gravity anomalies used to compute geoids, etc.) is also worth noticing. The incorporation of fast and precise space methods for positioning gravity stations in a global reference frame has had a considerable impact that seems most likely to increase, on scientific and engineering applications of physical geodesy. An example of this is the work now being done, combining precise levelling with GPS, local gravity data and global gravity field models (obtained from satellite tracking), to calculate directly orthometric heights, or else geoid undulations.

High Degree Potential Coefficient Models. To date, some satellite-derived models in the form of spherical harmonic expansions have been obtained complete to degree 36, with models to degree 50 on the horizon. The addition of surface gravity data and altimeter data enables the expansions to be carried to a higher degree. If area means at 1 degree are used, the expansions can be carried to degree 180, while 30' data enables expansions to degree 360. This mixing of tracking and terrestrial measurements results in models that combine the well-determined long wavelength information of the satellite data, with the high frequency information that is inherent in the surface data. Unfortunately, rigorous accuracy estimates for these high degree solutions are not available and clearly need to be developed.

The estimation of these high degree models has created new interest in the appropriate potential representation(spherical or ellipsoidal harmonics, for example), the role of terrain in the calculation of certain anomaly correction terms; and the various methods to estimate large parameter models. The high degree fields have proven useful in a number of applications in which either a fine resolution is needed, or where it is necessary to remove high frequency gravity signals, in a consistent fashion, from various data types (such as 1 degree gravity anomalies, or altimetric sea surface heights).   The extension of these expansions to higher degrees is doubtful because of data requirements and computational costs. It may be more appropriate to add special, regional terms to the high degree global expansion in order to describe the local structure. Such procedures are not now used in practice.

Mapping the Gravity Field from Space with Conventional Tracking Data. Present modeling efforts largely rely on the analysis of tracking data (range, range rate), acquired on a large number of diverse satellites. Much of these data have been taken to support the orbit determination needs of the individual satellite missions themselves. The distribution of data is limited, with many gaps in orbital inclinations and altitudes for optimal field recovery. These data provide accurate long wavelength and largely unambiguous geopotential information from the perturbations induced by the Earth's field on the orbits of near-Earth satellites; these perturbations appear as discrepancies between the observed and the computed observations of the spacecraft tracked.

Since the gravitational signal attenuates rapidly with altitude, tracking data acquired on satellites above 700 km altitude provide little intermediate and virtually no short wavelength field resolution. The improvement in ranging precision, especially with advances in satellite laser tracking, has provided the means for determining the long wavelength gravity field to high accuracy levels. Critically needed, however, are additional laser satellites at a variety of orbital heights and inclinations, to obtain a better sampling of the gravity field and, thus, a better separation of the individual spherical harmonics describing the field than currently possible. Lageos II and III, which will be at Lageos I altitude but at a very different inclinations, will be especially important for improved long wavelength geopotential modeling, and improved estimation of tidal and other time-varying gravitational effects.

Critical analysis of the present satellite-only geopotential models shows that above degree 10, few spherical harmonics, mostly associated with satellite resonances, are truly well-determined. The stability of least squares estimation decreases rapidly with increasing degree, due to the combined effects of signal attenuation and incomplete global data coverage. Most modeling efforts have utilized an estimation technique known as least squares collocation to numerically stabilize the field recovery through constraints applied to the size of the adjusting coefficients, but this method provides little information on the actual value of each high degree geopotential coefficient. The best way to improve the gravitational field models is to use new data that complements existing data in quality and coverage. Such data can be provided by missions like Lageos II.

Taking into account those missions for the next decade that have been approved or are under serious consideration, a reasonable projection of future gravitational modeling capabilities can be developed. What follows maps out the progress that may take place in the absence of a dedicated gravity mapping mission that uses novel techniques, such as satellite gradiometry, in studying the static part of the field (geopotential) and its temporal variations (tides, etc.).

Spherical harmonics are a natural choice for modeling the long wavelength gravitational field and provide an efficient means for upward continuing the model to calculate the gravitational accelerations experienced by an orbiting object. For satellites orbiting at 600 to 1500 km altitude, a gravity model complete to about 50 x 50 in harmonics plus selected resonances is sufficient to accommodate all significant gravitational perturbations that can be sensed by current techniques. These models are quite large, containing in excess of 2300 terms.

Geopotential models have become more accurate as data and computer capabilities have improved. Consequently, the basic assumption about the overall description of the gravitational model as a time invariant set of parameters has been substantially modified in recent solutions. For example, the calculation of GEM-T1 included estimation of additional spherical harmonic terms describing the temporal variation of the field at the main tidal frequencies. However, other temporal changes in the field are presently unmodeled like the secular change in the zonal harmonics due to post-glacial rebounding of the lithosphere, motions of the figure axis exhibited by secular and dynamic polar motion, and semi-annual, annual and long period variations in global mass transport (which may have variable amplitudes from one year to the next). It is important to to try to estimate these changes, as long as they have a significant signature in the data, not only because of their scientific value, but to avoid their subtle aliasing into other quantities of interest.

Each new field model has usually been based upon previous ones, with the new field primarily including the use of new data (in some cases replacing older, less precise measurements to the same satellites), and data from satellites previously unutilized. The restrictions on available satellite data sets has dictated that models develop in this way. In the past, special purpose geopotential models have also been developed which were tuned to accommodate the main orbital perturbations of a given satellite, to fulfill the precision orbit needs of a particular mission, but of no general application elsewhere (for example, PGSS3, for

Seasat). As general-purpose models become more accurate and detailed, reliance on tuned models will be less necessary.

General purpose gravity models should have a geoid adequate for geophysical/ oceanographic applications, and be suitable for the precise orbit determination needs of various geodetic missions. One of the best models recently developed, GEM-T1, gives better satellite orbits than those of older fields that were tuned specifically for the same spacecraft.

In addition to more and better data, the following questions require further work, in order to obtain significantly better results:

(a) Improved definition and calculation of the reference frame. This should include the full implementation of relativistic models, improved nutations, and better resolution of Earth orientation parameters. Station/plate tectonic velocity models must be included in the data analyses. Tidal motions at local tracking stations must be refined, to include frequency dependent Love numbers and ocean loading effects. Aspects of the gravitational interactions of celestial bodies, like indirect oblateness, require re-evaluation, given the demands of millimeter level measurement systems. Even small effects, like body tides in the moon, and the 18.5 year ocean tide, are now of significance if we are to understand better the long-term evolution of the lunar orbit. These effects should be considered for possible introduction in future geopotential solutions, to avoid their subtle aliasing into the results.

(b) Improved models of nongravitational forces, such as atmospheric drag, and solar and Earth re-radiated electromagnetic wave pressure.

(c) Development of more stable numerical estimation techniques. Solution stabilization through eigenvector constraints need to be assessed as an alternative to conditioning by such methods as least squares collocation.

Developments in Conventional Tracking Data Sources. Satellite observables and tracking systems are becoming more accurate. For example, satellite laser ranging is expected to reach the 2-3 mm normal point accuracy level within the next few years. TDRSS and GPS satellite-to-satellite tracking systems promise new dense and accurate sets of observations, that may extend considerably the resolving power of those already available. Over the past twenty years, data of many kinds have been accumulated and used for field modeling: optical (photographs against a star background), minitrack, S-band two- and three-way range rate, C-band range and range rate, TRANET Doppler, satellite-to-satellite range rate, altimetry, laser and deep resonance satellite passage observations, among others, from over thirty satellites of geodetic interest.

Two Important Data Types. Over the past decade, satellite-to-satellite tracking from high (geosyncronous) satellites to lower orbiting spacecraft, and satellite altimetry have been collected over large areas of the world not covered (or patchily covered) by conventional, land-based tracking systems. Difficulties in modeling these data properly, among other reasons, have kept their use to a minimum until recently. During the last two years, new efforts, particularly in connection with altimetry, involving more complex and faithful mathematical formulations, more thorough preprocessing, etc., have been started at various research centers (including Goddard Space Flight Center, The Ohio State University, and The University of Texas at Austin, in the USA) aimed at making full use of this valuable information.

Satellite-to-Satellite Tracking. There are now two main observation sets available, consisting in Doppler measurements between geostationary satellites and lower spacecraft. These are: the ATS-6/GEOS-3 data acquired during 1975-8, and the data available from TDRSS tracking a number of different spacecraft at present.

The altimeter-bearing GEOS-3 satellite was tracked from the geosynchronous ATS-6. GEOS-3 at the same time, was being tracked by ground-based laser systems. These data have already provided important local gravity measurements over South America, Africa and the Atlantic and Pacific Ocean Basins, where terrestrial data are scarce or unavailable.

TDRSS is a geosynchronous satellite system meant to replace most of NASA's ground tracking stations. TDRSS permits measuring inter-satellite range-rate data of 0.1 to 0.3 mm/sec precision. A second TDRSS satellite is soon to be launched, and will greatly increase geographical coverage, particularly over some continental regions that remain almost totally unmapped.

Satellite Altimetry. The spaceborne radar altimeter maps a sea surface which is complex, but which largely conforms to the equipotential gravitational surface known as the geoid. The ocean surface departs significantly (at the meter level) from the geoid, and this departure is of great interest to the oceanographers, meteorologists and geophysicists, for it reflects the global circulation patterns and also, in the form of tides and changes in sea level, fundamental climatic and solid Earth phenomena. For this reason, altimetry has become a major scientific tool for investigating these phenomena on a global scale. Furthermore, altimetry provides approximate estimates of the geoid itself, in the form of a mean sea surface obtained by averaging measurements taken over many satellite passes over the same places. Alternatively, altimetry can be described as vertical tracking from a satellite to the geoid. In either way, data from three altimetric missions have already been used to obtain geopotential models, over the last decade.

GEOS-3 was the first geodetic satellite to carry an altimeter, accurate to 45 cm. This satellite lacked an onboard memory system so its data acquisition was restricted by the geographical location of telemetry sites. However, over the two year time period that its intensive mode altimeter functioned, telemetry receivers were moved to permit nearly complete GEOS-3 mapping of the ocean surface between ± 65 degree latitudes, corresponding to the inclination of the GEOS-3 orbital plane.

Seasat was launched two years after GEOS-3 in the early summer of 1978. Seasat carried an altimeter accurate to 10 cm and an onboard data recording system and continuously tracked the ocean surface. It was placed into two distinct repeating groundtrack modes, the first every 17 days, the second every 3 days. A catastrophic power failure ended this mission after only four months. The data set for Seasat has extensive geographical coverage with dense mapping of the entire ocean surface between ± 72 degree latitudes. However, since the temporal distribution of the data was poor, much of the data from beyond 65° S latitude was contaminated by sea ice (it was winter in the southern hemisphere at this time, July 8 through October 10, 1978).

GEOSAT is an ongoing U.S. Navy satellite mission furnished with a Seasat-type altimeter. Since being placed in a repeating groundtrack orbit, coincident at the equator with the 17 day repeat orbit of Seasat, altimeter data from this mission have been made available to the civilian scientists. GEOSAT is still performing well and has been providing continuous data for more than three years. Approximately 18 months of ocean data are available, as of now, for analysis from this Extended Repeat Mission (ERM). And by being operational, it is continuously gathering more data, which extends the time scales open to detailed oceanographic investigation. The single greatest weakness of the ERM portion of the GEOSAT mission is the minimal tracking network available for orbit determination. The tracking is being done with five poorly geographically distributed Doppler sites.

TOPEX and ERS-1 are other altimeter satellites which are scheduled to reach orbit within the next five years. Data from these missions will also be used to understand the dynamics of the ocean, its temporal stability, and to strengthen future geopotential models.

For many applications of altimetry, good orbits and geoids must be calculated to interpret the data properly. These needs have driven a good deal of practical and theoretical research in physical and space geodesy in recent years. Current goals are for decimeter accuracies, or better, both for the geoid and the vertical component of the orbits. This requires refinements in existing techniques, including the incorporation of altimetry itself as a data type in calculations.

The nongeoidal signal, although small compared to the geoid, is a significant source of long wavelength aliasing if it is not properly modeled in orbit and gravity field determination with altimetry. The dominant source of this aliasing is to due that part of the oceanographic signal that is quasi-stationary (Q) in character, arising mostly from the dynamic sea surface topography (SST) caused by global geostrophic flow. This QSST component in the altimeter mapped surface is difficult to separate from that of the geoid. At the same time, from a scientific point of view, the QSST signal is of major interest to oceanographers.

To use altimetry without aliasing the gravity field and the orbits with the QSST while obtaining an accurate model of the QSST at long wavelengths, one can represent both gravity and QSST with spherical harmonics and solve for the coefficients of both expansions simultaneously. Such solutions, combining altimetry with data sets rich in gravity signal from LAGEOS and other laser satellites, have already yielded preliminary models for the QSST complete to degree and order 10, together with gravity field models complete to degree and order 36. Such solutions are only approximate, and only accommodate the largest source of modeling error when including altimeter data in gravitational solutions. However, absent a global, dedicated, precise geopotential research mission, altimeter data and surface gravimetry remain the major sources of information for all future gravitational models representing the field with a resolution beyond degree and order 40, or with a wavelength less than 1000 km.

New Satellite Data Systems. New spaceborne tracking systems like DORIS and GPS are being developed for future satellite missions. DORIS is a French one-way Doppler tracking system operating at dual frequencies (400 and 2000 MHz) which is to be flown on Topex/Poseidon. The transmitters are on the ground, and a Doppler receiver is carried in a spacecraft. DORIS is scheduled to be tested as part of the SPOT 2 satellite mission, in 1989. A very dense global network of about 50 tracking stations (DORIS transmitters) is now being deployed.

GPS (Global Positioning System) is a USA military system of spacecraft in very high orbits carrying dual-frequency radio beacons transmitting in the L-band, using very stable atomic clocks. GPS has been originally designed for fast positioning of combat units on the battlefield, and of ships, planes and other craft in motion. This system has been used with surprisingly good results by civilian geodesists, surveyors and scientists for very precise position determination, using its beat carrier phase signal (used in a mode similar to VLBI measurements at a single known frequency), and its pseudorange, when possible (given the possible classified nature of this information). The system is now in the final stages of its experimental phase, and the USA Air Force is about to begin deployment of up to 24 new spacecraft of advanced design. Preliminary studies indicate that the instantaneous position fixes of a spacecraft carrying a GPS receiver may constitute a data set of remarkable value, given its accuracy and potentially complete global coverage. Moreover, the availability of GPS tracking to a lower satellite, combined with data collected simultaneously by a terrestrial network of receivers, can greatly strengthen the measurement of long baseline vectors, of interest in studying plate tectonics. In this way, two important geophysical endeavors (charting the geopotential and mapping crustal deformations) can benefit at the same time from the use of an orbiting GPS receiver and will provide roughly 80% tracking coverage enabling an improvement of the gravity field thanks to this very good coverage.

The Study of Temporal Variations in the Field. The orbits of near-Earth satellites are perturbed by temporal departures of the gravity field from the static geopotential represented by most field models. These departures are due to tidal effects, largely driven by the sun and the moon, and to redistributions of mass inside the Earth, and in the oceans and atmosphere: glacial rebound, changes in ice cover, global fluctuations in sea level, changes in Earth rotation, etc. Some of these phenomena, linked to the complex dynamics of our planet, which is far from being a simple, rigid body, have been studied by space geodesists primarily with special spacecraft carrying laser retroreflectors, such as LAGEOS, Starlette, and (more recently) Ajisai. More satellites of this type are planned for the near future: LAGEOS II and III, Stella, etc. The gravitational effects in question are much weaker than those of the main, stationary field, and can be detected largely through resonant perturbations of the orbits, which selectively magnify some of their signatures in the tracking. These perturbations happen at certain very low frequencies (periods from weeks to months) and depend on the spacecraft orbit height and inclination. Two different orbits will be resonant to different kinds of field changes. To have a reasonably complete picture, several spacecraft in a variety of heights and inclinations are needed. The planned laser satellites mentioned above, in new orbits, will reveal new aspects of the changes in gravity. To go further, a worthwhile idea proposed some years ago by C. F. Yoder and others is to build small, spherical, dense laser targets (similar to LAGEOS, but smaller), and launch them together with other spacecraft whose orbits have desirable characteristics, but whose purpose may be quite removed from gravitational studies (spacecraft of opportunity). This could be a relatively inexpensive way of gathering new information on gravity variations over very long periods of time, as these "mini Lageos" (as they are called sometimes) will experience little air drag because of their small area-to-mass aspects, and are likely to remain in their orbits for centuries. Table 1 lists several important causes of changes in the gravitational field of the Earth, together with estimates of the magnitudes of the corresponding changes in the second and the third zonal harmonics of the geopotential.

Gravity Missions. Tracking a relatively low spacecraft from several high ones, such as the proposed Gravity Probe-B satellite (600 km altitude) from the GPS spacecraft (25000 km), should contribute significantly to our knowledge of features larger than 300 km across, of interest in determining mean global circulation from altimetry, upper and whole mantle convection, etc., as well as for computing satellite orbits, and other geodetic applications. Interim gravity missions based on this idea could be relatively inexpensive (if the cost of the low spacecraft is part of an already funded mission). Valuable preliminary experience may be gained from the GPS experiments planned on board TOPEX and other satellites scheduled for the early 1990's.

Fine Resolutions, Dedicated Gravity Missions. In these missions, the spacecraft would be placed at an altitude of less than 200 km, in near polar orbit, to provide complete global coverage. After several months, the spacing between adjacent passes would be of a few tens of kilometers at the equator [approx. 20000 km / (15 * length of mission in days)] and closer elsewhere; the separation of measurements along track would be also of a few tenths of kilometers (approx. 8 km * sampling time in seconds). Furthermore, the orbit could have a ground-track repeating precisely once over the whole length of the mission, for the most uniform geographic coverage possible in that time. The smallest feature that could be resolved, at least in principle, would be twice the sampling interval in space (in the North-South direction determined by the data-rate, and in the East-West direction by the mission length).

     The analyses of the large data sets to be gathered in these missions, to the fine resolution that such data will allow, involves the simultaneous estimation of up to hundreds of thousands of unknowns from millions of observations. The result will be uniformly accurate, self-consistent gravity models, used as reference fields in geodetic and geophysical work, refined further, if possible, by means of local solutions (which could combine them with surface data).

     Such analyses present special challenges that can only be met by close integration of all main aspects of mission planning and implementation. This is unusual: until now, data analysis has rarely been considered before the data became available (often having been collected for a completely unrelated purpose, such as tracking for orbit determination).

     The new techniques for estimating the gravity field parameters globally (or worldwide) will require special orbit selection (near circular, repeating precisely every several months, etc.), and almost uninterrupted data gathering.

     Many questions concerning global techniques, like the choice of base functions, numerical stability, effect of certain simplifying assumptions made on results, re-computation of the orbit at each iteration when solving for very many gravitational parameters, etc., are only beginning to be addressed.

     A number of preliminary error analysis studies suggest that the broad scientific objectives of the advanced gravity missions could be met with the instrument accuracies envisioned (1 micron per second for low-low sat. sat. tracking, 10^{-4} E to 10^{-2} E for gradiometers). More has to be done to establish just how relevant these costly experiments can be to specific key questions in solid Earth physics.

Low-Low Satellite-Satellite Tracking. This technique employs two satellites following each other along the same orbit, a few hundred kilometers apart. An advanced two-way tracking system would be used to measure their relative velocities with extreme accuracy (in the order of one micron per second). The irregular variations of this velocity convey gravitational information. Alternatively, a main spacecraft (such as an orbiting platform) may track two or more small orbiters at the same altitude and orbit as itself. The most developed concept is the first one above; it has been considered for the GRAVSAT-GRM project by NASA, where both spacecrafts were to be drag-free thanks to a compensation device based on the DISCOS system

Table 1. Some Recent Estimates of Temporal Variations in Zonal Harmonics of the Earth's
Gravitational Field

Source	Reference	Variation	ΔC_{20}	ΔC_{30}
• earthquakes	Chao & Gross (1987)	nonperiodic w/long period trend	$\pm 5 \times 10^{-13}$/yr	$\pm 2 \times 10^{-13}$/yr
• deglaciation rebound of crust	Yoder et al. (1983)	secular (observed)	-3.0×10^{-11}/yr	n.a.
	Rubincam (1984)	on LAGEOS	-2.6×10^{11}/yr	n.a.
• snow cover	Chao et al. (1987)	periodic: annual	1×10^{-10} amp	6×10^{-11} amp
		semiannual	3×10^{-11} amp	1×10^{-11} amp
• continental drift Greenland moving at 10 cm/yr in latitude with a depth of immersion of 50 km	Sconzo (1980)	secular	$\pm 2 \times 10^{-14}$/yr	n.a.
• tidal breaking	Paddack (1967)	secular	$< -5 \times 10^{-13}$	n.a.
• earth, ocean tides	Christodoulidis et al. (1988) and others	periodic (observed)	-- variable -- nontidal contributions lumped in tidal recoveries at forcing frequencies	-- variable --
• air pressure & groundwater	Gutierrez & Wilson (1988)	periodic: annual	1×10^{-9} amp (shows atmosphere/oceans to be ~10 times water storage at annual and ~3 times at semiannual periods)	n.a.
		semiannual	1.5×10^{-10} amp	n.a.
• changes in sea due to ice cap/ glacial melting	Peltier (1988) Yuen et al. (1987)	secular	2×10^{-11}/yr 2 to 8×10^{-12}/yr	n.a. 2 to 7×10^{-12}/yr
• growth of the Antarctic ice sheet equivalent to drop in sea level of 0.3 mm/year	Yuen et al. (1987)	secular	5 to 10×10^{-12}/yr	6 to 11×10^{-12}/yr
• continental water storage, aquifers, lakes	Chao (1988)	periodic: annual semiannual secular	1.5×10^{-10} amp 5×10^{-11} amp 1×10^{-12}/yr	1.4×10^{-10} amp 4×10^{-11} amp n.a.

of the US Navy. The tracking was to be ultimately between two proof masses, one in each satellite, being kept in constant free-fall by that system. A magnetometer was to be included in one of the two satellites, carried at the end of a boom. Both spacecrafts were to be deployed during a Shuttle mission.

Gravity Gradiometry. A pair of accelerometers aligned along their sensitive axes measure the spatial rate of change of the gravity force in the direction of those axes, which is proportional to the second gradient of the gravity potential. Gravity is the combination of gravitational and rotational forces, so to get information on the former it is necessary to eliminate the effects of rotation. Also the orientation of the instrument in space with respect to an Earth-fixed system, and the positions of the spacecraft where the measurements are taken (i.e., the orbit) need to be known with sufficient accuracy. Finally, this type of instrument is very sensitive to all gravitational forces, including the attractions of the components of the spacecraft itself. All this, and the extreme accelerometer accuracies required (of the order of a part per trillion to a part per quadrillion of the normal acceleration of gravity), makes the successful construction of such a device a major engineering challenge. The instrument must be also mechanically and electronically stable, as well as physically rugged, to withstand the vibrations during launch, and to operate uninterruptedly for many months. A number of attempts in the past have failed to produce such a device. At present, there are two main projects under way: one at the University of Maryland, USA, sponsored by NASA (this is a cryogenic device operating in a bath of liquid helium); the other project, sponsored by ESA, is taking place at the research and engineering facilities of ONERA near Paris (this instrument operates at room temperature, using an improved version of the French CACTUS accelerometer, previously used to measure radiation pressure and drag on spacecraft). The instruments should be able to measure all nine second derivatives (full-tensor device) or some subset of them (in fact, only five convey independent signals, the others could be measured to increase redundancy). As in the case of the satellite-satellite tracking signal, the gradients would be averaged over a few seconds, and sampled at intervals equal or longer than the averaging time (typically, every 4 seconds). The orbit would also be low, near polar, and repeating, ensuring the most uniform global coverage possible within the duration of the mission (approx. six months).

Gravity gradiometers can be expected to perform as well or better in space probes orbiting other bodies of the solar system, than they would on near-Earth satellites, in terms of scientific returns. Very little information on the structure of most of those bodies can be obtained by other means from space; little is known about them at present, so much that is new can be learned from data that might provide only a marginal gain in knowledge on Earth; other planets, moons and the asteroids have very thin or virtually no atmospheres, so the main limiting factor for altitude and mission duration around the Earth (air drag) is absent there. Being able to measure from closer up, for longer periods of time, the gravitation of geological features that are often larger than those found on Earth (consider Olimpus Mons in Mars), such features may be studied better than those of our planet using gradiometers.

Gravity gradiometers require the development of extremely precise and stable accelerometers that can be used in applications unrelated to gravity missions. In this way, valuable technological spinoffs may come as a consequence of the resources invested.

4. ORBITAL ANALYSIS

4.1 INTRODUCTION

Accurate orbit determination for the benefit of applications in space geodesy depends on observational data accuracy, their temporal and spatial distribution, physical modeling accuracy, computational approach and spacecraft engineering.

1. Provide satellite orbit determination accuracies of 1 part in 10^9 above an altitude of 10,000 km and a uniform accuracy of less than or equal to 1 cm below.
2. Support the ocean sciences in the determination of variable and mean topography to an accuracy of 1 cm over basin scales.
3. Support geophysics in the determination of baseline vectors to an accuracy of 1 mm at all scales.
4. Support solid Earth dynamics in the determination of the Earth's gravity field (including temporal variations) and Earth orientation.
5. Support planetary and lunar missions.

One of the inherent requirements for satellite geodesy is the precise determination of the position of the satellite with respect to a defined reference frame. To meet this task there are a wide variety of space techniques including Satellite Laser Ranging (SLR), Lunar Laser Ranging (LLR), Very Long Baseline Interferometry (VLBI), Global Positioning System (GPS), Global Navigation Satellite System (Glonass), Doppler Orbitography and Radio Positioning Integrated by Satellite (DORIS), Precise Range and Range Rate Equipment (PRARE), Altimeter (ALT), Geodynamics Laser Ranging System (GLRS), and spaceborne gradiometers, which are currently or will be soon in use. All of these technologies will be fully deployed and become operational in the next decade. While the measurement precision of some of the instruments is currently at the 1 to 5 centimeter level, it is anticipated that the precision will approach the millimeter level over this time frame. A stringent requirement is now placed on the physical force and measurement models, the Earth orientation models and the fundamental reference systems to keep pace with the rapid advances in observation accuracy and system developments. The need to compute orbits, some extending with dynamic continuity to decade lengths, will further tax the performance of these models. The current state-of-the-art being achieved in orbital accuracies and the concurrent performance in precise surface positioning which result when utilizing various space techniques are given in Table 2. To date, the advancements in orbit determination accuracy have been achieved through a dynamic modeling approach. It is now evident that this approach will be complemented with a geometric approach as GPS and similar technology becomes available.

Table 2. State-of-the-Art Orbital and Surface Positioning Accuracies

Satellite (Altitude)	Radial Orbit Accuracy (cm)	Surface Positioning Accuracy (mm)
GPS (20000 km)	30	5-10 (up to 200 km) 10-40 (up to 2000 km)
LAGEOS (6000 km)	3	10-40 (global, system dependent)
STARLETTE (800 km)	15	N/A
Seasat (800 km)	30	N/A

Current computational approaches within existing major software systems include least squares and numerical integration techniques, semi-analytical methods and stochastic processes. Computational techniques exploiting existing and future computer technologies are required for orbit determination and data processing applications. Computational algorithms to utilize vectorization and parallel processing capabilities of modern and future computers are currently in progress and will be needed in the future. Mini-supercomputers in spaceborne environments may become routine for real time navigation and geodetic product generation. Depending on the applications, large-scale orbit determination software systems may require modification to support the science objectives of the lunar and planetary missions. The continued development of analytic orbital perturbation theory and computation of mean orbital elements is desirable to enhance our understanding of orbital dynamics and to advance our understanding of long-term orbital evolution.

4.2 TRACKING SYSTEMS

Radiometric Techniques. Four radiometric tracking systems now under development will be useful in addressing a variety of Precision Orbit Determination (POD) requirements over the next few decades. These are (1) the U.S. Defense Department's Global Positioning System (GPS), (2) the Soviet Union's Global Navigation Satellite System (Glonass), (3) the Federal Republic of Germany's Precise Range and Range Rate Equipment (PRARE), and (4) the French system for Doppler Orbitography and Radio Positioning Integrated by Satellite (DORIS). In addition, for precise tracking of geosynchronous and other high Earth orbiters, very long baseline interferometry (VLBI) and analogous techniques have been successfully demonstrated and are well suited. Several other tracking systems, including NASA's Tracking and Data Relay Satellite System (TDRSS) and the U.S. Defense Mapping Agency's Tranet Doppler system, are now routinely used for many operational orbit determination requirements. Tranet, however, will be taken out of service in the early 1990s.

PRARE and DORIS. PRARE and DORIS are ground-based systems developed specifically for precise tracking and geodetic applications. DORIS is a one-way Doppler system featuring a global network of approximately 50 ground beacons which transmit continuously at two frequencies: 401 and 2036 MHz. The global network should provide approximately 80% tracking coverage. Satellites equipped for DORIS tracking carry a receiver that observes individual ground beacons in sequence (current receivers can observe only one beacon at a time) and measures the Doppler frequency of the received signals over successive 9 second intervals. The Doppler measurements are then transmitted to the ground for processing. PRARE, by contrast, broadcasts dual frequency signals (2200 and 8500 MHz) from a transmitter onboard the satellite to receiver-transponders on the ground. The signals are modulated by pseudonoise ranging codes to permit both range and range-rate measurements. The 8500 MHz signal is coherently transponded back to the satellite (at 7200 MHz) for onboard range and range rate extraction. The 2200 MHz signal is received and tracked on the ground to provide an ionospheric correction. With both PRARE and DORIS, orbit determination requires conventional dynamic techniques which in turn require highly accurate physical models to achieve high orbit accuracies. To approach the 1 cm level of orbit accuracy, enormous improvements in the models for both gravitational and nongravitational forces are necessary, as described in Section 4.2. It is unlikely that 1 cm dynamic orbit determination will ever be realized for satellites orbiting at altitudes below about 1000 km, because of the difficulty in modeling shorter wavelength gravity effects and atmospheric drag. Indeed, given the problems of modeling the effect of solar radiation pressure on satellites of complex shape, 1 cm dynamic orbit determination accuracy at any altitude may remain beyond reach for all but the simplest satellite forms.

GPS and Glonass. GPS and Glonass are very similar general purpose positioning systems consisting of constellations of 18—24 satellites in roughly 12-hour orbits, continuously broadcasting dual frequency (roughly 1200 and 1600 MHz) navigation signals. The carriers are modulated with pseudonoise ranging codes to permit both Doppler and one-way range measurements (also known as "pseudorange" measurements). The signals from each satellite blanket the Earth, extending to roughly 3000 km beyond the Earth limb. At an altitude below 3000 km, any user of GPS or Glonass will have four or more transmitters continuously in view. This combination of dual data types (phase and pseudorange) and continuous three--dimensional coverage will enable a quasi-geometric positioning technique, which combines instantaneous geometric positioning from pseudorange measurements with long-term kinematic smoothing against carrier phase, as a powerful alternative to purely dynamic orbit determination. (Dynamic orbit determination will of course always be an option.) Because kinematic tracking is not subject to dynamic modeling errors, it offers positioning performance that is virtually independent of altitude for satellites below 3000 km. With a network of 6-10 GPS receivers on the ground, which observe GPS in concert with the orbiter and serve as precise reference points, kinematic tracking is expected to readily achieve sub-decimeter accuracy regardless of orbiter dynamics. Such a system will be flight tested on Topex/Poseidon in early 1992. Simulations and covariance studies have indicated that with emerging GPS receiver technology offering low system noise and effective multipath suppression, and with expected improvements in the determination of ground reference networks, continuous 1-2 cm accuracy in all three dimensions will be achievable for satellites below 3000 km by 1995.

The optimal technique for applying GPS and Glonass in satellite tracking is in fact a synthesis of the dynamic and kinematic techniques, which, when properly tuned, must produce a result superior to either the kinematic or dynamic technique alone. This so-called "reduced dynamic" technique employs an optimal weighting of dynamic and geometric information. Instantaneous geometric position measurements from pseudorange are smoothed both kinematically against the empirical record of position change obtained (geometrically) from carrier phase, and dynamically against a physical orbit model. As dynamic models are improved over time, greater weight can be given to the dynamic component of the solution, reducing the effective number of solution parameters and yielding an overall improvement in orbit accuracy. In a world of perfect physical models, the optimum orbit determination strategy would be purely dynamic. If, as is likely, the power of GPS and Glonass are combined in a dual function receiving system, it is not unreasonable to expect that within 20 years, sub-centimeter accuracy will be achievable for a broad class of remote sensing missions. Such accuracies will be highly beneficial for improved modeling of geopotential and nonconservative forcing effects.

We should not conclude, however, that GPS and Glonass will answer all needs. There is great appeal in a "Super-GPS" optimized for geodetic and POD applications. Such a constellation, broadcasting a well chosen set of tones, would vastly simplify data analysis and would permit instantaneous geometric determination of user position with sub-centimeter accuracy. We strongly recommend that the world geodetic community investigate the potential of such a system and, if indicated, organize a multi-nation development program.

Interferometry. All of the tracking systems discussed thus far—DORIS, PRARE, GPS, and Glonass—can be applied at any altitude, but all, except possibly PRARE, will lose effectiveness at geosynchronous altitude. (Doppler, for example, is useless for a geostationary satellite.) A tracking technique well suited to POD for geosynchronous and other high altitude satellites is VLBI, of which there are a number of variations. VLBI produces differential range and phase measurements between two or more widely separated points on Earth. This can be achieved either with active (two-way) range and Doppler measurements or with passive observations from the ground. PRARE range measurements, for example, can be formed into interferometric observables by differencing. The most promising passive approaches make use of reference sources to calibrate systematic errors, a powerful technique that doesn't have a ready analogue in two-way ranging. Candidate reference sources include extra-galactic radio sources and other precisely known satellites, such as the GPS and Glonass satellites. An attractive variation is to fly a GPS-like beacon on the high orbiter and track it along with the other GPS and Glonass satellites using suitable receivers on the ground. Such techniques promise few decimeter accuracy at geosynchronous altitude.

Laser Techniques. Satellite laser ranging is expected to continue to be a strong contributor to advances in precision orbit determination. Laser range accuracy has reached the 1 cm level and continues to improve. An ultimate accuracy of 1 mm is the projected goal.

The major difficulty with laser tracking data are the outages due to weather and the nonglobal distribution of tracking sites. Although weather will continue to be a problem further deployment of fixed and mobile tracking stations is encouraged so that a more complete global coverage can be accomplished. In particular, development of laser tracking programs in countries not currently active in this area is very desirable.

The data outages due to weather and a limited number of tracking sites will become more critical as the number of satellite targets increases in the coming years. Currently, LAGEOS is the prime target and the acquired data has resulted in significant contributions to the study of Earth orientation, temporal variations in the gravity field and plate tectonic motion. In addition to LAGEOS, the Starlette and Ajisai satellites have been tracked to the extent possible. These satellites will be joined in the near future by LAGEOS II and III, Topex/Poseidon, ERS-1 and others. The question of tracking priorities will then become quite critical with respect to the scientific goals of each of these space missions. Thus, it is recommended that techniques be investigated for obtaining the maximum amount of tracking data from multiple targets at any given site. Due to the rapid data rate that current systems achieve it is proposed that methods for a station to quickly move between targets be studied. By interleaving targets it should be possible to obtain nearly complete tracking coverage of satellites that simultaneously pass a station.

The gaps in the acquired laser range tracking data sets typically necessitates a dynamic orbit determination approach. Although, geometric techniques can be applied over short time intervals where multiple station coverage is obtained. Further work in this area is encouraged in hopes of providing rapid measurements of surface displacement.

Physical Modeling. The physical models required to propagate the motion of an orbiting satellite have grown in number and complexity as observations of the motion have steadily improved. In order to meet the objectives of future space geodesy continual improvement will be required. Need is great in both the modeling of gravitational and nongravitational effects. Currently, success is being achieved in the area of modeling the temporal and stochastic variations of gravity. However, advancement in the modeling of nongravitational effects is slower due to the poorly defined satellite environment. It should be emphasized that future efforts will require an interdisciplinary approach to reach the desired accuracy levels.

Nongravitational Forces. Nongravitational forces includes atmospheric drag, solar radiation pressure, Earth albedo radiance forces and anisotropic thermal radiation forces and perturbations associated with specific spacecraft, such as due to active thermal control. Nongravitational forces can be accommodated through essentially three approaches. First, a drag compensation mechanism can be flown for those satellites requiring explicit removal of nongravitational effects in order to fulfill their science objectives [e.g., Gravity Probe-B (GP-B) and Geopotential Research Mission (GRM)]. Secondly, on-board measurement of the effects, either directly through accelerometers or indirectly by measuring the total momentum of flux impacting the spacecraft, can be used to separate gravitational signal from other effects. Lastly, most satellites require precise modeling of these effects through models developed on the basis of physical principles or through an empirical determination.

Of these approaches, the development of improved physical models for nongravitational forces will improve the performance of the greatest number of missions although it is the least accurate of the three alternatives. It is recognized that the capability for modeling these effects has inherent limitations. When such limitations are reached it is recommended that the residual effect be modeled through an appropriate stochastic process.

Atmospheric Density Model. The limitation in modeling atmospheric drag accelerations on planet orbiting artificial satellites is knowledge of the physical properties of the upper atmosphere. The characteristic of the Earth's upper atmosphere is associated with physical phenomena such as the thermal diurnal bulge which is caused by the solar heating; the solar activities which include an approximate 11 year solar cycle, a 27 day variation caused by the solar rotation and the geomagnetic activities; the seasonal latitudinal effect due to the seasonal migration of helium; and the semi-annual variation whose origin is debated.

The current commonly used semi-empirical thermospheric density models include the Drag Temperature Model (DTM), the ESRO-4 Models, Jacchia Models (1965, 1971, and 1977 versions), and the Mass Spectrometer-Incoherent Scatter Models (MSIS). None of these models are reliable above 1,000 km. These models will therefore be limited for the commutation of the effect of drag acceleration for satellites like Topex/Poseidon (1,300 km) and Ajisai (1,500 km).

Without an improved density model, techniques to accommodated drag errors through empirical parameterization of unmodeled errors are required so that these effects can be removed without absorbing desired physical signals. Atmospheric density correction models have been shown to be effective while further detailed analysis is required to test the limits of this approach.

Solar Radiation Pressure. The solar radiation pressure force acting on a satellite is a complicated function of specularly and diffusely reflected and emitted solar incident fluxes which can produce variable momentum flux caused by temperature differences on the surfaces of a composite satellite. The conventional radiation pressure models assume a lumped effect of specular and diffuse refection and the unbalanced thermal radiation, which are inadequate to account for the solar radiation forces for satellites which require decimeter orbital accuracy, such as the Topex/Poseidon. It is recommended that generalized models for composite satellites be studied to account for appropriate reflectivity and emissitivity effects of the solar radiation pressure. More studies are required to improve the modeling of Earth shadowing effects and to develop improved techniques to better handle the force discontinuity when crossing shadowing boundaries.

Earth Radiation Pressure. The Earth radiation pressure is the radiation pressure exerted on a satellite due to (1) sunlight reflected off the Earth (shortwave), and (2) heat emission of the Earth (longwave). The reflected solar radiation pressure, the albedo and radiance, can be categorized into the isotropic effect caused by the diffused reflection (Lambertian effect), and the anisotropic effect caused by specular reflection. Current Earth radiation pressure models include the variation of latitude and seasonal effects for albedo and longwave emissivity, longitudinal albedo effects represented by spherical harmonic expansion, diurnal and anisotropic albedo effects. It is recommended that Earth albedo data collected by current and future remote sensing satellites be used to improved the Earth radiation pressure model.

Neutral and Charged Particle Drag. The effect of charged particle drag appears to explain part of the decay observed in the LAGEOS semimajor axis. The effect of anisotropic thermal emission caused by the anisotropic temperature distribution on the surface of LAGEOS can also explain part of secular decrease of the LAGEOS semimajor axis. It is recommended to continue the study of the neutral and charge particle drag, the dust drag and the so-called "Poynting-Robertson" drag, to better understand the satellite environment and to improve models of the drag force.

Gravitational Forces. Accurate dynamic modeling of satellite motion requires a precise description of the gravitational field through which the satellite is moving. Modeling the gravity influence of the Moon, Sun and planets has not posed a major problem. The gravity field of the Earth though has been found to be a limiting factor in accurately predicting the dynamic motion of a satellite. The Earth's gravitational field is typically modeled as a mean field with a variability component (e.g., due to tides) then added to this mean. The variable component has been found to have both predictable and nonpredictable constituents.

Gravity and Tides. Determination of the mean field of the geopotential has been carried out using a combination of observation methods. These include surface gravity measurements, satellite tracking data and satellite altimeter data. Gravity models based on various combinations of these data have been produced by a number of research groups. These independent efforts have been one of the important reasons for the success and continual improvement in gravity modeling. However, gravity model accuracy is still the main limiting factor in producing accurate dynamic orbits. Further work in this area, through incorporating more data and improved methodology, is of critical need.

Modeling the predictable variations of the gravity field also requires much further development. Particularly with regard to the ocean tides. Ocean tidal models more complete in the number of modeled constituents, in the degree and order of their expansion and their respective accuracy are needed.

The unpredictable variations of the gravity field at seasonal and interannual frequencies are only beginning to be understood. Statistical representations of these effects should be determined so that their influence may be accommodated in the precision orbit determination process. These variations are caused, in part, by earthquakes, viscous relaxation in the mantle following the Pleistocene deglaciation, water and air pressure redistribution, mass transport between the atmosphere and the ocean and the ocean and the bathymetry, ice-cap melting/growing and snow cover distribution. These problems are very much of an interdisciplinary nature. Research programs that attempt to span the disciplines required to fully understand the effects of these phenomenon should be strongly supported.

General Relativity. In light of current ranging accuracies, the modeling of general relativistic perturbation effects, which include nonlinear Schwarzschild Field effects of the Earth, the geodesic precession, the Lense-Thirring effect and light time corrections due to the Earth's mass will be required for computation of near Earth satellite orbits in the geocentric reference frame. Recent work has demonstrated the equivalence of the Solar system barycentric and geocentric reference frames to within some limiting accuracy.

Solar-Planetary. The knowledge of the precise position, masses and other physical characteristics of the planets and their natural satellites are required for planetary missions, and to a lesser extent, for near Earth satellite missions. The current ephemerides used for most orbit determination purposes is the DE-200 Planetary Ephemerides. It is envisioned that the accuracy of the planetary ephemerides will be driven by improved knowledge of planetary gravitational fields, and other physical characteristics of the planets will also be improved to support the scientific objectives of planetary missions such as the Mars Observer.

Planetary and Lunar Applications. All of the models developed for accurate orbit determination about the Earth will require parallel counterparts for support of missions about the Moon and planets. Additionally, it can be expected that alternative approaches and algorithms will be necessitated in instances where the environment differs significantly from that prevailing for missions about Earth.

The data required for the development of models to support these missions can be expected to come from the missions themselves. In order to maximize the usefulness of such missions it is recommended that plans for obtaining accurate and optimally distributed tracking data be implemented. Further, it is recommended that dedicated geodetic missions to the moon and planets be flown so that knowledge of their physical models can be advanced.

Spacecraft Engineering. Spacecraft of complex shape and functionality present additional problems in accurate determination of their orbital position. These problems arise due to highly variable area projection, uneven heating of surfaces, and the number of surfaces with differing absorptivity/emissivity properties and flexure which require modeling. Accurately modeling of atmospheric drag and radiation forces can be prohibitive from a computational perspective, for spacecraft that cannot be represented by a combination of simple, rigid, geometric shapes. Additionally, uneven heating of surfaces, from the intercepted flux or internally generated sources, produces accelerations that are extremely difficult to model without extensive onboard measurements. Also, the ability to maintain very accurate knowledge of the position and attitude of nonrigid structures of the spacecraft presents additional limitations.

In view of these difficulties and the potential limitations they impose for obtaining 1 cm level orbital accuracies, these concerns should be directly addressed in the design stage of any satellite mission requiring this level of precise orbit determination.

Dynamic Orbit Determination. Dynamic orbit determination and physical parameter recovery software systems employ numerical integration techniques and least squares methods to estimate orbital and physical parameters. Such software systems usually require large-scale computers which include multi-processor vector computers. Efforts to vectorize the computer algorithms has greatly increased computational efficiencies. Success in this area has been a significant factor in the recent progress being made in refining estimates of the Earth's gravity field based on satellite tracking data. It is recommended that efficient algorithms continue to be developed in consort with the present and future computer technologies for the benefit of rapid orbit and physical parameter determination.

Semi-analytical Methods. Semi-analytical techniques are both efficient and computationally inexpensive for orbit determination purposes where the orbit accuracy requirement is not so stringent. These techniques have additional advantages for addressing a specific class of physical problems associated with orbital dynamics. These techniques are currently used mostly for planetary mission analysis and simulation studies for Earth satellite missions. It is recommended that studies of these techniques continue in the areas of general perturbation and stability theory, and provide improved theoretical tools to understand long--term linear and chaotic dynamical systems.

Stochastic Modeling. At the 1 cm level, deterministic modeling of the forces acting on a satellite will not always be possible. This is due to either the complexity of the physics involved or the existence of true stochastic processes perturbing the satellite. In this event, it is recommended that these effects not be left unmodeled but that appropriate stochastic models be applied. To this end, it is necessary that the statistics of the various stochastic phenomenon be known. These statistics comprise the power of the process and their decorrelation times. Successful implementation will enable the improved separation of signal from noise. Such models have already been developed and applied in some existing orbit determination applications. However, this approach has not been widespread and further work in this area is highly recommended.

5.   CONCLUSIONS AND RECOMMENDATIONS

It is clear that the major long-term goal over the next decade will be positioning at the Earth's surface within the 1 mm level. Significantly improved gravity models for orbital determinations, improved modeling of optical and radiometric signals through the ionosphere and troposphere, proper interpretation of Earth surface motions, and properly designed and implemented space missions will be required to achieve this goal. Subsurface mass distribution models will improve concurrently as missions will appropriate gravity sensors (including gradiometers) are approved and implemented. This will enhance greatly our knowledge of the gravity field which will in turn benefit other satellite missions. While improvement in each of these areas is required, only with a coordinated effort in all areas will our understanding of planet Earth evolve in a coherent fashion. Specifically, recommendations in each of the three major areas follow.

Measurement/Environmental Modeling. As demands approach 1 mm in site positioning and 1 cm in the determination of the height of the sea surface, measurement modeling from environmental sources becomes one of the major or is the major limiting factor when using space technologies. To improve on our capabilities to (1) better model signal transmissions through the atmosphere and (2) to correct site motions for transient environmental signals, it is recommended that:

1) Improved models of the propagation delay effects of signals passing through the atmospheric medium be developed:

For the tropospheric effects:
(a)   Overall improvement in the refraction model and modeling algorithms must be advanced;
(b)   Incorporation of new measurements into the models either through the use of advanced technologies (i.e., multi-colored laser systems), operational synergy among systems (e.g., shared targets for radio and optical signals), improved regional meteorological inputs (e.g., extension of models to accommodate gradients in

pressure, temperature and humidity found through regional meteorological models) and improved model testing;

(c)   Improved understanding of the physics of refraction due to water vapor, atmospheric chemistry over a range of signals, and better understanding of the horizontal gradients and turbulence affecting refraction modeling;

(d)   Increased interaction and collaborative efforts with atmospheric scientists; and

(e)   Deployment of improved measuring systems, such as dual or multi-colored laser trackers with reduced sensitivity to tropospheric refraction errors.

For ionospheric effects:

(a)   Overall improvement in higher order ionospheric refraction modeling is required.

(b)   Instruments with three or more widely spaced frequencies in higher frequency bands are needed to provide data for higher order modeling investigations.

(c)   Additional models of the local magnetic field are required.

(d)   Increased interaction and collaboration with ionospheric physicists are required.

(e)   The use of higher and multiple frequencies will improve the ability of different measurement systems to actively correct the effects of the ionosphere, to include or minimize the higher order effects. The inclusion of higher frequencies on the GPS satellites would be a major step in this direction. The continuous global coverage provided by multiple GPS satellites would permit improved ionospheric data to be gathered and more accurate ranging information; and

(f)   Other systems such as DORIS and PRARE which already operate at higher frequencies and can provide information from a variety of orbits, could materially aid global coverage.

2) Station modeling at the 1 mm level requires extensive knowledge of the stations motions due to environmental factors at this level. These motions can be categorized as those which are largely periodic due to tides and tidal loading, those which are nonpredictable, like those resulting from loading due to atmospheric, thermal and other sources, and other motions which are related to the reference frame and the motion exhibited by the entire Earth in inertial frame. Therefore, it is recommended that;

(a)   Improved models be developed for body tides and tidal loading which takes into account anelastic Earth responses, lateral inhomogeneities within the Earth and more complicated ocean loading effects. This will require extension of existing tidal model to include motion modeling as a function of latitude, longitude and frequency. Space geodetic networks will have to be expanded to include other motion sensitive instrumentation such as tidal gravimeters and tiltmeters to improve regional monitoring of body tidal displacements and tidal loading.

(b)   Improved understanding and monitoring of local loading effects must be undertaken. This includes measurements of local groundwater and groundwater topography, regional atmospheric pressure, temperature and wind stress. Regional models will have to be developed and tested in tectonically quiescent areas to improve the monitoring and correction algorithms for station displacements attributable to these regional effects. Dense and precise deformation networks should be established around each fiducial station marker. Frequent precise surveys within these networks will be required on an ongoing basis to determine local motions, monument integrity, and the relative isolated motions of the station markers.

3) With the growing number of different instrumentation systems with higher precision, the standardization of timing systems and the time reference is becoming increasingly more important. Timing has manifested itself in geodetic systems heretofore as frequency stabilities necessary to perform the basic measurements. In future systems the intercomparison of different measurement types will require a common time reference for accurate integration of the different data types. It is therefore recommended that, in the near term, the time scale of choice should be UTC since it is directly available and can be used by all the different geodetic measurement systems. The GPS can already disseminate UTC

and is capable of high precision and accuracy. When GPS is operational the availability of UTC will be greatly enhanced and independent of navigation denial. The possibility of new and higher frequencies to GPS, discussed elsewhere in the report, would also increase the time transfer capability of the system into the subnanosecond range.

Gravity. Considering the need for improving our knowledge of the gravity fields of the Earth and planets for scientific and technical applications; the consequent need to collect more and better quality data both on the ground and from space, with as extensive and fine a coverage as possible; the need for improving existing techniques and developing new ones, both for instrumentation and for data analysis; recognizing the importance of cooperation, particularly international cooperation, in this essentially global enterprise of great complexity, cost, and logistical difficulty, it is recommended that:
1) That all nations distribute their gravity data in the interest of scientific and technical advancement.
2) The inclusion in some of the currently planned missions (ERS-1, Topex/Poseidon) of a nonrepeat phase, to achieve the fine sampling needed to distinguish features as small as 10 km across in the ocean geoid.
3) The development of advanced terrestrial methods, such as air-born gravimetry and gradiometry, strapdown inertial navigation combined with precise space navigation (GPS),etc., which are recognized as likely to become economically viable options for the detailed study of gravity at resolutions finer than 50 km, at the milligal accuracy level.
4) The launching of new laser spacecraft (Lageos II, Lageos III, and Stella), because of the further increase in knowledge of the gravitational field they may contribute. (Stella, for example, will be unique among laser satellites,given its high orbital inclination of 98 degrees, and its low altitude of 800 km.)
5) The deployment in space of a number of small, simple, passive laser targets,similar to Lageos but smaller, if this could be done relatively cheaply by sharing launch vehicles with other missions, to be placed on a variety of orbits to sense, over long periods of time, slow changes in terrestrial gravitation (caused by variation in ice-cover, post-glacial rebound, mass transport between the oceans and the atmosphere, tides, etc.)
6) The establishment of new permanent and mobile laser stations in areas of the world which have small or no previous tracking coverage. These areas include Asia, South America, Africa, etc.
7) The installation of GPS and other radio-tracking, in suitable spacecraft of opportunity, such as Gravity Probe-B (about 600 km height, in near polar orbit, drag-compensated), and likely to map details of the geoid with a resolution of 300 km. Implementation of this approach should not interfere with the pursuit of the advanced missions under consideration, which should give a resolution of 50-80 km.
8) That, in planning the advanced missions for the mid-1990's and beyond, care be taken to ensure a very close integration of their main aspects (orbit selection, instrument design, spacecraft design, data analysis), because this is essential to the fulfillment of their goals.
9) The undertaking of further studies on how to analyze the data from advanced gravity missions, to the intended resolution and accuracy. This is a large and difficult task requiring innovative approaches; new developments in computers should be exploited as fully as possible.
10) The support of advanced gravity missions such as ESA's Aristoteles and NASA's GRM, as they will have significant engineering applications, as well as scientific ones.
11) Recommend that work towards developing sensitive devices, such as gravity gradiometers, be done with a view to their long-term use in space probes for studying the terrestrial planets and major moons in the solar system.

Orbital Analysis. To support the ocean, geophysical, atmospheric and planetary/lunar scientific objectives of space geodesy it is desirable to provide orbital position accuracies of 1 centimeter or better up to altitudes of 10,000 km and 1 part in 10^9 of altitude above, it is recommended that:

1. Pursue precise geometric positioning techniques by exploiting the technology of GPS/Glonass and other similar systems with a geometric positioning capability. For the long-term development of these techniques, including the capability of instantaneous sub-centimeter orbit position accuracy, study the development of a satellite constellation (e.g.,"Super-GPS") dedicated and optimized to the purposes of precise surface and orbital positioning.
2. Encourage an interdisciplinary approach to the development of physical models needed for precise dynamical orbit positioning. These models encompass temporal gravity variations from ocean tides and other meteorological and geological effects, atmospheric pressure and other disturbances such as nongravitational effects due to atmospheric drag, solar and albedo radiation and other components as well as other physical models necessary to support the lunar/planetary applications.
3. Provide operational services for the international dissemination and archival of precise and mean orbital ephemerides. The ephemerides should be referenced to adopted standards to the extent possible and departures from the standard should be documented and disseminated. These services should be capable of providing near time distribution and updated ephemerides as they become available. Technical support of these services should be provided by a wide range of international research groups. Particular emphasis should be placed on the development of algorithms to adapt to vectorization and parallel processing technologies of present and future computers, and for the efficient production and upgrading of orbital ephemerides and geodetic data products.

REFERENCES

Abshire, J.B. and C.S. Gardner, 1985, *IEEE Transactions on Geoscience and Remote Sensing*, Vol. GE-23, (4), pp. 414.

Budden, K.G., 1985, *The Propagation of Radio Waves! The Theory of Radio Waves of Low Power in the Ionsphere and Magnetosphere*, Cambridge University Press.

Chao, B.F. and R.S. Gross, 1987, *Geophys. J.R. Astr. Soc.*, 91, 569.

Chao, B.F., 1988, *EOS*, 69, No. 16, 326.

Chao, B.F., W.P. O'Connor, A.T. Chang, D.K. Hall and J.L. Foster, 1987, *J. Geophys. Res.*, Vol. 92, 89, 9415.

Christodoulidis, D.C., D.E. Smith, R.G. Williamson, and S.M. Klosko, 1988, *J. Geophys. Res.*, Vol. 93, B6, 6216.

Davis, J.L., T.A. Herring, I.I. Shapiro, A.E.E. Rogers and G. Elgered, 1985, *Radio Sci.*, Vol. 20, (6), 1593.

Goad, C.C., 1980, *J. Geophys. Res.*, Vol 85 , B5, 2679.

Guinot, B. and P.K. Seidelmann, 1988, *Astron. Astrophys.* 194, 304.

Gutierrez, R. and C.R. Wilson, 1988, in press.

H.M. Nautical Almanac Office, 1961, *Explanatory Supplement to the Astronomical Ephemeris and the American Ephemeris and Nautical Almanac*.

Janssen, M.A., 1985, *IEEE Transactions on Geoscience and Remote Sensing*, Vol. GE-23, (4), 414.

Lichten, S.M. and J.S. Border, 1988, in press.

Munk, W.H. and G.J.F. MacDonald, 1960, *The Rotation of the Earth: A Geophysical Discussion*, Cambridge University Press.

Paddack, S.J., 1967, *J. Geophys. Res.*, Vol. 72, 22.

Peltier, W.R., 1988, *Science*, Vol. 240, 895.

Rubincam, D.P., 1984, *J. Geophys. Res.*, vol. 89, B2, 1077.

Sconzo, P., 1980, *Cel. Mech*, 22, 61.

Van Dam, T.M. and J.M. Wahr, 1987, *J. Geophys. Res.*, Vol. 92, (B2), 1281.
Yoder, C.F., J.G. Williams, J.O. Dickey, B.E. Schutz, R.J. Eanes, B.D. Tapley, 1983, *Nature*, 303, No. 5920, 757
Yuen, D.A., P. Gasperini, R. Sabadini, 1987, *GRL*, V. 14, 812.

Chapter 7

REFERENCE COORDINATE SYSTEMS

Ivan I. Mueller

1. INTRODUCTION

Geodynamics has become the subject of intensive international research during the last decades, involving plate tectonics, both on the intraplate and interplate scale, i.e., the study of crustal movements, and the study of earth rotation and of other dynamic phenomena such as the tides. Interrelated are efforts improving our knowledge of the gravity and magnetic fields of the earth. A common requirement for all these investigations is the necessity of a well-defined coordinate system (or systems) to which all relevant observations can be referred and in which theories or models for the dynamic behavior of the earth can be formulated. In view of the unprecedented progress in the ability of geodetic observational systems to measure crustal movements and the rotation of the earth, as well as in the theory and model development, there is a great need for the definition, practical realization, and international acceptance of suitable coordinate system(s) to facilitate such work. Manifestation of this interest has been the numerous specialized symposia organized during the past decade or so, such as those held in Stresa (Markowitz and Guinot, 1968), Morioka (Melchior and Yumi, 1972; Yumi, 1971), Torun (Kołaczek and Weiffenbach, 1974), Columbus (Mueller, 1975b, 1978, 1985), Kiev (Fedorov, Smith and Bender, 1980), San Fernando (McCarthy and Pilkington, 1979), Warsaw (Gaposchkin and Kołaczek, 1981), and Coolfont (Wilkins and Babcock, 1988). There seems to be general agreement that only two basic coordinate systems are needed: a Conventional Inertial System (CIS), which in some 'prescribed way' is attached to extragalactic celestial radio sources, to serve as a reference for the motion of a Conventional Terrestrial System (CTS), which moves and rotates in some average sense with the earth and is also attached in some 'prescribed way' to a number of dedicated observatories operating on the earth's surface. In the latter, the geometry and dynamic behavior of the earth would be described in the relative sense, while in the former the movements of our planetary system (including the earth) and our galaxy could be monitored in the absolute sense. There also seems to be a need for certain interim systems to facilitate theoretical calculations in geodesy, astronomy, and geophysics as well as to aid the possible traditional decomposition of the transformations between the frames of the two basic systems.

There is an understanding on how the two basic reference systems should be established; operational details are part of a recent international agreement. There are still, however, a number of open questions which have to be discussed further. These include the type of interim systems needed and their connections to be CIS and CTS, the type(s) of observatories, their number and distribution, whether all instruments need to be permanently located there or only installed at suitable regular intervals to repeat the measurements; how far the model development should go so as not to become impractical and unmanageable; and how independent observations should be referenced to the CTS, i.e., what kind of services need to be established for the user of the systems.

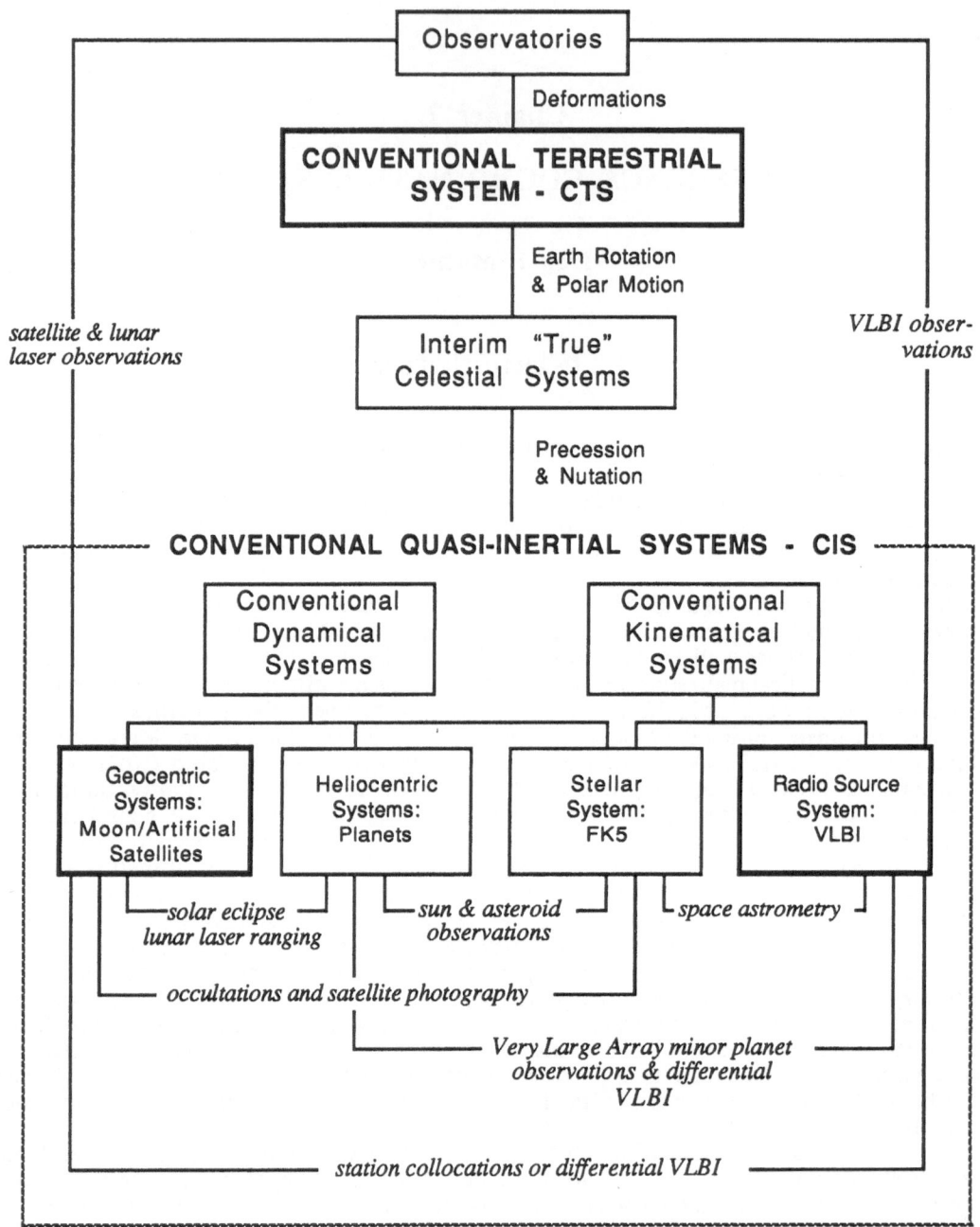

Fig. 1   Conventional terrestrial and quasi-inertial systems of reference with some possible connections.

2. CONVENTIONAL INERTIAL SYSTEMS (CIS) OF REFERENCE

The actual availability of the systems is obtained through their realization in the form of reference frames. This materialization can be done in two different ways so that one can distinguish between two kinds of reference frames (Kovalevsky and Mueller, 1981):

(a) *Conventional kinematic frames.* The fiducial points are presently stars or extragalactic radio sources. In case of the latter, it is necessary to provide connection to stellar catalogues, so that the celestial system can be made available to optical instrumentation.

(b) *Conventional dynamic frames.* In such frames, one or several moving objects are used as the materialization of the system. The theory supporting the corresponding reference system provides the apparent ephemeris of the objects (satellites or planets) as a function of time and the observed successive positions are the fiducial points needed to refer the observations to the system.

Current kinematic and dynamic frames are shown in Fig. 1. It is to be noted that there is not necessarily a bi-univocal correspondence between the two types of frames and quasi-inertial systems. For instance, the FK4 or FK5 stellar systems are dynamical (due to the method of determination of the equinox), while their frame is stellar.

Extragalactic Radio Source System. This system is attached to radio sources which generally either are quasi-stellar objects (quasars) or galactic nuclei. Very long baseline interferometers rotating with the earth determine the declinations of these sources with respect to the instantaneous rotation axis of the earth (see Section 4.2), as well as their right ascension differences with respect to a selected source [3C273, NRAO 140, Persei (Algol), etc.]

There are at present several catalogues of extragalactic radio sources in the J2000.0 system. They vary considerably in number of sources, distribution of sources, and precision. See Table 1 for a summary (Ma, 1989).

Table 1   J2000.0 Catalogues of Extragalactic Compact Sources

Organi-zation	Instru-ment	Baseline Length (km)	No. of Sour-ces	Uncer-tainties mas	Reference
NRAO	CEI	35	36	20-40	Wade & Johnston, 1977
NRAO	CEI	35	16	10	Kaplan et al., 1982
JPL	Mark II	8000-11000	836	300	Morabito et al., 1986
NSF	VLA	<27	700	20-100	Perley, 1982
JPL	Mark II	8000-11000	117	1-5	Fanselow et al., 1984
NASA	Mark III	800-6000	85	0.3-13	Ma et al., 1986
NGS	Mark III	800-6000	26	0.5	Robertson et al., 1986
NASA	Mark III	800-11000	101	0.2-9	Ma, 1988
JPL	Mark III	8000-11000	128	0.5-7	Sovers et al., in press

Stellar System. This system is attached to stars in the FK5 catalogue, i.e., the adopted right ascensions and declinations of the FK5 define the equator and the equinox and thus the frame of the Stellar-CIS. The FK5 is the fifth fundamental catalogue in a series which began with the FC in 1879 (Fricke and Gliese, 1978).

As far as the accuracy of the FK5-CIS is concerned, the question again is meaningful only in the sense of how precise the star positions in the FK5 are. The 1535 *Basic FK5* stars have a mean precision of 0''.02 and 0.7 mas per year in proper motion. The 980 stars in the bright (magnitudes 5.5–7.0) *FK5 Extension* are precise to about 0''.03, while the 2150 additional faint (magnitudes 6.5–9.5) stars to about 0''.06.

An important extension of the FK5 is the International Reference Star (IRS) catalogue which is almost completed and will include about one star per square degree.

Dynamical Systems. The dynamics expressed in the equations of motion define a number of nonrotating planes which could be the basis of reference frames. Considering the observable planes that could be the basis of such a Dynamic-CIS, there are the planetary (including the earth-moon barycenter) orbital planes, the equator, the lunar orbital plane, and the orbital planes of certain high flying, thus only slightly perturbed, artificial earth satellites (e.g., Lageos or GPS). In more practical (observational) terms one can distinguish between Planetary, Lunar and (artificial) Satellite CIS's, each frame defined, in theory, by two of the above-mentioned planes, and in practice by the available ephemerides.

In the case of the *planetary systems*, the defining planes are the equator and the ecliptic, their intersection being the line of the equinoxes. In the case of the *lunar system*, the main references are the orbital plane of the moon and the equator of the earth.

Data types to which modern planetary and lunar ephemerides are adjusted are listed in Table 2. The post-fit rms residuals indicate the accuracy of the data. The values listed without brackets are the units of the original observations; those within brackets give the comparable values for comparison purposes.

Table 2   Data in Modern Lunar and Planetary Ephemerides (Williams and Standish, 1989)

Type of Observation	Time Span	Post-Fit Rms (km)	Residuals	No. of Observ.
Radar Ranging				
Mercury	1966-	1.5	[0''002]	500
Venus	1965-	1.5	[0''002]	1000
Mars	1967-	2.2	[0''003]	40000
Mars Closure	1969-82	0.15	[0''0002]	200
Spacecraft Ranging				
Ma9 Orbiter (Mars)	1972-73	0.040	[0''0002]	600
Viking Lander (Mars)	1976-80	0.007	[0''000003]	900
	1980-82	0.012	[0''000006]	400
Spacecraft Tracking (Range, Doppler)				
Pion&Voy (Jup,Sat)	1973-80	[200, 400]	[0''05]	20000
Lunar Laser Ranging	1969-70	0.00100	[0''0005]	10
	1970-75	0.00030	[0''00016]	1700
	1976-85	0.00015	[0''00008]	3000
	1985-	0.00006	[0''00003]	600
Radio Astrometry				
Jupiter, ..., Neptune	1983-	[100, ..., 600]	0''03	10
Ring Occultation				
Uranus	1978-	[1500]	0''1	14
Optical Transits (Manual)				
Sun, Mercury, Venus	1911-	[700]	1''0	37000
Mars, ..., Neptune	1911-	[150, ..., 10000]	0''5	18000
Optical Transits (Photoelectric)				
Mars, ..., Neptune	1982-	[100, ...,6000]	0''3	1000
Astrolabe				
Mars, ..., Uranus	1961-	[100, ..., 4000]	0''3	1500
Astrometry				
Pluto	1914-	[15000]	0''5	1600

In the case of *satellite systems*, the problem is compounded by additional modeling problems related to the force field in which the satellite moves and by the fact that nowadays there are no direct connections to other frames of reference. Modern satellite tracking techniques (laser, GPS, etc.) all basically observe ranges or range differences and contain no direct directional information. The main reference planes, the orbital plane of the satellite and the equator, intersect along the line of nodes, the initial orientation of which therefore must be defined more or less arbitrarily.

Discussion:

1. The most accurate, long-term CIS is the one attached to extragalactic radio sources. It is accessible through VLBI observations. Other systems can be accurately connected to it by station collocation, space astrometry, VLA and differential VLBI observations (Fig. 1). The number of primary radio sources must be increased, especially in the Southern Hemisphere, to achieve isotropy.

2. The CIS attached to the FK5 is somewhat less accurate. Direct access to it is through optical star observations, which by nature are generally less accurate than VLBI observations. Its main value is in defining the fundamental mean system of coordinates and thereby providing a direction (the FK5 equinox) for the time (UT1) definition, and for the possible orientation of the Radio Sources-CIS. The latter function, however, stems from more of a traditional requirement and not from theoretical needs.

3. Of the Dynamical-CIS's, the accuracy of the planetary (including lunar) system is better than the FK5, with a rotational stability of 3 mas per century. The satellite systems by themselves are suitable for medium-term to short-term work only. The rotational stability can be extended by connections to the Radio Source-CIS through accurate and continuous observations at collocated stations or differential VLBI. Lunar laser ranging provides the best connection between the lunar and planetary systems.

4. If a *dynamical system* is based on the motion of planets, the ecliptic plays a privileged role and, naturally, the ecliptic is used in the definition of coordinates. Since equatorial coordinates are preferred to ecliptic ones for obvious instrumental reasons, the intersection of the ecliptic and the equator, the vernal equinox, becomes the natural origin of right ascensions. When the dynamical system is geocentric, the natural reference plane is the Laplace plane whose position depends upon the relative magnitude of the perturbations. For the moon, the solar effects are dominant and, practically, the Laplace plane is the ecliptic and, again, the equinox is the natural origin of equatorial coordinates. In the case of artificial satellites the perturbations due to the earth flattening are predominant so that the Laplace plane is the equator. The equator is, therefore, the natural fundamental plane, but the origin may be arbitrary.

Similarly, the choice of the equinox in the stellar systems is justified by the fact that they are partially dynamical systems based upon planetary theories. However, in the construction of the corresponding stellar frame, the difficulty of maintaining the theoretical origin is so serious that one is led to distinguish between the dynamical equinox which defines the origin of the system and the catalogue equinox which is the origin of the frame. In practice, the actual origins of the stellar reference frames are purely conventional and are not the dynamical equinox.

5. The situation will become even more conspicuous for frames derived from conventional *kinematic systems*. Even if, for the sake of continuity, the origin and the fundamental plane of such a system should be close to the equinox and the equator, they should be conventional points defined only by the realization of the corresponding frame. Otherwise, it would be necessary to introduce a complex dynamical model to define the origin at the expense of introducing inaccuracies in the system and an uncertainty in its realization by the frame. In practice, the solution might be analogous to the present situation for the terrestrial reference frame. One would establish an international organization that would provide the coordinates of radio sources in the conventional kinematic frame, taking into account eventual changes in the number and position of the reference sources, due, for instance, to the disappearance or motion of quasars or better measurements, in such a way that the changes should not introduce a rotation (or translation) of the system in the average statistical sense. It is an almost

unavoidable conclusion that for geodetic and geodynamic applications the most useful CIS is just such a system (Kovalevsky and Mueller, 1981; Guinot, 1986).

3. CONVENTIONAL TERRESTRIAL SYSTEMS (CTS) OF REFERENCE

There seemed to be general agreement that the new CTS frame conceptually be defined similarly to the earlier used CIO-BIH system (Bender and Goad, 1979; Guinot, 1979; Kovalevsky, 1979; Mueller, 1975a), i.e., it should be attached to observatories located on the surface of the earth. The main difference in concept is that these can no longer be assumed motionless with respect to each other. Also they must be equipped with advanced geodetic instrumentation like VLBI or lasers, which are no longer referenced to the local plumblines. Thus the new transformation formula, again assuming a common origin, may have the form

$$[\vec{OBS}]_j = \vec{L}_j' + [\vec{CTS}]_j = \vec{L}_j' + \mathbf{SNP} \ [\vec{CTS}]_j \tag{1}$$

where $\vec{L}_j'$ is the vector of the 'j' observatory's movement on the deformable earth with respect to the CTS, computed from suitable models; $\mathbf{NP}$, the nutation and precession matrices computed with the new 1976 IAU constants and the 1980 IAU series of nutation; and $\mathbf{S}$, the rotation matrix between the CTS and the true frame for which the nutation is computed.

In the above equation the coordinates of the observatory 'j' $[\vec{CTS}]_j$, in the Conventional Terrestrial Reference Frame, are related to the coordinates determined by the technique 'o' $[\vec{OBS}]_j{}^o$, in its own reference frame, through the well-known transformation equation:

$$[\vec{CTS}]_j + \vec{\delta}^o + \mathbf{R}_1(\beta_1{}^o) \ \mathbf{R}_2(\beta_2{}^o) \ \mathbf{R}_3 \ (\beta_3{}^o)[\vec{CTS}]_j$$
$$+ c[\vec{CTS}]_j + \vec{L}_j' = [\vec{OBS}]_j{}^o + \vec{v}_j \tag{2}$$

where $\vec{\delta}^o$ contains the three translation components between the CTS frame and that inherent in the technique 'o', $\vec{\beta}^o$ are the three (usually very small) rotations, and c a differential scale factor. $\vec{L}_j$ is the vector of deformation, not containing global rotations nor translations, and $\vec{v}_j$ is the residual vector.

Another set of equations derived in (Zhu and Mueller, 1983) relate other parameters in eq. (1), specifically the earth rotation parameters (ERP) in the matrix S determined by the technique 'o' in its own frame of reference, to those referring to the CTS frame:

$$x_p - \beta_2{}^o + \alpha_1{}^o \sin\theta + \alpha_2{}^o \cos\theta = x_p{}^o + v_{x_p}$$
$$y_p - \beta_1{}^o - \alpha_1{}^o \cos\theta + \alpha_2{}^o \sin\theta = y_p{}^o + v_{y_p} \tag{3}$$
$$\omega_d \ UT1 + \beta_3{}^o - \alpha_3{}^o \qquad\qquad = \omega_d \ UT1^o + v_{UT1}$$

where $x_p{}^o$, $y_p{}^o$ and $UT1^o$ are the observed ERP's; x_p, y_p and UT1 are those referenced to the CTS; ω_d is the conversion factor, $\vec{v}$ is the residual vector of the observed ERP's and θ the sidereal time. Finally, $\vec{\alpha}^o$ are the small rotations between the Conventional Inertial Reference Frame and the Inertial Frame inherent in the technique 'o', i.e.,

$$[\vec{CIS}]^o = \mathbf{R}_1(\alpha_1) \ \mathbf{R}_2(\alpha_2) \ \mathbf{R}_3 \ (\alpha_3) \ [\vec{CIS}] \tag{4}$$

For each ERP series 'k' of 1-1.2 years length (or longer), generated by the technique 'o', one can fit the following type of circular model:

$$a_1^{o,k} + a_2^{o,k} \cos A + a_3^{o,k} \sin A + a_4^{o,k} \cos C + a_5^{o,k} \sin C = x_k^o + v_{x_k}$$

$$a_6^{o,k} - a_2^{o,k} \sin A + a_3^{o,k} \cos A - a_4^{o,k} \sin C + a_5^{o,k} \cos C = y_k^o + v_{y_k}$$

$$(5)$$

where A is the annual frequency and C the Chandler frequency, x_k^o, y_k^o are the observed ERP's in the series k, and v the residuals.

The coefficients $a^{o,k}$ allow the computation of the amplitude of the annual motion

$$\sqrt{(a_2^{o,k})^2 + (a_3^{o,k})^2} \ ,$$

that of the Chandler motion

$$\sqrt{(a_4^{o,k})^2 + (a_5^{o,k})^2} \ ,$$

as well as the coordinates of the center of the polhode $a_1^{o,k}$ and $a_6^{o,k}$.

Equations (2) - (5) can be used as observation equations by an international service to determine the parameters defining and maintaining the CTS and providing the relationship versus the terrestrial frame of each technique ($\vec{\delta}^o$, $\vec{\beta}^o$, c), versus the CIS (x_p, y_p, UT1), and the latter's relationship to the technique inertial frame ($\vec{\alpha}^o$):

$[\vec{CTS}]_j$ and $\vec{L}_j$ for each observatory 'j', to define the CTS

$\vec{\delta}^o$, $\vec{\beta}^o$, c, $\vec{\alpha}^o$ and $\vec{a}^{o,k}$ for each technique 'o' and ERP series 'k'

x_p, y_p, UT1   for the service to provide the S matrix in eqs. (1) and (9)

As far as the origin of the CTS is concerned, it could be centered at the center of mass of the earth, and its motion with respect to the stations can be monitored either through observations to satellites or the moon, or, probably more sensitively, from continuous global gravity observations at properly selected observatories (Mather et al., 1977).

The IAU and IUGG recently made practical recommendations on the establishment of such a (or very similar) Conventional Terrestrial System, including the necessary plans for supporting observatories and services by establishing the International Earth Rotation Service, effective 1 January 1988 (Wilkins and Mueller, 1986). The goal of the service is the determination of the total transformation between the CTS and CIS. Thus the service will publish not only ERP determined from the repeated comparisons (the past situation), but also the models and parameters described above in eqs. (1) - (5), i.e., the parameters defining the whole system.

4. REFERENCE FRAME TIES

Ties Between the CIS Frames. 'Measurements are inherently simpler to make and generally more accurate in their "natural" frame and hence should always be reported as such.

However, to benefit from the complementarity of the various techniques, knowledge of the frame interconnections (both the rotation and the time-variable offset) is essential' (Dickey, 1989). These are summarized in Figs. 1 and 2.

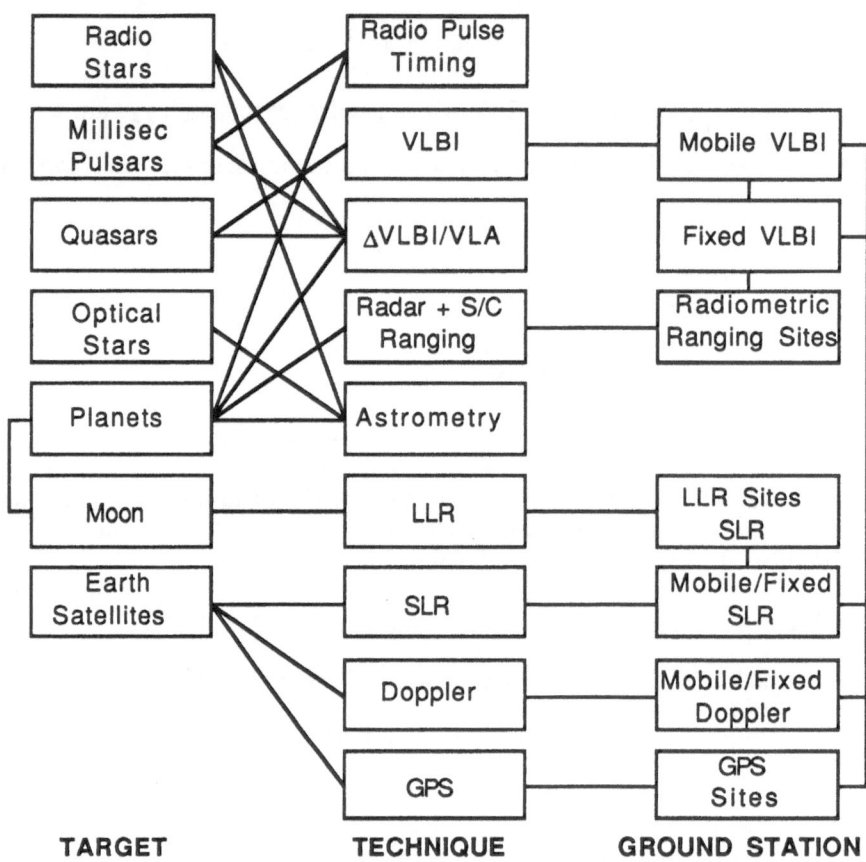

TARGET   **TECHNIQUE**   **GROUND STATION**

Fig. 2 Reference frame connections (Dickey, 1989).

Dickey (1989) also outlines the future with ongoing and planned efforts in several areas: Improved ephemeris-radio frame ties can be accomplished by VLBI observations of pulsars, additional VLA observations of the outer planets and satellites, and future differential VLBI experiments (such as that with orbiting spacecraft around Jupiter and Saturn). The millisecond pulsar PSR1937+214, having a period of 1.6 ms, has exceptionally low timing noise. Its position in the ephemeris frame can be measured to ~1 mas. This will allow a radio-planetary frame tie, limited only by the accuracy of an interferometric position measurement. Roughly, a factor of five improvement (down to 0.''01) is expected here with the full implementation of VLBI observations. An initial experiment of this type has been executed by R. Linfield and C. Gwinn.

As already mentioned, for optical astrometry, Hipparcos will measure a network of stars over the entire sky with accuracies of ~2 mas (Kovalevsky, 1980), while the Space Telescope will measure small fields with similar differential accuracy. However, the Space Telescope can observe much fainter objects (Jeffreys, 1980) and could observe the optical counterparts of extragalactic radio sources, all but possibly one of which are too faint for Hipparcos. A joint program would produce an accurate stellar network linked to the quasar radio frame by the

Space Telescope. The occultations of stars by planets and planetary rings can provide an additional link between the optical and ephemeris frames. Also, optical interferometry offers exciting possibilities with the potential resolution being two or three orders of magnitude finer than that of VLBI (Reasenberg, 1986). More details are given in (Dickey, 1989).

Ties Between the CTS Frames. Boucher and Altamimi (1987) established relationships between a number of Conventional Terrestrial Reference Frames based on collocated observation stations and eq. (2). The selected sets of station coordinates defining each CTS are as follows:

(i) CTS (VLBI). Three sets of station coordinates have been selected:
CTS(NGS) 87 R01. The coordinate data are derived from a composite set of Mark III VLBI observations collected under the aegis of project MERIT, POLARIS, and IRIS and conducted between September, 1980, and January, 1987. Westford coordinates were fixed to their initial values. The IRIS terrestrial frame is made more nearly geocentric by applying the BTS 1985 translations (Carter et al., 1987).
CTS(GSFC) 87 R01. The data acquired since 1976 by the NASA Crustal Dynamics Project and since 1980 by the NGS POLARIS/IRIS programs. The terrestrial frame is defined by the position of the Haystack 37-M antenna and the BIH Circular D values for 1980 October 17 (Ma et al., 1987).
CTS(JPL) 83 R05. The coordinate data are from the JPL Time and Earth Motion Precision Observations (TEMPO) project, using the DSN radio telescopes. The reference frame solution is tied to the BIH on 20 December 1979 (Eubanks et al., 1985).

(ii) CTS (Lunar Laser Ranging). The coordinate data are from the JPL solution: SSC(JPL) 87 M01 containing four stations, two at Fort Davis, one at Haleakala (Maui), and one at Grasse. The nominal planetary and lunar ephemeris DE121/LE65 was used in the reduction. The ephemeris uses the equator and equinox of B1950.0. It is on the dynamical equinox and has a zero point consistent with the FK5 catalogue (Newhall et al., 1987).

(iii) CTS (Satellite Laser Ranging). Two sets of station coordinates have been selected:
CTS(CSR) 86 L01. The solution is based on Lageos ephemeris from May, 1976, to September, 1986, using the model Lageos Long Arc 8511. The force model, referred to as the CSR 8511 system, adheres closely to the MERIT standards. The tectonic plate motion model AM1-2 of Minster and Jordan (1978) was used and the epoch of the derived station coordinates is 1983 January 1. The GM value is 398600.4404 km^3/s^2 (Schutz et al., 1987).
CTS(DGFI) 87 L01. The solution is computed from Lageos observations covering the period 1980 to end 1984 and based on five yearly solutions. By the rates of change of the yearly solutions, the station coordinates then were related to the same reference epoch 1984.0. The reference frame was defined by the three coordinates (longitude, latitude of Yaragadee (7090) and latitude of Wettzell (7834)) which were held fixed in the five solutions. The GM value is 3.98600448 E + 14 m^3 s^{-2}, initial ERP series were from homogeneous BIH series and other constants from MERIT Standards (Reigber et al., 1987).

(iv) CTS (Doppler). Station coordinates are from DMA Doppler project SSC(DMA) 77 D01 solution, and other Doppler campaigns containing more than 100 station positions. They are determined in the NSWC9Z2 datum by point positioning using Precise Ephemerides.

A combination of all above data has been performed incorporating 51 collocated sites and making use of the plate tectonic absolute motion model AMO-2 derived from the global RM-2 model (Minster and Jordan, 1978).

Table 3 lists the transformation parameters of the individual systems with respect to a global one whose origin is constrained to that of JPL 87M01 (LLR) and CSR 86L01 (SLR), the scale to CSR 86L01 (SLR), and the orientation to NGS 87R01 (VLBI). Some conclusions about the origin, scale and orientation of the individual CTS's with respect to the global one: Knowing that the origin of the adjusted system is from SLR and LLR, the origin of all VLBI solutions remains arbitrary. The level of consistency of the scale factor is 10^{-8} for the different solutions. Some variations for VLBI and LLR solutions are due to a relativistic bias in the definition of the terrestrial system (Hellings, 1986; Boucher, 1986). The orientation of the individual terrestrial systems is usually realized through BIH values. The differences in orientation of the different solutions are arbitrary and of some mas level.

Table 3   Transformation Parameters from the Individual 1984.0 CTS Systems to the 'Global' CTS (Boucher and Altamimi, 1987) (the uncertainties are given on the second line)

CTS	δ_1 m	δ_2 m	δ_3 m	$(c-1)10^{-6}$	β_1	β_2	β_3
NGS 87 R01	-0.009	-0.111	-0.112	0.023	0".000	0".000	0".000
	0.035	0.036	0.035	0.004	0.000	0.000	0.000
GSFC 87 R01	-1.696	0.862	-0.463	0.020	0.001	0.000	0.003
	0.029	0.034	0.032	0.004	0.001	0.001	0.001
JPL 83 R05	-0.062	0.234	0.140	0.015	0.001	0.011	0.000
	0.032	0.036	0.035	0.005	0.002	0.002	0.001
JPL 87 M01	0.000	0.000	0.000	0.020	-0.004	0.009	0.004
	0.000	0.000	0.000	0.017	0.005	0.005	0.005
CSR 86 L01	0.000	0.000	0.000	0.000	0.003	0.005	0.008
	0.000	0.000	0.000	0.000	0.002	0.001	0.002
DGFI 87 L01	-0.015	0.021	-0.053	-0.015	-0.010	0.014	-0.115
	0.041	0.041	0.040	0.006	0.002	0.002	0.002
DMA 77 D01	0.302	0.096	4.645	-0.605	-0.030	-0.005	0.797
	0.219	0.206	0.195	0.026	0.009	0.009	0.006

5. REFERENCE FRAME REQUIREMENTS

5.1 EXPECTED CHANGES IN THE ADOPTED CONSTANT OF PRECESSION AND SERIES OF NUTATION

Recent analysis of modern highly accurate observations (e.g., VLBI) indicates significant departures from the IAU 1980 nutation series. None of the existing theories based on various Earth models can adequately explain these departures from Wahr's model. Apparently more efforts are required both in theory and in observations to arrive at a resolution.

Assuming that the CTS is to be maintained unchanged, corrections to the nutation terms in longitude ($\delta\Delta\psi$) and obliquity ($\delta\Delta\epsilon$) would theoretically change the polar motion components and GAST, utilized in the transformation equation (1), i.e., in the matrix S, as follows (Zhu and Mueller, 1983):

$$\Delta x_p \quad = \quad \delta\Delta\varepsilon \sin\theta + \delta\Delta\psi \sin\varepsilon \cos\theta$$

$$\Delta y_p \quad = \quad -\delta\Delta\varepsilon \cos\theta + \delta\Delta\psi \sin\varepsilon \sin\theta$$

$$\Delta(GAST) \quad = \quad \delta\Delta\psi \cos\varepsilon$$

where θ is the sidereal time. As it is seen, the theoretical effects on polar motion are diurnal terms ($\delta\Delta\psi$ and $\delta\Delta\varepsilon$ being long periodic).

Williams and Melbourne (1982) and Zhu and Mueller (1983) investigated the effects of a change in the constant of precession. The effect on polar motion is a diurnal periodic term with an amplitude increasing linearly in time; on the GAST it is a linear term.

5.2 INTERMEDIATE REFERENCE FRAME ISSUES

The complete transformation from the CIS to the terrestrial frame CTS is given by eq. (1). In geodetic applications generally only the complete transformation **SNP** is needed. Changes in the 'intermediate' reference frame defined by the **NP** transformation must either by 'absorbed' in the **S** matrix by changing appropriately x_p, y_p and GAST (UT1), or the CTS must change its orientation. There are seven options to choose from, and they are a matter of preference (Zhu and Mueller, 1983). One of these which would neither change the CTS orientation nor the UT1 is probably preferred by geodesists. It would however change the definition of the Greenwich Mean Sidereal Time by referring it to a point on the equator insensitive to precession. A similar option has been advocated by Guinot (1979) during the past decade but for different reasons. A recent proposal by Capitaine and Guinot (1988) is based on the observation that the classical definition of GAST representing the rotation of the Earth (i.e., CTS) is not satisfactory mainly for two reasons:

(i) It is referred to the true equinox of date which is an inadequate and unnecessary intermediate reference point because modern observations of the CTS's orientation in space (especially VLBI) are practically insensitive to the orientation of the ecliptic and consequently to the position of the equinox.

(ii) The presently adopted expression converting GAST to UT1 (Aoki et al., 1982) neglects some cross-terms between precession and nutation which are of the order of $0\!''\!001$ and should now be considered.

The definition advocated would thus be better adapted to the new methods of observation and would provide an accuracy of the order of $0\!''\!0001$. It would also result in a new definition of Universal Time which would remain valid even if the adopted model for the **NP** transformation is revised (see also Capitaine et al., 1986). The proposal is not without its critics. See (Aoki and Kinoshita, 1983; Aoki, 1988).

Related to the above issue is the definition of the third axis of the intermediate frame as defined by the transformation model **NP**, specifically, by the adopted theory of nutation. This pole, the Celestial Ephemeris Pole (CEP), conceptually has no diurnal motion with respect to an Earth-fixed or a space-fixed reference frame. Some of the modern observational techniques, however, are not very sensitive to this axis and, in fact, on the level of $0\!''\!001$ accuracy, define a variety of technique dependent conventional poles. Capitaine et al. (1985) and Capitaine (1986) point out that clarification of this issue is necessary in order to intercompare and interpret polar motion coordinates determined at the level of $0\!''\!001$ accuracy, by means of a variety of techniques ranging from VLBI to superconducting gravimetry.

5.3 THE RADIO SOURCE REFERENCE FRAME

Current definitions of conventional inertial reference frames at radio wavelengths make use of the extragalactic references provided by suitable objects, primarily quasars. Such references, when combined with interferometric astrometry, are *nonpareil* when applied to the determination of Earth Rotation Parameters (ERP) and Crustal Dynamics (CD) studies. Indeed, these programs have made the major contribution to the various radio reference catalogues currently available. It is now necessary to extend the use of such systems in general. This must include extensions in wavelength, in applications, and in the classes of objects included in such systems. There are, of course, obstacles to be overcome if the full potential of this conceptually straightforward approach is to be realized. However, if the phenomena of precession, nutation and polar motion as well as the concepts of the ecliptic and the vernal equinox can be disconnected from the realization of a reference frame, and be regarded as simply describing various aspects of the Earth's complicated motions, then a great simplification will have been achieved. Of course all of the above phenomena and concepts are basic, and a knowledge of them is absolutely necessary. This knowledge will continue to be supplied by the classical, dynamical observations, radio astrometry and pulsar observations. However, it is now possible to consider these items in their proper context and to define a reference frame which is independent of them. Such independence will benefit not only the reference frame, but also aid in the study of the very phenomena from which the concept of a reference frame will have been freed. Essentially, observations will have been decoupled from the observing platform. As a result of this, the accuracy of the reference frame will become primarily dependent upon the precision and accuracy of the underlying measurements, and will have a minimal, noncritical dependence upon any companion theories.

An extragalactic reference frame which will serve as the initial system of the International Earth Rotation Service (IERS) was compiled on the basis of four individual catalogues from the Goddard Space Flight Center, the Jet Propulsion Laboratory and the U.S. National Geodetic Survey. The compilation was carried out at the IERS (Arias et al., 1988) and includes 228 extragalactic, compact sources divided into primary, secondary and complementary sources depending upon geometrical and physical considerations as well as observational histories. Unfortunately, this reference frame contains no sources south of −45°, and of the 23 primary sources which define the directions of the axes, eight are in the Southern Hemisphere between the equator and −29°. This points up the fact that even with the excellent ERP results, the distribution of well-observed radio sources and radio interferometry baselines is far from ideal for the purposes of a global reference frame. Nevertheless great improvements are taking place. Indeed, dozens of new sources have been observed most recently many in the Southern Hemisphere. Thus the problem of sufficient coverage on a global basis, whatever coverage that may ultimately prove to be, is being addressed. At present the density and the distribution of radio sources necessary to provide an acceptable transformation between radio and optical systems depend primarily upon the homogeneity or isotropy of the optical system. If, for example, one had an optical catalogue with relative coordinates of the stars at some epoch with the same accuracy as in radio catalogues, then merely applying a correction to the zero points could serve as the transformation for that epoch. The forthcoming HIPPARCOS catalogue is intended to approach this ideal and will provide an excellent example regarding the matter of radio/optical transformations. If the HIPPARCOS system is successfully referred to an extragalactic frame, then the extension of this frame to magnitudes intermediate to HIPPARCOS stars and the quasars will follow through the use of astrographs and Schmidt telescopes. Imperfect proper motions complicate the situation of course, but the while point of improving the reference frame is to provide a better standard coordinate system within which improved stellar motions, for example, can be determined. It is important to note that given an accessible extragalactic reference frame, optical reference frame positional observations would be freed of the burden of simultaneously determining the zero points of a dynamical system, the improvements to the assumed planetary orbits and the individual star positions. The emphasis could then be upon achieving isotropy and observing to fainter magnitudes. Questions involving source structure and any evolution thereof can only be resolved by repeated and carefully programmed observations.

6. THE INTERNATIONAL EARTH ROTATION SERVICE

6.1 THE MERIT-COTES PROGRAMS

The acronyms MERIT and COTES refer to two international programs that were started independently, but which developed together. MERIT refers to an international program to monitor the earth's rotation and intercompare the techniques of observation and analysis with a view to making recommendations about the form of a new international service. On the other hand, the objective of the COTES program program was to provide a basis for recommendations on the establishment and maintenance of a new conventional terrestrial reference system for the specification of positions on or near the earth's surface. The two programs were linked when it became clear that the observational campaign planned for MERIT and the new earth rotation service would provide results that could be used for COTES. In particular, in order to determine the earth rotation parameters to high accuracy, it is necessary to establish the positions of the observing sites (or 'stations') in a worldwide network that provides a suitable basis for a new terrestrial reference system. The observational data and results that have been obtained in the course of these programs have been collected together for further analysis and for use in current and future scientific studies and practical applications.

Project MERIT was conceived in 1978 at IAU Symposium No. 82 on "Time and the Earth's Rotation." The Symposium recommended the appointment of a "working group to promote a comparative evaluation of the techniques for the determination of the rotation of the earth and to make recommendations for a new international program of observation and analysis in order to provide high quality data for practical applications and fundamental geophysical studies." Two years later, in 1980, the participants in IAU Colloquium No. 56 on "Reference Coordinate Systems for Earth Dynamics" recommended the setting up of a working group "to prepare a proposal for the establishment and maintenance of a Conventional Terrestrial Reference System." Information discussions at the First MERIT Workshop in 1981 were followed eventually by the merging of the two groups and the production of a Joint Summary Report (Wilkins and Mueller, 1986). This report describes briefly the development of the programs of observation and analysis and gives recommendations for new terrestrial and celestial reference systems and for the setting up of a new International Earth Rotation Service (IERS); this report also includes references to earlier reports that describe the techniques used, the organizational arrangements and the programs of the activities, and that give the principal results and references to relevant papers.

The MERIT and COTES programs have been very successful in stimulating the use and development of new techniques of observations using laser ranging and radio interferometry; they also led to improvements in the results from optical astrometry and the Doppler (radio) tracking of satellites, which were in regular use before 1978. Coordinators were appointed for each technique and for certain associated activities, such as the operation of a Coordinating Center for the combination and dissemination of results, the preparation of MERIT Standards, and the collocation of equipment of different techniques.

The quantities measured by each of the techniques that were used in the programs are as follows:

Doppler tracking of satellites: The Doppler shifts (range rates) in the radio transmissions from Transit navigation satellites.

Satellite laser ranging: The time for pulses of laser light to travel to and from geodetic satellites carrying retroreflectors.

Lunar laser ranging: Time of flight for pulses of laser light to travel to and from retroreflectors on the surface of the moon.

Optical astrometry: Directions to stars measured with respect to local reference frames.

Connected-element radio interferometry, and Very long baseline radio interferometry: Differences between the travel times of the radio emission from quasars to two or more radio telescopes.

Organizational arrangements for the regular transmission and processing of data already existed for optical astrometry and Doppler tracking, but for the other techniques it was necessary to set up both operational centers and analysis centers. The operational centers coordinated the observations, collected the observational data, computed earth rotation parameters on a rapid-service basis from 'quick-look data', and distributed the observational data (perhaps after some processing) to the analysis centers, which determined both earth rotation parameters and station coordinates from all the available data.

There were several designated periods when all stations were requested to make observations and send them as quickly as possible to the operational centers. The first was the MERIT Short Campaign from 1 August to 31 October 1980. This was primarily a test of the technical and organizational arrangements, but it also produced much valuable data and showed clearly the potential of the new techniques. The MERIT Main Campaign covered the 14-month period from 1 September 1983 to 31 October 1984 and included the first COTES Intensive Campaign, which ran from 1 April until 30 June 1984. The data were analyzed independently at two or more analysis centers for each technique, and many excellent series of earth rotation parameters and sets of station coordinates were obtained. These data are still being studied to determine, for example, the systematic differences between the reference systems of the various techniques. The results have established beyond doubt the very close correlation between the short-period variations in the length of day and in the angular momentum of the atmosphere. The pole of rotation has been shown to move much more smoothly than had earlier been thought, but there is still controversy about the sources of excitation of the 14-month term in the motion.

6.2 THE INTERNATIONAL EARTH ROTATION SERVICE

By the end of the MERIT Main Campaign it had become clear that laser ranging and radio interferometry were able to provide more precise estimates of polar motion, universal time and length of day than could optical astrometry and the Doppler tracking of satellites, which were the prime contributors to the international services in 1978. This conclusion has since been substantiated by the more detailed analyses of the data that have been reported at the MERIT Workshop and Conference held at Columbus, Ohio, on 29 July - 2 August 1985 (Mueller, 1985). The accuracy of the regular determination of the coordinates of the poles by SLR and VLBI is about 5 cm, compared with 30 cm by optical astrometry and Doppler tracking, while for UT and excess length of day the accuracy is about 0.2 ms and 0.06 ms, compared with 1 ms and 0.2 ms.

It must be realized, however that other factors besides precision had to be taken into account before recommendations about the future international services could be formulated. Perhaps the most important factor was whether it is reasonable to expect that the organizations concerned are likely to continue to make and process observations at an appropriate level and to make the results available to the international community without restriction. The MERIT Main Campaign was a period of special activity, and it cannot be assumed that any technique would provide results of the same high quality (as judged by the combination of precision, accuracy, frequency, reliability and promptness) on a long-term basis.

The International Latitude Service was initially set up a a set of five dedicated stations, but it was eventually replaced by the International Polar Motion Service which relied on receiving data from a much larger number of instruments which provided local services and data for other scientific purposes as their prime justification. It is to be expected that any new International Earth Rotation Service will also have to depend largely on the use of observations and results that are obtained for other national and international programs.

In particular it must be recognized that an important application of the Service will be the establishment and maintenance of the new conventional terrestrial reference system. The permanent stations used for monitoring earth rotation will comprise a primary geodetic network of large scale and high precision that will be densified, partly by the use of mobile systems using the same techniques, but mainly by the use of other geodetic techniques, such as the use in radio interferometric mode of signals for navigation satellites.

The choice of the techniques to be used in the new service depends on the subjective evaluation of many factors and not merely on a comparison of the potential quality of the determination of each rotation parameters. Although it is conceivable that a single VLBI network could provide an adequate international earth rotation service, the general conclusions of the discussions in the MERIT and COTES working groups is that the new service should be based on both laser ranging and VLBI and should also utilize any other appropriate data that are made available to it.

The three recommendations given in Appendix 1 were adopted at a joint meeting of the MERIT Steering Committee and the COTES Working Group that was held at Columbus, Ohio, on 3 August 1985. Earlier drafts had been subject to critical review at the MERIT Workshop on 30 July and by interested participants in the Conference on Earth Rotation and Reference Systems held 31 July to 2 August. The joint meeting also adopted a draft resolution for consideration by a Joint Meeting of the IAU Commissions 19 and 31 on 22 November 1985 during the XIXth General Assembly of the IAU at New Delhi. Amended versions of this resolution were adopted by the Joint Meeting and subsequently by the Union on 28 November 1985. A further recommendation concerning the assignment of responsibility within the IAU for matters relating to the celestial and terrestrial reference systems was adopted by the MERIT/COTES meeting on 3 August and served to stimulate a discussion within the IAU, but no decision was announced.

The final version of the IAU resolution on the MERIT/COTES program and recommendations is given in Appendix 2. In effect the resolution endorsed this report and the principal recommendations on concepts, organization and interim arrangements. As a consequence the MERIT and COTES Working Groups were replaced by a Provisional Directing Board for the new International Earth Rotation Service which was to come into operation on 1 January 1988. The IAU resolution was endorsed by the Executive Committee of the International Association of Geodesy in March, 1986 (Mueller and Wilkins, 1986). The recommendations of the Provisional Directing Board were considered and adopted by the IUGG during its XIXth General Assembly in Vancouver, B.C., in August, 1987 (Appendix 3).

With this last action, after ten years of preparation the new International Earth Rotation Service became a reality.

Organization of the Service. For each technique of observation (VLBI, SLR and LLR), prospective host organizations were invited to submit proposals for participation in one or more of the following ways:
- as a coordinating center,
- as an observing station or a network of stations,
- as a data collection (and distribution) center for quick-look and/or full-rate observational data. Such a center could, if appropriate, also process the data to form normal point data for use in analyses, or the task could be carried out by separate centers,
- as a quick-look operational center that would provide rapid service results,
- as a full-rate analysis center that would determine ERP's, station coordinates and other parameters to a regular schedule.

Several of these activities might be carried out by one center, and the actual organization would differ according to the number of observing stations and networks and to the nature of the processing required. There will be nod need for associate analysis centers in the formal structure, although it is expected that many groups will wish to analyze data provided by the Service. Offers of the deployment of mobile systems for use in improving the terrestrial reference system would be welcomed.

The principal tasks of the Central Bureau are specified in Recommendation B in Appendix 1, and some of them would be carried out by sub-bureaus. There is a need also for separate centers for relevant data from other fields, such as data on atmospheric angular momentum (AAM) and appropriate geodetic data (e.g., GPS results). The former might prove to be useful in predicting the variations in the rate of rotation of the earth, while the latter would be useful in the establishment and maintenance of the terrestrial reference system.

Kovalevsky and Mueller in their 1980 review of the Warsaw Conference listed a number of actions required to assure that the reference system issue be resolved "early and that the uniformity is assured by means of international agreements." There were the following:

Re CTS:
1. Selection of observatories whose catalogue will define the CTS.
2. Initiation of measurements at these observatories.
3. Recommendation on the observational and computational maintenance of the CTS (e.g., permanent versus temporary and repeated station occupations, constraints to be used).
4. Decision on how far and which way the earth deformation should be modeled initially.
5. Plans and recommendations for the establishment of new international service(s) to provide users with the appropriate information regarding the use of the CTS frame.

Re CIS:
6. Selection of extragalactic radio sources whose catalogue will define the CIS.
7. Improvement of the positions of these sources to a few milliseconds (arc).
8. Final decision on the IAU series of nutation and to assure that it describes the motion of the Celestial Ephemeris Pole.
9. Early completion of the FK5 and revision of astronomical equations due to the changed equinox (e.g., transformation between sidereal and Universal times).
10. Extension of the stellar catalogues (FK5 and later Hipparcos) to higher magnitudes.
11. Connection of the FK5, and later Hipparcos, reference frames to the CIS frame.

Eight years later it is gratifying to note that significant progress has been made on all items.

REFERENCES

Aoki, S., 1988, Relation between the celestial reference system and the terrestrial reference system of a rigid Earth, *Celes. Mechan.,* in press.

Aoki, S. and Kinoshita, H., 1983, *Celes. Mechan.,* **29**, 335.

Aoki, S., Guinot, B., Kaplan, G., Kinoshita, H., McCarthy, D. and Seidelmann, P., 1982, *Astron. Astrophys.,* **105**, 359.

Arias, E., Lestrade, F. and Feissel, M., 1988, in BIH Annual Report for 1987, Paris.

Bender, P. and Goad, C., 1979, in *The Use of Artificial Satellites for Geodesy and Geodynamics, Vol. II,* G. Veis and E. Livieratos (eds.), National Technical Univ., Athens.

Boucher, C. and Altamimi, Z. , 1987, Intercomparison of VLBI, LLR, SLR, and GPS Derived Baselines on a Global Basis, IGN No. 27.450, France.

Boucher, C., 1986, GRGS Tech. Rep. No. 3, IGN, France.

Bureau International de l'Heure, 1987, BIH Annual Rep. for 1986, Paris.
Capitaine, N. and Guinot, B., 1988, in Wilkins and Babcock (eds.), Reidel , 33.
Capitaine, N., 1986, *Astron. and Astrophys., 162*, 323.
Capitaine, N., Guinot, B. and Souchay, J., 1986, *Celes. Mechan.,* **39**, 283.
Capitaine, N., Williams, J. and Seidelmann, P., 1985, *Astron. Astrophys.,* **146**, 381.
Carter, W., Robertson, D. and Fallon, F., 1987, in BIH Annual Rep. for 1986, Paris, p. D-19.
Dickey, J., 1989, in Kovalevsky, Mueller and Koľaczek (eds.), Kluwer.
Arias, E., Lestrade, F. and Feissel, M. (eds.), Kluwer.
Eichhorn, H. and Leacock, R. (eds.), 1986, *Astrometric Techniques*, Reidel.
Eubanks, T., Steppe, J. and Spieth, M., 1985, in BIH Annual Rep. for 1984, Paris, p. D-19.
Fanselow, J. et al., 1984, *Astron. J.,* **89**, 987.
Fedorov, E., Smith, M., and Bender, P. (eds.), 1980, *Nutation and the Earth's Rotation,* IAU
    Symp. 78, Reidel.
Fricke, W. and Gliese, W., 1978, in Prochazka and Tucker, 421.
Gaposchkin, E. and Koľaczek, B. (eds.), 1981, *Reference Coordinate Systems for Earth
    Dynamics*, Reidel.
Guinot, B., 1979, in McCarthy and Pilkington (eds.), 7.
Guinot, B., 1986, in Eichhorn and Leacock (eds.), Reidel.
Hellings, R., 1986, *Astron. J.,* **91**, 650.
Jeffreys, W., 1980, *Celestial Mech.,* **22**, 175.
Kaplan, G. et al., 1982, *Astron. J.,* **87**, 570.
Koľaczek, B. and Weiffenbach, G. (eds.), 1974, *On Reference Coordinate Systems for Earth
    Dynamics,* IAU Colloq. 26, Smithsonian Astrophys. Obs., Cambridge, Mass.
Kovalevsky, J. and Mueller, I., 1981, in Gaposchkin and Koľaczek (eds.).
Kovalevsky, J., 1979, in McCarthy and Pilkington (eds.), 151.
Kovalevsky, J., 1980, *Celestial Mech.,* **22**, 153.
Kovalevsky, J., Mueller, I. and Koľaczek, B.(eds.), 1989, *Reference Frames*, Kluwer Publ.
Ma, C., 1988, in Wilkins and Babcock (eds.).
Ma, C., 1989, in Kovalevsky, Mueller and Koľaczek (eds.).
Ma, C., Clark, T., Ryan, J., Herring, T., Shapiro, I., Corey, B., Hinteregger, H., Rogers, A.,
    Whitney, A., Knight, C., Lundquist, G., Shaffer, D., Vandenburg, N., Pigg, J.,
    Schupler, B. and Ronnang, B., 1986, *Astron., J.* **92**, 1020.
Ma, C., Himwich, W., Mallama, A. and Kao, M., 1987, in BIH Annual Rep. for 1986, Paris,
    p. D-11.
Markowitz, W. and Guinot, B. (eds.), 1968, *Continental Drift, Secular Motion of the Pole, and
    Rotation of the Earth,* IAU Symp. 32, Reidel.
Mather, R. et al., 1977, Uniserv G 26, Univ. of New So. Wales, Australia.
McCarthy, D. and Pilkington, J. (eds.), 1979, *Time and the Earth's Rotation,* IAU Symp. 82,
    Reidel.
Melchior, P. and Yumi, S. (eds.), 1972, *Rotation of the Earth,* IAU Symp. 48, Reidel.
Minster, J. and Jordan, T., 1978, *J. Geophys. Res.,* **83**, 5331.
Morabito, D., Preston, R., Linfield, R., Slade, M., Jauncey, D., 1986, *Astron. J.,* **92**, 546.
Mueller, I., 1975a, *Geophys. Surveys, 2*, 243.
Mueller, I. (ed.), 1975b, *Dept. of Geod, Sci,. Rep.* **231**, Ohio State Univ., Columbus.
Mueller, I. (ed.), 1978, *Dept. of Geod, Sci,. Rep.* **280**, Ohio State Univ., Columbus.
Mueller, I. (ed.), 1985, *Proc. Int. Conf. on Earth Rotation and the Terrestrial Reference Frame,"*
    publ. Dept. of Geodetic Sci. and Surveying, Ohio State Univ.
Mueller, I. and Wilkins, G., 1986, *Adv. Space Res.,* **9**, 5.
Newhall, X, Williams, J. and Dickey, J., 1987, in BIH Ann. Rep. for 1986, Paris, p. D-29.
Perley, R., 1982, *Astron. J.,* **87**, 859.
Prochazka, F. and Tucker, R. (eds.), 1978, *Modern Astrometry,* IAU Colloq. 48, Univ. Obs.
    Vienna.
Reasenberg, R., 1986, in Eichhorn and Leacock (eds.), 789.
Reigber, C., Schwintzer, P., Mueller, H. and Massmann, F., 1987, in BIH Ann. Rep. for
    1986, Paris, p. D-39.

Robertson, D., Fallon, F., Carter, W., 1986, *Astron. J., 91*, 1456.

Schutz, B., Tapley, D. and Eanes, R., 1987, in BIH Ann. Rep. for 1986, Paris, p. D-33.

Sovers, O., Edwards, C., Jacobs, C., Lanyi, G., Liewer, K., Treuhaft, R., *Astron. J.*, in press.

Wade, C. and Johnston, K., 1977, *Astron. J., 82*, 791.

Wilkins, G. and Babcock, A., 1988, *The Earth's Rotation and Reference Frames for Geodesy and Geodynamics*, Reidel.

Wilkins, G. and Mueller, I., 1986, *EOS, Trans. Am. Geophys. Union*, **67**, 601.

Williams, J. and Melbourne, W., 1982, in *High-Precision Earth Rotation and Earth-Moon Dynamics*, O. Calame (ed.), Reidel Publ., Dordrecht, 293.

Williams, J. and Standish, E., 1989, in Kovalevsky, Mueller and Kołaczek (eds.).

Yumi S. (ed.), 1971, *Extra Collection of Papers Contributed to the IAU Symp. No. 48, Rotation of the Earth*, International Latitude Obs., Mizusawa, Japan.

Zhu, S.Y. and Mueller, I., 1983, *Bull. Geodes.*, **57**, 29.

APPENDIX 1

PRINCIPAL RECOMMENDATIONS OF THE MERIT AND COTES WORKING GROUPS

A. Technical Recommendation on Concepts

The IAU/IUGG MERIT and COTES Joint Working Groups recommend that the following concepts be incorporated in the operation of an international earth orientation service:

(1) The Conventional Terrestrial Reference System (CTRS) be defined by a set of designated reference stations, theories and constants chosen so that there is no net rotation or translation between the reference frame and the surface of the earth. The frame is to be realized by a set of positions and motions for the designated reference stations.

(2) The Conventional Celestial Reference System (CCRS) be defined by a set of designated extragalactic radio sources, theories and constants chosen so that there is no net rotation between the reference frame and the set of radio sources. The frame is to be defined by the positions and motions of the designated radio sources. The origin of the frame is to be the barycenter of the solar system.

(3) This international service should provide the information necessary to define the Conventional Terrestrial Reference System and the Conventional Celestial Reference System and relate them as well as their frames to each other and to other reference systems used in the determination of the earth rotation parameters. The information should include, but not be limited to, pole positions, universal time, precession, nutation, dynamical equinox, positions of the designated reference stations and radio sources, and crustal deformation parameters.

B. Recommendation for the Organization of a New International Earth Rotation Service

The IAU/IUGG MERIT and Cotes Joint Working Groups recommend that IAU and IUGG establish a new international service within FAGS for monitoring the rotation of the earth and for the maintenance of the Conventional Terrestrial Reference System to replace both the International Polar Motion Service (IPMS) and the Bureau International de l'Heure (BIH) as from 1 January 1988.

The new service will be known as the International Earth Rotation Service (IERS) and will consist of a Directing Board, a Central Bureau, coordinating centers and observatories. The Central Bureau, the centers and the observatories will be hosted by national organizations.

The Directing Board will exercise organizational, scientific and technical control over the activities and functions of the Service including such modifications to the organizational structure and participation in the Service as are appropriate to maintain an efficient and reliable service while taking full advantage of advances in technology and theory. The voting membership of the Directing Board will consist of one representative each of the IAU, the IUGG, the Central Bureau, and each of the coordinating centers. Additional nonvoting members may be appointed to advise the Board on complex technical and scientific issues.

The Central Bureau will combine the various types of data collected by the Service to derive and disseminate to the user community the earth rotation parameters in appropriate forms, such as predictions, quick-look and refined solutions, and other information relating to the rotation of the earth and the associated reference systems. The Central Bureau will conduct research and analysis to develop improved methods of processing and interpreting the data submitted. The Central Bureau may include sub-bureaus that carry out some of the specific tasks of the Central Bureau.

Coordinating centers will be designated for each of the primary techniques of observation to be utilized by the Service as well as for other major activities which the Directing Board may deem appropriate. Initially, there will be three centers for (1) very long baseline interferometry (VLBI), (2) satellite laser ranging (SLR), and (3) lunar laser ranging (LLR). Additional coordinating centers may be designated for the improvement of the determination of the earth rotation parameters and the maintenance of the conventional reference system by other techniques and to ensure that relevant data on the atmosphere, oceans and seismic events are available.

The coordinating centers will be on the same level as the Central Bureau in the organizational structure o the Service and will be responsible for developing and organizing the activities by each technique to meet the objectives of the Service. Associated with the coordinating centers there may be network centers for subsets of observatories that may, for reasons of geometry or system compatibility, work more efficiently as autonomous units. There may also be associated analysis centers to process the observational data regularly or for special applications and studies. These centers may submit their results directly to the Central Bureau.

National Committees for the International Unions for Astronomy and for Geodesy and Geophysics will be invited to propose before 1 January 1987 national organizations and observatories that will be willing to host the Central Bureau or one of the centers and/or to provide observational data for use by the Service.

It is essential that the new service have redundancy throughout the organizational structure to insure the uninterrupted timely production of consistent, accurate, properly documented earth orientation and reference frame parameters, even in the event that one of the host national organizations should terminate its participation. A widespread distribution of observatories that regularly make high precision observations by one, or preferably more, modern space techniques by fixed and/or mobile equipment will be needed for this purpose, and national organizations are urged to provide appropriate resources.

APPENDIX 2

RESOLUTION OF INTERNATIONAL ASTRONOMICAL UNION (1985)

The following resolution was adopted by the XIXth General Assembly of the International Astronomical Union at New Delhi on 28 November 1985.

The International Astronomical Union

recognizing the highly significant improvement in the determination of the orientation of the earth in space as a consequence of the MERIT/COTES program of observation and analysis, and

recognizing the importance for scientific research and operational purposes of regular earth orientation monitoring and of the establishment and maintenance of a new Conventional Terrestrial Reference Frame,

thanks all the organizations and individuals who have contributed to the development and implementation of the MERIT and COTES programs and to the operations of the International Polar Motion Service and the Bureau International de l'Heure,

endorses the final report and recommendations of the MERIT and COTES Joint Working Groups;

decides

(1) to establish in consultation with IUGG a new International Earth Rotation Service within the Federation of Astronomical and Geophysical Services (FAGS) for monitoring earth orientation and for the maintenance of the Conventional Terrestrial Reference Frame; the new Service is to replace both the IPMS and the BIH as from 1 January 1988,

(2) to extend the MERIT/COTES program of observation, analysis, intercomparison and distribution of results until the new service is in operation,

(3) to recommend that an optical astrometric network be maintained for the rapid determination of UT1 for so long as this is recognized to be useful,

(4) to set up a Provisional Directing Board to submit recommendations on the terms of reference, structure and composition of the new service, and to serve as the Steering Committee for the extended MERIT/COTES program,

invites National Committees for the International Unions for Astronomy and for Geodesy and Geophysics to submit proposals for the hosting of individual components of the new service by national organizations and observatories, and

urges the participants in Project MERIT to continue to determine high precision data on earth rotation and reference systems and to make the results available to the BIH until the new service is in operation.

APPENDIX 3

RESOLUTION 1 OF THE INTERNATIONAL UNION OF GEODESY AND GEOPHYSICS, XIX GENERAL ASSEMBLY VANCOUVER, 21 AUGUST 1987

The International Union of Geodesy and Geophysics

Noting that the improved determination of the Earth's orientation parameters resulting from the MERIT and COTES programs of observation and analysis is highly significant,

considering the importance for scientific research and operational purposes of regularly monitoring the Earth's orientation and of establishing and maintaining a new conventional terrestrial frame of reference,

approving the replacement of the International Polar Motion Service (IPMS) and of the Bureau International de l'Heure (BIH) by the International Earth Rotation Service (IERS) which will be responsible both for earth rotation and for the associated conventional frames of reference, and

recognizing that organizations in many countries have indicated their willingness to participate in such a new service,

endorses the recommendations of its Provisional Directing Board on the terms of reference, structure and composition of the new service,

decides to establish, in cooperation with the International Astronomical Union, the International Earth Rotation Service within the Federation of Astronomical and Geophysical Data Analysis Services (FAGS) a from 1 January 1988,

thanks all organizations and individuals who have helped to develop and implement the MERIT and COTES programmes, all who have operated IPMS and BIH in the past and all who have indicated their willingness to participate in the new Service.

Chapter 8

EDUCATION

A. Ashour, G. Birardi, M. Caputo, D. Christodoulidis, C. Harrison,
J. Rais, B. Schutz, S. Tatevian, G. Veis and S. Zerbini

1. INTRODUCTION

Geodesy is concerned with the study of the size, the shape and the gravity field of the Earth and other celestial bodies, together with their temporal variations through the definition, realization, maintenance and interrelation of the appropriate reference systems. Today this is accomplished to a large extent through the use of advanced space techniques which are foreseen to be the predominant methods in the near future. This new approach is referred to as "space geodesy."

The Williamstown Report (NASA, 1970) expressed an early vision of the interdisciplinary role of modern space geodesy. Although the report did not specifically address the educational aspects, there was an implication throughout the report of such a need. The need for specialists in geodesy to be cognizant of the basic principles and problems associated with the disciplines making use of geodetic results and, conversely, the need for those disciplines to be familiar with geodetic science is an implicit theme of the Williamstown Report.

In the two decades since the Williamstown Report was written, educational institutions have introduced either specialized courses in space geodesy or added selected topics to existing courses. According to the U.S. National Academy of Sciences (1978) and Brandenberger (1976), there are about 250 departments at higher institutions throughout the world specializing in surveying and mapping. There are, however, no dedicated departments of space geodesy. Instead, space geodesy is incorporated into the program of geodetic science, engineering or other academic disciplines.

2. INTERACTION WITH OTHER DISCIPLINES

The capability of present-day geodesy in the realization and maintenance of four-dimensional reference frames has reached accuracy levels beyond our expectations in the early 1960's. Through the use of space techniques, geodesy is currently producing positional accuracies, in a consistent global reference frame, of better than two centimeters. The inertial reference systems offered by the orbits of near-Earth satellites provide the means for the analysis of altimeter data to levels of unprecedented resolution. Determination of the coefficients in the gravitational potential has reached a level of accuracy that allows for the resolution of its temporal variations across a wide band of frequencies.

It is evident that at these accuracy levels of reference frame definition and realization, the interaction between the various physical processes is not only very rich, but also very complicated. It appears to the analysts that the number of interactions between these processes increases exponentially with the required accuracy of reference frame definition. At low accuracies these processes can be considered uncoupled to a large extent, whereas at the millimeter level, coupling becomes the predominant limiting factor in the analysis process.

Geodesy's capability to provide a wealth of information over a wide range of Earth and planetary science results in a natural interaction with these disciplines. It has recently become evident that these new types of information could further advance our knowledge of the Earth through substantial interaction among the various disciplines. This is by no means a one-way interaction.

Many disciplines are expected to gain from geodesy and vice versa. Among them are included solid earth geophysics, geodynamics, astronomy, atmospheric sciences and meteorology, oceanography, planetology, and a number of engineering disciplines such as surveying mapping, remote sensing, navigation and aeronautics. In fact, as the geodesist's requirements become more demanding in terms of accuracy and in terms of spatial and temporal resolution, the more physical processes that are unveiled thus require a closer cooperation with other disciplines. On the other hand, the more the geodesist provides the other sciences with measures of relevant processes, the more these disciplines advance the state of the art within their own domain of research. So, it is apparent that only through cooperative efforts can the ever more demanding requirements be met.

In order to meet the requirements of the expected higher demands in the field of space geodesy for the 1990's, it is apparent that an appropriate effort should be placed on the proper education of those involved both in the research development as well as in the implementation through practice of space geodesy.

The fact that interaction among astronomy, earth sciences, engineering and space geodesy is vital for research indicates that a similar interchange must take place in the educational process. This means that those studying space geodesy should have an appropriate knowledge in other other disciplines, and those studying astronomy and earth sciences should be knowledgeable about the new space geodesy methods, since these methods will be important for their future research. It is also quite obvious that the studying "classical geodesy" or geodetic surveying should take courses in space geodesy, since this is the direction in which geodesy is moving.

Accordingly, it is believed that the following will improve the advancement of the field of space geodesy through education:

1. Include dedicated courses of space geodesy in the existing programs of geodesy and geodetic surveying at first university degree level, covering the following topics: reference systems and time, instruments for observations, orbital mechanics, positioning, gravity field determination and combination solutions.

2. For an advanced university degree level in space geodesy, in addition to a strong background in mathematical and physical sciences, high-level courses in space geodesy as well as selected topics from astronomy, earth sciences and engineering should be included. Because of the rapid development in these fields, it is essential that the courses must include a strong element of principles, and strengthen the basic knowledge of the appropriate mathematical and physical sciences. The following subjects for an advanced degree should be considered:

Space Geodesy
- Reference systems, orbital mechanics, space geodesy instrumentation (including lab)
- Positioning and gravitational fields determination

Science
- Advanced adjustments and statistics
- Solid Earth physics (tectonic processes, Earth tides)
- Physical oceanography (temporal variations, tides)
- Astronomy (Earth orientation, radioastronomy, planets)
- Environmental physics (atmospheric effect on wave propagation)

Engineering
- Computers (data base management, advanced computer architecture)
- Electronics (digital systems)
- Navigation
- Applications to photogrammetry, mapping and remote sensing

It is considered important that geodesy and especially space geodesy as disciplines be better known to the general public and specifically to more elementary and high school students. This will facilitate the process of attracting more and better students to study space geodesy at the university level.   This can be accomplished by introducing notions of geodesy in the appropriate science courses and through the distribution of brochures, video tapes, specially prepared for this purpose, emphasizing the scientific problems to be solved.

Space geodesy is both interdisciplinary and international because of the global scale processes being investigated.   Therefore, the educational process should also have an international character.   This has the additional advantage of sharing the cost of establishing several complete university programs.

3. SUMMARY AND RECOMMENDATIONS

The space geodesy requirements of the 1990's will influence the educational process in a variety of ways. For example, the achievement of 1 mm laser range measurements will result in greater interdisciplinary understanding of the factors contributing to the measurement at that level. The space geodesists of the 1990's should be familiar with the disciplines that contribute to both the signal and the noise of those measurements and should be able to utilize the most modern techniques to support applications of space geodesy in a number of scientific disciplines. This familiarity should be obtained through structured university-level degree programs, especially at the graduate level, that address the fundamental interdisciplinary nature of modern space geodesy. The formal educational process should be augmented by short dedicated courses in space geodesy, of duration from a few days to a few weeks to address the interdisciplinary aspects. Furthermore, short courses in space geodesy should be prepared for the purpose of familiarizing individuals who are not specialists in this field with the techniques, methods and capabilities offered by space geodesy. General brochures and video cassettes which describe the modern techniques and applications for the nonspecialists should be developed.

Based on the foregoing considerations, space geodesy education should be enhanced and upgraded through:

- Introduction of interdisciplinary courses in space geodesy curricula.

- International summer schools and workshops on specific topics in the field of space geodesy.

- The exchange of professors, students, young scientists and engineers among universities and research centers from different countries through existing or newly developed exchange programs.

- The development of graduate exchange programs between universities, or "sandwich" programs.

- The initiation of joint research projects between universities for mutual benefit.

- The development of space geodesy sample data bases for educational purposes.

- The development of educational material in space geodesy, such as books and video cassettes, to educate the general public and to excite the interest of young people to the subject of space geodesy and to promote its introduction in high school curricula.

- With the cooperation of interested scientific institutions, a symposium or workshop on "Teaching of Interdisciplinary Fields" should be organized.

Problems facing teaching of geodesy in general and space geodesy in particular at universities in most developing countries are more severe, and they include: lack of textbooks, high cost of scientific journal subscriptions, lack of teaching aids and references, high cost of equipment or maintenance problems of purchased equipment, and minimal number of qualified teaching staff. In order to promote geodesy education in the developing countries, the following could be of help:

- Initiate regional centers for Earth sciences, including geodesy, on the line which UNESCO and ICSU have followed in establishing the network of biological sciences. Such centers would serve, among others, as data and documentation banks for the region.

- Support for attendance of students from developing countries at the International School/Summer School/Short Courses conducted in the developed countries or Centers.

- Technical assistance from the developed countries to strengthen the educational and research institutions of the developing countries.

Additional information about space geodesy, including international commissions devoted to specific aspects, can be obtained from the International Association of Geodesy. Contact:

International Association of Geodesy
140, rue de Grenelle
75700 Paris, France
Telex: 204989 F

Space geodesy research is reported in several journals, for example:

Bulletin Géodésique
manuscripta geodaetica
Journal of Geophysical Research
Navigation

Not all of these journals are specifically devoted to geodesy, a further indication of the interdisciplinary nature of space geodesy.

REFERENCES

National Academy of Sciences, 1978, Geodesy: Trends and Prospects, Committee on Geodesy, National Research Council.

National Aeronautics and Space Administration, 1970, The Terrestrial Environmental: Solid-Earth and Ocean Physics, NASA CR-1579, Washington, D.C., April.

Brandenberger, A.J., 1976, *Study on the Status of World Cartography*, United Nations Economic and Social Council, First United Nations Regional Cartographic Conference for the Americas, Panama.

APPENDIX 1

THE EARTH OBSERVATION ACTIVITIES OF THE EUROPEAN SPACE AGENCY

B. R. K. Pfeiffer

1. INTRODUCTION

The Earth Observation activities of the European Space Agency (ESA) were initially focused on a preoperational geostastionary meteorological demonstration Programme (Meteosat). This programme provided the European meteorological community with visible infrared and water vapour imagery from a geostationary orbit.

The first and second flight models as well as the prototype (P2) were launched in 1977, 1981 and June 1988 respectively.

The data products included in particular cloud maps, cloud heights, winds and sea surface temperatures.

The processed data were disseminated by two channels to the users through the broadcast capability of the satellite.

Following the success of the preoperational programme, the European Meteorological Satellite Organisation (EUMETSAT) was established to provide continuation of the meteorological data products. ESA procures and will operate the additional Meteosat spacecraft on behalf of EUMETSAT. The launch of the first operational meteorological spacecraft (MOP–1) is foreseen in February 1989, to be followed by the launch of the second spacecraft (MOP–2) scheduled approximately one year later.

In order to broaden its experience in other instrumentation and to provide the Earth observation users with microwave remote sensing data, ESA flew in 1983 on board of the first Spacelab mission, the Microwave Remote Sensing Experiment (MRSE) which was built and provided by the Federal Republic of Germany. This experiment was operated only in the passive mode but allowed ESA to gain valuable experience for future microwave missions.

Another activity of ESA was initiated, to give European Earth Observation users data from numerous Earth observation satellites. This Earthnet programme provides as a service the acquisition, preprocessing, dissemination, archiving and cataloguing of satellite data. The satellite missions covered include: Landsat, Seasat, Nimbus–7, NOAA (TIROS–N), MOS–1, SPOT and ERS–1.

The operation of three stations in Fucino (Italy), Kiruna (Sweden) and Maspalomas (Spain) is part of the Earthnet programme. For specific missions, additional stations are involved such as Tromsoe (Norway) in the MOS–1 mission and Gatineau (Canada) for the ERS–1 mission.

2. PRESENT PROGRAMME

The present activities of ESA approved by its member countries are:

-   the European Remote Sensing Satellite Programme (ERS-1)
-   the continuation of the Earthnet Programme
-   the Earth Observation Preparatory Programme (EOPP)
    preparing the future missions
-   the Meteosat Operational Programme in which ESA acts as
    a procurement and operational agent for EUMETSAT.

2.1 THE ERS-1 PROGRAMME

This programme consists of the development, launch and operations of the first European Remote Sensing satellite. Its launch is foreseen for September 1990. The preoperational science and applications mission aims to enhance the scientific knowledge of ocean, ice and coastal zones and to contribute to climate research in general. The programme is aimed to promote research and operational applications of microwave remote sensing data taken over the Earth's environment monitoring and the "global change" research programme.

The programme provides an "end to end" system with a number of regular data products, which are described in Table 1.

The instrument playload consists of the following core instruments:

-   an Active Microwave Instrumentation (AMI) which combines a C–Band Synthetic Aperture Radar and a C-Band Wind Scatterometer
-   a K–band Radar Altimeter (RA)
-   a Laser Retro–reflector (LRR).

This instrument package is complemented by nationally provided Announcement of Opportunity (AO) instruments. These are:

-   an Along Track Scanning Radiometer (ATSR) [Microwave and IR parts] provided by the United Kingdom and France
-   a Precise Range and Range Rate Experiment (PRARE) provided by the Federal Republic of Germany.

The spacecraft is to be launched by ARIANNE–4 from French Guyana into a near polar, sun–synchronous orbit of about 780 km altitude. The nominal lifetime is approximately 3 years.

The ERS-1 overall Ground Segment and the potential Ground Stations concerned are given in Figs. 1 and 2 respectively.

An Announcement of Opportunity for science and application experimentation proposals (including calibration and validation of geophysical parameters) was issued in May 1986.

Approximately 300 proposals from 23 countries and 4 international organisations, combining about 1000 investigators and 500 laboratories/institutes, were received. It is expected that 200 proposals have been selected for further implementation. A very large number of Earth Observation disciplines will benefit from the ERS–1 mission.

Table 1 Geophysical Measurements and ERS-1 Performance Parameters

Main Geophysical Parameter	Range	Accuracy	Main Instrument
Wind Field			
- Velocity	4–24 m/s	±2 m/s or 10% whichever is greater	Wind Scatterometer & Altimeter
- Direction	0°–360°	± 20°	Wind Scatterometer
Wave Field			
- Significant Wave-Height	1–20 m	± 0.5 m or 10% whichever is greater	Altimeter
- Wave Direction	0° – 360°	± 15°	Wave Mode
- Wavelength	50–1000 m	20%	Wave Mode
Earth Surface Imaging -Land/Ice/ Coastal Zones etc.	80 km (minimum swathwidth)	Geom./Radiom. Resolutions: a) 30 m/2.5 dB b)100 m/1 dB	SAR imaging Mode
Altitude -Over ocean	745–825 km	2 m absolute ± 10 cm relative	Altimeter
Satellite Range		± 10 cm	PRARE
Sea surface temp.	500 km swath	± 0.5 K	ATSR (IR)
Water Vapour	in 25 km spot	10%	μWSounder

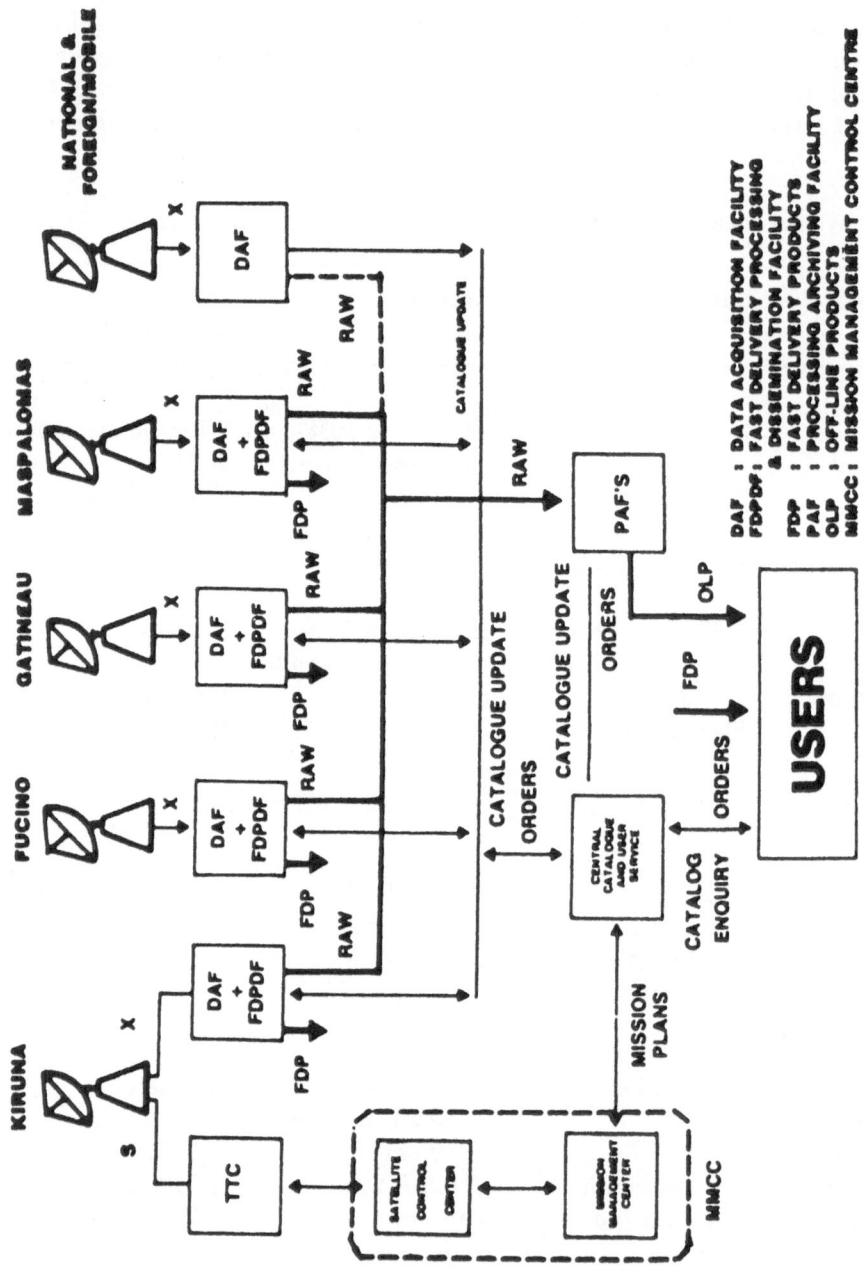

Fig. 1   ERS-1 overall ground segment and data flow

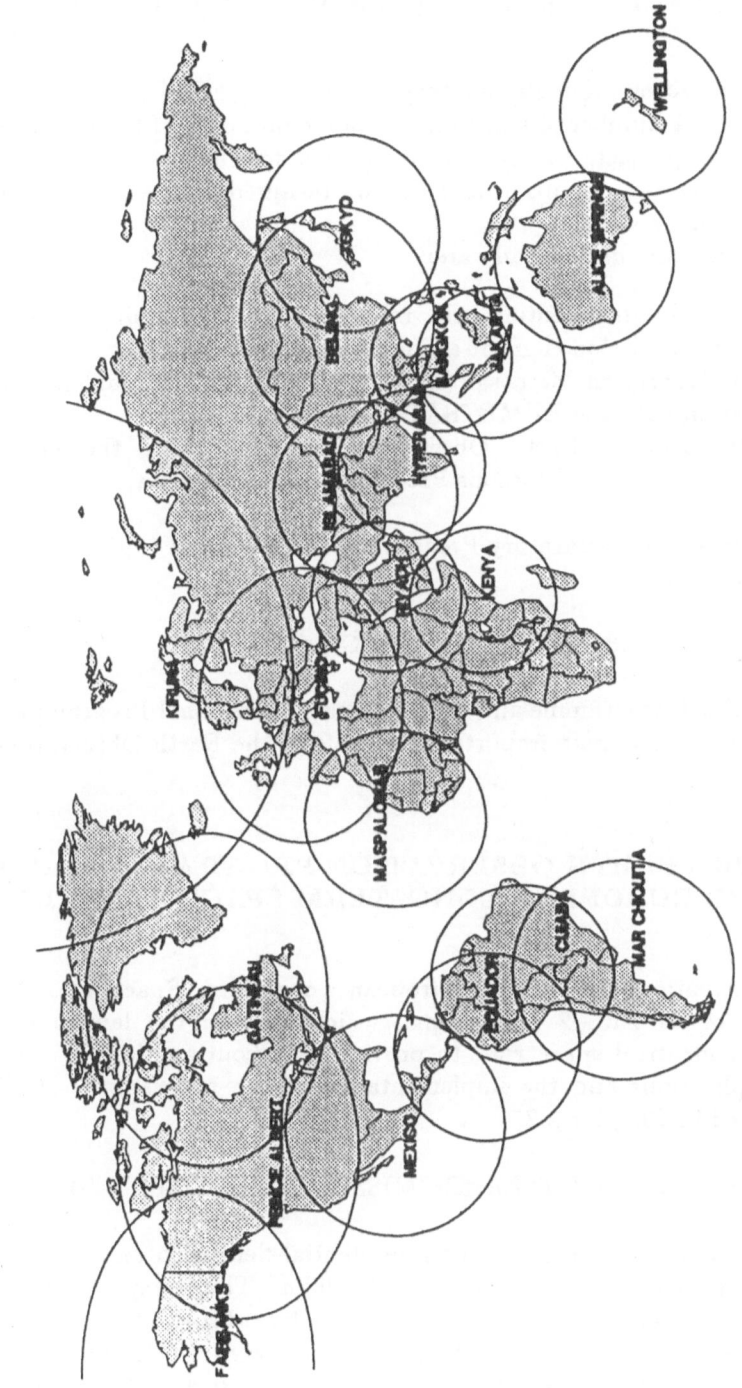

Fig. 2 ERS-1 potential stations coverage

2.2 THE EARTH OBSERVATION PREPARATORY PROGRAMME

In 1986 an Earth Observation Preparatory Programme (EOPP)was agreed with ESA member countries. A number of new programmes identified by ESA as potential future Earth Observation projects are studied to system feasibility (phase A) level. The preparation includes the definition and limited technology research for future Earth observation instrumentation.

The programmes under study are:

- a Solid Earth geopotential field mapping and positioning mission (with a magnetic field mission element as an option), ARISTOTELES
- a Second Generation Meteosat (MSG) programme to be implemented later on collaboration with the EUMETSAT organisation
- two polar orbiting Earth Observation missions using the polar platform developed within the Columbus Programme of the Agency.

The Earth Observation Preparatory Programme will last until 1991 with the intention of a further extension.

2.3 OTHER ONGOING ACTIVITIES

The Earthnet programme and the Meteosat Operational Programme (MOP) as described in Chapter 1 remain important elements of the Earth Observation activities of the Agency.

3. FUTURE EARTH OBSERVATION PROGRAMMES AS PART OF THE EUROPEAN LONG–TERM SPACE PLAN (LTP)

The Earth Observation part of the European Long–Term Space Plan foresees the implementation of an ERS–2 programme, which foresees the launch in 1994 and operations of an identical spacecraft, to provide data continuation for ocean and ice science and applications and the implementation of the programmes studied in the EOPP as outlined in Chapter 2.2.

3.1 THE SOLID EARTH MISSION (ARISTOTELES)

This programme aims at measuring the geopotential field with an accuracy of better than 5 mgal as a goal and a resolution of about 100 km. This is expected to be achieved by a gradiometer onboard of the satellite. Its near polar orbit will be at 200 km altitude.

The optional Earth magnetic field measurement accuracy is planned to be better than 3nT. After about 6 months in the low altitude dawn-dusk orbit the satellite's orbit will be lifted to approximately 700 km for a period of about 3 years for the positioning mission of this spacecraft.

3.2 THE SECOND GENERATION METEOSAT MISSION (MSG)

This programme is foreseen to be implemented as a joint venture between ESA and EUMETSAT. The mission objectives need still to be agreed in all details, but it is already obvious that the second Generation Meteosat instrumentation will provide data continuity to the present data products. However, higher spatial and spectral resolution will be required for the visible and infrared imaging.

The mission will include some instrumentation allowing sounding of temperature and humidity profiles of the atmosphere through IR and/or microwave measurements.

The mission will continue to provide an improved data circulation capability and include also a scientific instrument package of limited mass.

The launch of the first MSG spacecraft is foreseen by 1988.

3.3 POLAR ORBITING EARTH OBSERVATION PROGRAMME

The most significant future Earth Observation undertakings of the European Space Agency will be the polar platform developed by the Columbus Space Station programme.

The European Space Agency will contribute through this polar orbiting programme to the worldwide global Earth resources and environmental monitoring. Several polar platforms with a 4–5 year in–orbit lifetime are foreseen to provide data well into the next century.

The first and second such platforms are assumed to be launched by 1997 and 2000 respectively.

The instruments to be flown on this platform fall into four categories:

- Operational meteorological instuments provided by NOAA and EUMETSAT.
- Core/research facility instruments such as provided by ESA
- Announcement of Opportunity Earth Observation instruments provided nationally or by commercial entities
- Announcement of Opportunity Space Science instruments provided nationally.

Tables 2 and 3 give candidate operational and core research instruments under consideration for the European polar platform.

Table 4 shows the Earth Observation fields addressed by the candidate European instruments.

A tentative schedule for polar orbiting platforms operation is given in Fig. 3.

The planning of the future ESA mission is given in Fig. 4.

The polar orbiting Earth Observation missions have to be seen in the worldwide Earth Observation scenario. The most prominent worldwide missions are shown in Fig. 5.

Table 2   Candidate Operational Instruments for the Polar Orbiting Platforms

AMRIR	Advance Medium Resolution Imaging Radiometer
AMSU	Advanced Microwave Sounding Unit (A and B)
ARGOS	Data collection and location system direct broadcast
ERBI	Earth Radiation Budget Instrument (non-scanning)
S & R	Search and Rescue Package
SEM	Space Environment Monitor

Possible Providers - EUMETSAT and NOAA
Preferably on both morning and afternoon platforms
Mass is of the order of 500 kg

Table 3   Core/Research Facility Instruments (European Instruments)

ALT–2	Altimeter derived from the ERS-1 concept
ATLID	Backscatter laser (probably Nd-Yag)
ALADIN	Laser–based (probably CO2) Doppler wind lidar
HRIS	High Resolution Imaging Spectrometer
HRTIR	High Resolution Thermal Infra-red Radiometer
LISA	Interferometric limb sounder
MERIS	MEdium Resolution Imaging Spectrometer
MIMR	Multiband Imaging Microwave Radiometer
SAR-C	Synthetic Aperture Radar
SCATT-2	Wind Scatterometer
VHROI	Very High Resolution Optical Imager

core/research instruments - European interest
European interest reflected in eleven candidates under study by ESA (within EOPP)

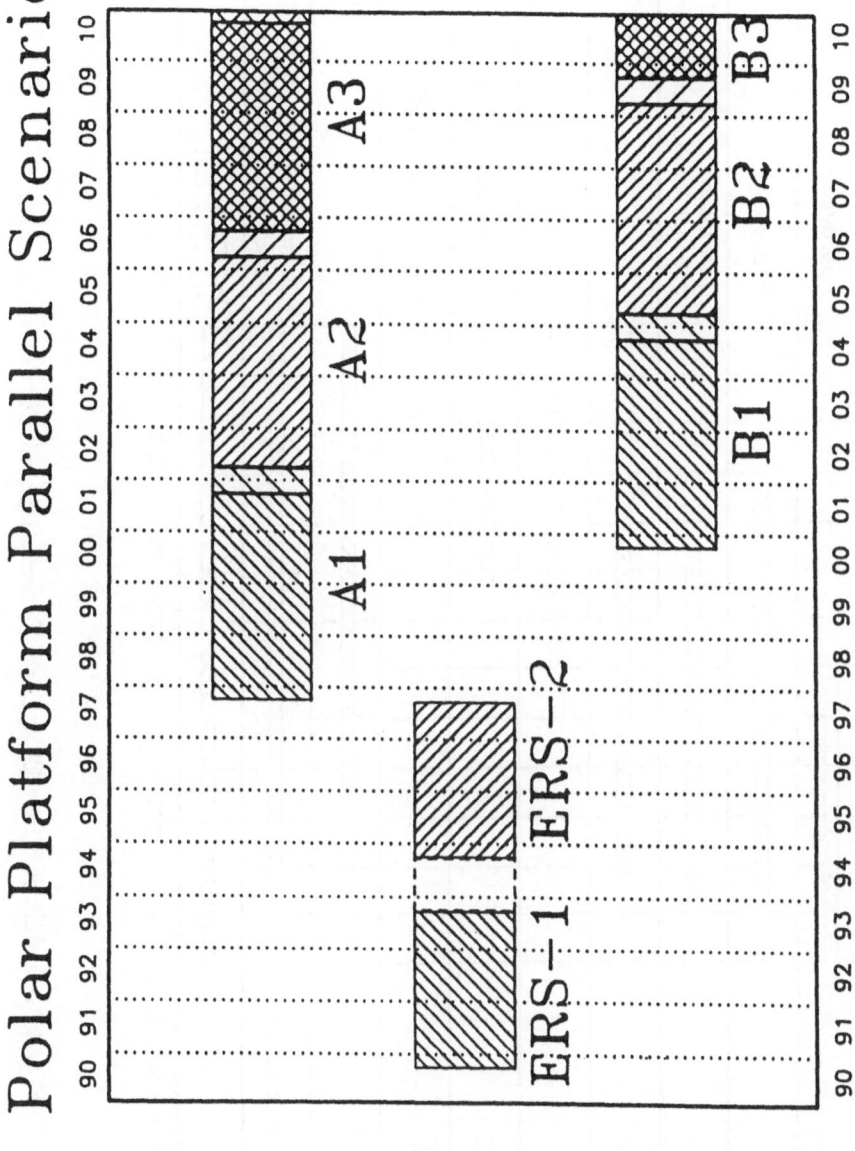

Fig. 3 Tentative polar platform operation schedule

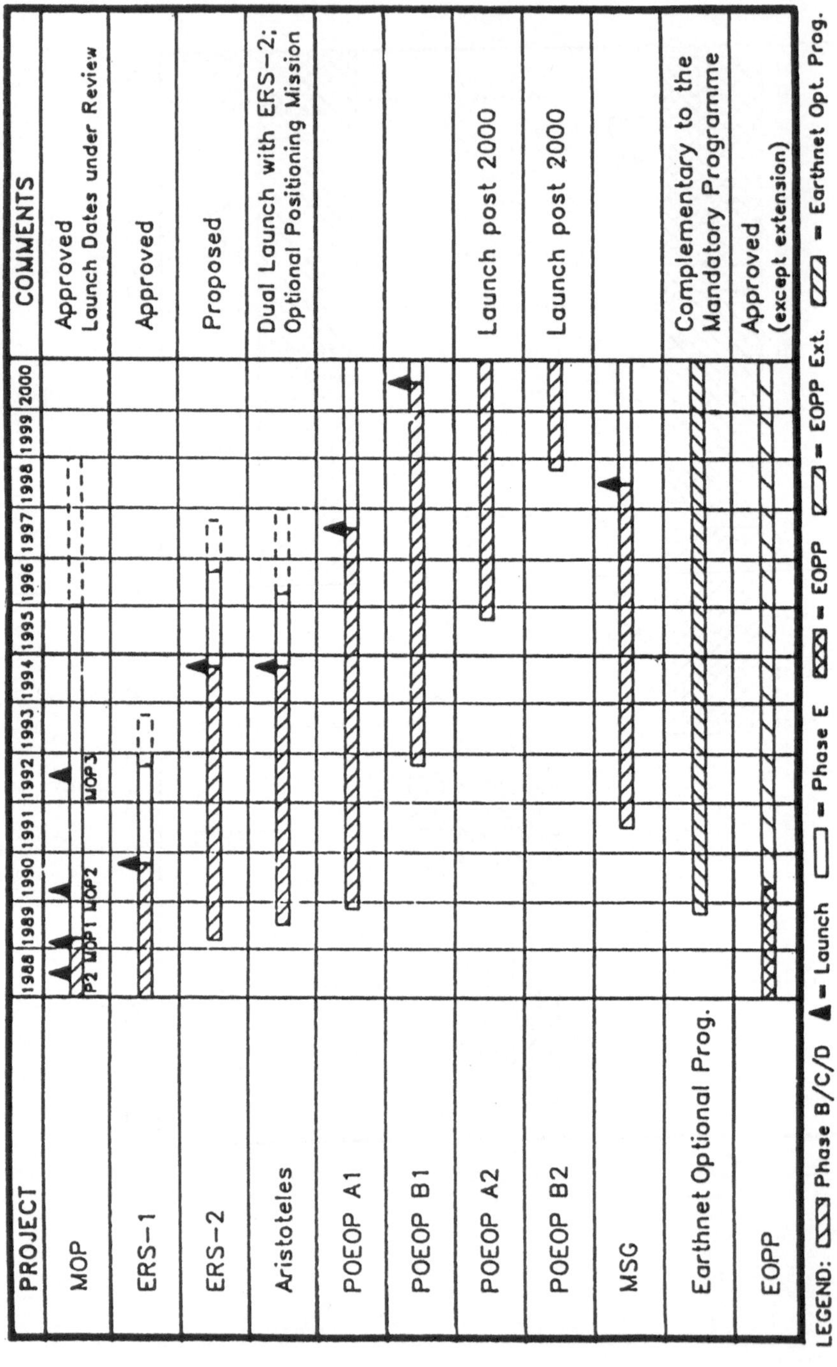

Fig. 4   Schedule of future ESA Earth Observations missions

AGENCY/ COUNTRY	MISSION	1987	1988	1989	1990	1991	1992	1993	1994	1995	1996	1997	1998	1999	2000
EUROPE (ESA AND EUMETSAT)	METEO (EUMETSAT)		X P2	* MOP-1	* MOP-2		* MOP-3 - - - -	*(tbd)					o MSG		
	ATMOS (ESA)														
	ENVIRONMENT MONITORING (ESA)				* ERS-1				o ERS-2			o EPOP-1			o EPOP-2
	RESOURCES MONITORING (ESA)								o ARISTOTELES	- - - o					
U.S.A. (NASA, NOAA, EOSAT)	METEO	X GOES-7 (a.m.) (p.m.)	NOAA-11 X	* NOAA-D	* GOES-I	* GOES-J NOAA-I *	* GOES-K * NOAA-J	NOAA-K *	* GOES-L * NOAA-L		* GOES-M NOAA-M *				
	ATMOS					* UARS			o TRMM						
	ENVIRONMENT MONITORING				* ATLAS	* TOPEX	* ATLAS	* ATLAS	o N-SCAT	o N-POP-1		o N-POP-2			
	RESOURCES MONITORING					* SIR-C o LANDSAT-6	* SIR-C		o LANDSAT-7 * SIR-C						
FRANCE	ENVIRONMENT MONITORING					* POSEIDON									
	RESOURCES MONITORING			* SPOT-2			* SPOT-3			o SPOT-4				o SPOT-5	
JAPAN	METEO			* GMS-4					* GMS-5						
	ATMOS								o TRMM						
	ENVIRONMENT MONITORING	X MOS-1			* MOS-1b				o ADEOS				o J-POP-1		
	RESOURCES MONITORING						* J-ERS-1								
INDIA			X INSAT-1C X IRS-1A		o INSAT-2A * IRS-1B		o INSAT-2B o IRS-1C								
CANADA									o RADARSAT						

X - Operational * - Under Development o - Planned

Fig. 5 World Earth Observations space programmes

Table 4   Fields Addressed by the European Candidate Instruments

	Atm	Climate	Land	Ocean	Ice	Sol.Earth
ALT–2	(x)		x	x	x	x
ATLID	x	x				
ALADIN	x		x		x	
HRIS	(x)	(x)	x	(x)		
HRTIR	(x)	x	x	(x)	x	
LISA	x	x				
MERIS	x	(x)	x	x		
MIMR	x	x	x	x	x	
SAR-C			x	x	x	
SCATT–2	x	x		x		
VHROI				x	(x)	

x denotes a field of primary application,
(x) indicates one of secondary application

4. CONCLUSIONS

It is the aim of the European Space Agency to provide, within existing overall budget
constraints, data and mission continuity to the many disciplines of Earth Observation
and to maintain also the necessary flexibility to cope with future Earth Observation
data needs. It is foreseen to achieve these objectives through a large contribution to
worldwide international activities and the collaboration of other international partners.

APPENDIX 2

THE ROLE OF NASA IN GEODYNAMICS RESEARCH IN THE DECADE 1991 – 2000

Edward A. Flinn

1. INTRODUCTION

NASA is now in the early stages of development of a long–term plan for its research activities in geophysics and geodynamics for the years beyond 1992. The purpose of this note is to discuss possible scenarios for NASA involvement in the solid Earth sciences during this period, as a basis for the planning that will take place in the next eighteen months.

In 1969 NASA held a conference in Williamstown, Massachusetts to determine what contributions space technology could make to geophysics and geodesy. This conference, attended by many domestic and foreign research workers, produced a report which defined a broad program of precise positioning studies of plate movement and deformation using Satellite Laser Ranging (SLR) and Very Long Baseline Interferometry (VLBI), as well as Lunar Laser Ranging (LLR), for measuring Earth orientation; ocean altimetry for physical oceanography; and for studies of the rheology of the oceanic lithosphere. The main recommendations concerned the use of space techniques to measure plate movements and deformation, satellite geodesy to measure the Earth's gravity field, and space flight missions to measure in–situ gravity and magnetic fields.

As a result of the Williamstown report, NASA established in 1972 an Earth and Ocean Dynamics Applications Program, which built several satellite laser ranging facilities (SLR) and operated them in California to measure displacements across the San Andreas Fault System, performing laser ranging to several satellites equipped with cube corner retroreflectors. This program successfully launched the GEOS–3 altimetry satellite, Seasat, the Laser Geodynamics Satellite (LAGEOS I), and the Magnetic Field Satellite (Magsat–A).

In 1978 physical oceanography was separated from geodesy and geodynamics, and solid Earth geophysics was made into a new program, the Geodynamics Program. The objectives and content of this program were decided upon after consultation with other Federal agencies – the National Science Foundation (NSF), the Defense Mapping Agency, the NOAA National Geodetic Survey (NGS), and the U.S. Geological Survey (USGS) – as well as with individuals and organizations in many other countries. A Geodynamics Program Plan was published (NASA, 1979), comprising three major elements: first, geopotential field studies; second, studies of the dynamics of the Earth's deep interior; and third, studies of movement and deformation of the lithosphere. The last was implemented through the Crustal Dynamics Project (CDP) at Goddard Space Flight Center (GSFC), which began in 1979. Lunar laser ranging (LLR), which was made part of the CDP, was concentrated at the University of Texas at Austin and the Jet Propulsion Laboratory (JPL). In 1982 development of the capability to make use of radio signals from the Defense Department's Global Positioning System Satellites (GPS) was assigned to JPL, with an emphasis on technology development and demonstration campaigns in Southern California and the Caribbean – Central American area. The CDP, which has the same status as space flight missions such as Voyager, has made research grants to nearly one hundred scientists in NASA Centers, universities, and other organizations in the U.S., and has cooperative arrangements with over twenty foreign countries. While building and deploying SLR, LLR, and VLBI facilities worldwide, the CDP has also developed technology to improve accuracy and reliability of the field measurements, and established a Data Information System to make the data easily accessible to scientific investigators.

NASA Projects have definite objectives, fixed budgets, and finite lifetimes; the CDP is scheduled to wind up at the end of 1991. NASA has chosen to handle the evident necessity to continue and improve on the work of the CDP by establishing a new "level–of–effort" scientific research and technology (SR&T) program whose organization and objectives will be very similar to those of the CDP. Thus the formal termination of CDP in late 1991 does not imply that NASA is stopping its activities in space geophysics and geodesy: quite the opposite is true. NASA looks forward to a long and fruitful cooperation with other Federal agencies and organizations in other countries as a participant in a global program of research in geodynamics, geophysics, and geodesy, for many years to come.

Thus 1991 marks a significant opportunity to re-think and reconsider the role NASA should play in international space geodynamics through the rest of this century. A detailed program plan is now being formulated, similar to the one produced in 1978 but taking into account the fact that the space geodynamics has changed in the past ten years, and that many other countries and other organizations now have or soon will have capabilities equivalent to those of NASA.

In this long–range plan, several tasks will have highest priority. The first is to work toward the establishment of an international organization for archiving and disseminating high–accuracy GPS baseline results, and for distributing GPS orbit information. Another high–priority activity is to continue to support and contribute NASA data to data and network centers of the International Earth Rotation Service (IERS). The strong emphasis NASA has placed on international cooperation, sharing of data, and cooperative projects since the beginning of its Geodynamics Program will remain at the heart of the new Geodynamics Program.

A third important activity is to continue to support the WEGENER Medlas project to measure crustal deformation in the Mediterranean, and help to establish similar international regional deformation projects in other tectonically interesting parts of the world.

Fourth, NASA intends to work energetically for approval of space flight missions to measure the in–situ geopotential fields – geomagnetic and gravity – preferably to be undertaken in cooperation with space agencies in other countries. Finally, in this short list of highest priority actions, NASA must develop, deploy, and operate the Geodynamics Laser Ranging System (GLRS), a downward ranging laser system to be flown on one or more polar platforms of the Earth Observing System.

Consistent with these priorities we can also state several principles that will govern the future Geodynamics Program.

1. NASA should strengthen the worldwide IERS network of observatories, both in geographical distribution and in technological capabilities, in order to obtain better polar motion, Earth, and plate motion data for the scientific community. For example, more observatories are needed in South America, Africa, and Antarctica – some NASA mobile stations can be moved from Western North America to fill in such voids in the IERS network.

2. As rapidly as possible, NASA should develop inexpensive and highly portable GPS receivers, with antennas that minimize multipath effects.

3. NASA should increase the quality of the geophysical data and the accuracy of the baseline length measurements by continuing to develop space geodetic technology, and to provide this improved technology to the world scientific community.

4. There should be closer cooperation between the NASA Geodynamics and Geology Programs. The latter uses remote sensing (SAR, multispectral optical sensing) to study the geological structure and history of the lithosphere. Both programs need global topographic mapping of the continents, which can most effectively be done by a joint space flight mission.

5. In geopotential field measurements, NASA should press strongly for an early start on the Magnetic Field Explorer mission in cooperation with the CNES Magnolia mission; MFE/Magnolia will be a five–year flight of scalar and vector magnetometer. The objective of MFE/Magnolia will be to measure secular changes in the geomagnetic main field.

6. NASA should also press strongly for approval of its proposed Earth Probe line of low–cost space flights, and use this resource as soon as possible to augment the extremely austere ESA Aristoteles mission (which will fly part of a gravity gradiometer at an altitude greater than 200 km, and carry neither magnetometers nor a GPS receiver). The international science community has been pointing out strongly since 1969 the urgent need for gravity and magnetic field measurements with high accuracy and high spatial resolution. The lamented NASA Geopotential Research Mission would have achieved 2–3 mgal accuracy at 100 km horizontal resolution; Aristoteles is now planned to achieve about 5.5 mgal at this same resolution. However, with an inner stage, drag compensation, magnetometers and GPS receiver supplied by NASA, the full–component Gradio instrument and magnetometers could almost certainly achieve all the objectives of GRM.

7. To further improve the accuracy and spatial resolution of gravity field models, NASA should vigorously pursue the development of the Superconducting Gravity Gradiometer.

A number of questions must be answered in formulating this long–range program for NASA Geodynamics. For example, what should be done about the existing NASA networks of VLBI, SLR, and LLR stations? Should the present program of measuring only relatively long baselines for studies of plate motion and regional deformation with VLBI and SLR be continued? How should the activities of cooperating U.S. Federal agencies be handled in the post–1991 era? Similar questions exist about cooperation with ESA and with space agencies in France, Italy, Japan, and other countries.

One point must be emphasized: NASA will not simply terminate operation of its fixed and mobile SLR and VLBI stations on grounds that the international IERS networks would be adequate for geophysics and geodesy. However, to the extent that other organizations are now gathering data for polar motion, Earth rotation, and global plate motion, NASA can divert its activities to other fields of interest – in other words, the more that other countries do, the more freedom NASA has to redirect its activities without adversely impacting space geophysics.

2. INTERNATIONAL COOPERATION

A nongovernmental international agency is needed to provide coordination and general scientific oversight of the many bilateral and multilateral research projects that are now or soon will be undertaken by individual research organizations in different countries. A major step in this direction was taken in the early 1980's by a joint commission of the International Association of Geodesy, the International Astronomical Union, and the Committee on Space Research (Cospar) of the International Council of Scientific Unions (ICSU). This commission organized two international observation campaigns in 1982 and 1984 (MERIT – "Monitoring Earth Rotation and Intercomparison of Techniques"), and subsequently established the International Earth Rotation Service (IERS), based on the Project MERIT results. Global networks of stations, and international analysis centers, are now operating to gather, archive, and distribute data, and to carry out and distribute analysis of the laser ranging and VLBI data from all countries. The IERS centers were formally established in January 1988 and are now functioning well. NASA believes that a similar procedure should be followed for geodetic GPS measurements, with the objective of establishing centers for data archiving and distribution, as well as for standardized GPS orbit calculations. Without such centers the availability of high–quality GPS data for research in geodesy and geophysics will be inadequate.

Regional international cooperation in measurement of crustal deformation was first done under the WEGENER Medlas Project in the Mediterranean region. Beginning as a loosely organized group of NASA scientific investigators in Europe, WEGENER has now grown into a well coordinated multinational project, and has received the endorsement of the European Council of Governments, CSTG, and the Inter–Union Commission on the Lithosphere, an ICSU inter–union (International Union of Geodesy and Geophysics, and International Union of Geological Sciences). WEGENER is beginning to expand in several ways – by including the use of GPS and VLBI measurements as well as the original satellite laser ranging measurements, and in considering other regions to study.

Informal cooperative GPS measurements have already been carried out in the Caribbean, Andean South America, and the Southwestern Pacific; detailed planning for similar campaigns is now taking place in other regions, such as the Himalayas and Eastern Asia. Other countries are now developing mobile facilities and observatories, and these should be in operation within the next few years, contributing to IERS and to regional deformation measurements.

3. NEW TECHNOLOGY

Since the NASA Geodynamics Program began, the relevant technology has changed profoundly. The accuracy of VLBI and SLR in 1978 was not even ten centimeters, and there were known systematic errors as well; in contrast, the accuracy of all systems is now better than 1 cm, and the goal of the CDP is to be routinely achieving accuracies better than 1 mm in all systems, by the mid–1990's. This will require substantial funding for the technology development within the next few years.

Comparing space geodynamics in 1978 with the prospects for the 1990's, we are on the brink of having available two revolutionary new measurement systems: GPS – which must still be regarded as an experimental technology not yet fully operational – and the Geodynamics Laser Ranging System (GLRS), to be available late in the 1990's. The first and second generation GPS receivers developed in the early 1980's made it possible to use facilities for measuring crustal deformation which were highly mobile and which cost a few hundred thousand dollars per unit instead of the several million for the much more cumbersome mobile VLBI and SLR facilities. The third generation GPS facilities now being developed cost less than $100K and will use much simpler data processing algorithms; within 2–3 years GPS systems will be placed on a single VLSI chip housed in a unit that could be transported by a small child, and which will cost less than $15K. At this price it is possible to think of semi–permanent deployment of arrays of such GPS receivers instead of continually moving a few units from site to site.

GLRS, which puts the laser in space and the retroreflectors on the ground, will reduce the cost of measuring crustal deformation drastically, even below that of GPS operations. Once the reflectors are in place, no further movement of field crews is necessary. Data analysis will be done by the Earth Observing System Project, and baseline length and length changes provided directly to scientific investigators, perhaps even in the form of maps and tables transmitted via FAX machines.

4. GRAVITY AND MAGNETIC FIELD MISSIONS

A surprisingly large number of space missions with geophysical and geodetic objectives are now approved or planned by various space agencies for flights in the 1990's – these include TOPEX–Poseidon, ERS-1, MFE/Magnolia, Aristoteles, the Superconducting Gravity Gradiometer Shuttle Experiment, Gravity Probe-B, Mini-LAGEOS (small laser ranging satellites, including the CNES Stella), magnetometers and laser altimeters on the Earth Observing System polar platforms, a topographic mapping mission, and the Superconducting Gravity Gradiometer Mission.

To take advantage of the common theme of these missions, the NASA Geodynamics Program is considering proposing an "International Decade of the Gravity Field" as a method of focusing attention on the opportunity for obtaining definitive measurements of the magnetic and gravity fields.

The broad outline of NASA's activities for the next decade seem reasonably clear at present; the difficult work of planning specific projects in cooperation with other organizations and other U.S. Federal agencies will be undertaken over the next year. Whatever specific form the international projects in space geophysics and geodynamics take, NASA looks forward to contributing to major new advances in these scientific disciplines.

REFERENCE

NASA, 1979, Application of Space Technology to Crustal Dynamics and Earthquake Research, *Technical Paper 1464*, Washington DC, U.S.A.

THE GLONASS SATELLITE NAVIGATION SYSTEM

Glavcosmos

1. SYSTEM DESCRIPTION

The GLONASS satellite navigation system currently under development is intended for determining civil aviation aircraft position coordinates and airspeed. It could also be used to serve marine and fishery needs.

The system will consist of twenty–four satellites (three of them in a standby mode) positioned on three orbital planes, each plane accommodating seven–eight satellites.

The GLONASS system satellites will be positioned on circular orbits with the following parameters:

Period   – 11 hours, 15 minutes;
Altitude   – 19100 km;
Inclination – 64°.8.

The GLONASS system user equipment would perform non–interrogation measurements on up to four satellites' navigational parameters (pseudo–range and radial pseudo–speed). The user equipment would operate in a passive mode.

A navigation message transmitted from each satellite would consist of information on satellite ephemeris position and correction relative to the GLONASS system–time–scale, as well as information concerning all satellites' condition.

Based upon measurements the user three–dimensional coordinates and speed vector components are determined and its time scale is referenced to that of the system.

The GLONASS system satellites would emit navigation signals within the frequency band 1600 MHz. The nominal carrier frequency on the j–th satellite's navigation signal is determined as follows:

$$f_j = f_1 + (j-1)4f, \qquad j = 1, 2, 3, ..., 24,$$

$$f_1 = 1602.5625 \; MHz, \quad \Delta f = 0.5625 \; MHz$$

The satellites are identified by their navigation signal nominal carrier frequency.

2. SYSTEM PERFORMANCE PARAMETERS

Satellites	Twenty–four, including 3 standby satellites; circular orbits rotation period 11 h 15 min.; altitude 19100 km; I= 64°.8; 3 orbital planes.
Number of users	Unlimited
Frequency band	$(1602.5625 + 1615.5) \pm 0.5$ MHz
Navigational determination technique	Pseudo-range and radial pseudo-speed finding, the user in passive mode

Parameter output accuracy

Plane coordinates	100 m;
Altitude	150 m;
Speed vector components	15 cm/s;
Time	1 mcs

Signal detection time	Signal detection time depends largely on the user specific equipment performance. The satellites transmit informations for navigational purposes during 30 seconds and satellite's condition during 2.5 minutes.
Coverage	Global
Integrity	A message transmitted to the user from each satellite would contain data on troubles concerning that satellite as soon as they occur. Such information would appear in the content of a navigation message of all satellites not later than 16 hours after the trouble occurred.
Implementation schedule	Approximately 1989-1990 - ten–twelve satellites 1991-1995 - twenty-four satellites
Applicability to comunication	The system would not be used for retransmission of any signals or additional messages
System upgrading	The system accuracy can be significantly increased when user operation is in a differential mode.

User radio link performance using an isotropic antenna.

APPENDIX 4

SPACE GEODESY IN FRANCE

Anny Cazenave

1. INTRODUCTION

The French Space center (Centre National d'Etudes Spatiales) has initiated some twenty years ago a national program in Space Geodesy. It includes the launch of several satellites for geodesy (the Diademe satellites, Peole, Castor and Starlette), the development of the satellite Laser station in Grasse (France) the international program of Doppler tracking of the Transit satellites (MEDOC) and with the help of the Institut National des Sciences de l'Univers, the development of the Lunar Laser Ranging station also located in Grasse.

Research activities in Space Geodesy are mostly developed in the Groupe de Recherches de Geodesie Spatiale (GRGS). GRGS which includes six different teams is in charge of the initiation and preparation of Space Geodesy missions, the participation in international tracking campaigns, the operations of the tracking stations, the data analysis and the associated scientific research. For the next few years, several new projects are presently under development either as part of the national space program or as a cooperative program with other countries.

2. NATIONAL PROGRAM IN SPACE GEODESY

2.1 THE DORIS PROGRAM (decided)

A new system for precise orbit determination and high–accuracy beacon location has been developed by the French Space Agency. This system called DORIS is based on the measurement of Doppler shifts in radio signals transmitted by ground beacons over two frequencies (401 and 2036 MHz) and received by the onboard package. The DORIS onboard package comprises: a receiver, an ultrastable crystal oscillator and an antenna. The two frequencies transmission will allow elimination of errors due to ionospheric propagation delays. The ground beacon package is composed of two transmittors emitting every 10 s, an ultrastable oscillator, a battery for power back-up, an antenna and meteorological sensors.

The DORIS system will be placed onboard the SPOT2 French satellite, dedicated to Earth observation and to be launched early 1989, and later on onboard the TOPEX/POSEIDON and future SPOT (3, 4, etc..) satellites. The first objective of DORIS concerns very high precision orbit determination: a worldwide network of about fifty beacons evenly distributed around the world is currently implemented. The DOPPLER measurements of this global network will be used to determine the satellite orbit with a radial accuracy of about 10 cm. These fifty beacons will stay in position for at least a decade for the purpose of precise orbit determination of all satellites carrying the DORIS system.

The DORIS system has also the potential for high accuracy positioning of ground beacons. The positioning can be either absolute or relative and takes advantage of the high accuracy orbit determination. The ground location beacons are functionally identical to the orbitography beacons. They will be packaged for field use in sealed housing with their own power system (generator, solar panel or battery).

Although the DORIS system does not offer the on–site processing ability of GPS, it enables, after a few days delay, ground location of automatic beacons dropped in unfriendly environments. The DORIS system will be useful for various geodetic applications as well as for monitoring natural phenomena such as tectonic deformations movements of glaciers, ground sliding, or for surveying ground deformation during major civil engineering projects. The DORIS location system is particularly suitable for permanent survey of areas of high tectonic activity. Thanks to its very great tracking coverage, DORIS will be also very useful in the gravity field improvement.

2.2 THE STELLA SATELLITE (decided)

Stella is a passive Laser satellite, identical to Starlette (12 cm in radius, 47 kg in weight covered by 60 reflectors) to be launched on a sun–synchronous circular orbit of 98°.03 inclination at an altitude of 790 km. Stella will be launched together with the SPOT3 satellite, the sooner at the end of 1990, nominal launch date being 1992. As for the Starlette satellite, the scientific objectives will concern:

(1) Improvement of the Earth gravity field:
 • the Laser data on Stella will be used in the GEM and GRIM global Earth models.
 • zonal harmonics of the gravity field will be determined by analyses of long series of orbital elements of several geodetic satellites.
(2) Temporal changes of the geopotential coefficients.
 Previous investigations with Starlette have shown that seasonal variations in J_2 due to seasonal changes in the atmospheric load could be recovered. A new secular variation determination of first zonal harmonics has been recently performed at Texas University. To obtain a good separation of the various parameters, Starlette data should be analyzed jointly with the Stella data.
(3) Ocean tides coefficients.
(4) Relativistic effects.
 Relativistic effects give rise to a secular motion of the argument of perigee of the satellite orbit. Accurate determination of the motion of perigee should permit a determination of the relativistic coefficient (2 + 2 Gamma – Beta).

3. BILATERAL COOPERATION IN SPACE GEODESY

3.1 THE TOPEX/POSEIDON PROJECT (decided)

The objective of the project is to measure with onboard altimeters, the sea surface topography for the determination of the global ocean circulation. TOPEX/POSEIDON is a bilateral cooperation between USA and France. NASA will provide a dual–frequency altimeter, able to measure the satellite altitude with 2 cm precision, as well as a microwave radiometer, Tranet beacons and the GPS receiver placed onboard the satellite. CNES will provide a solid–state altimeter and the DORIS tracking system, as well as the launch by Ariane, in 1992. TOPEX/POSEIDON will orbit at an altitude of about 1330 km with an inclination of 64°, and will fly on the same ground track every ten days. TOPEX/POSEIDON should allow major advances in the field of ocean dynamics.

3.2 THE MAGNOLIA EXPERIMENT (not yet decided)

Magnolia is a French–US project for measuring by satellite the magnetic field of the Earth core, in particular the secular variation. Two magnetometers measuring the three components as well as intensity of the internal magnetic field should fly as passengers onboard a satellite at about 600 km altitude. Two joint (US and French) phase A and phase B studies have been performed during the last three years. Magnolia is waiting a phase C decision, which should not occur before late 1989.

4.PARTICIPATION IN EUROPEAN SPACE AGENCY PROGRAMS

4.1 THE ARISTOTELES MISSION (not yet decided but in a study phase)

Aristoteles (previously called GRADIO) is a project dedicated to measure with high accuracy (< 5 mgal) and high resolution (< 100 km) the Earth gravity field at a global scale, with an onboard gradiometer. The project was initially proposed by GRGS, the gradiometer technology being currently studied at the Office National d'Etudes et de Recherches Aerospatiales (ONERA) in France. After several years of Phase A studies conducted at CNES and ONERA, Aristoteles has been proposed to ESA and is now part of the Earth science program of the European Space Agency. Aristoteles is waiting for a "new start" decision. It could be launched with ERS2 by the Ariane launcher in the middle of the next decade.

Other studies are being developed to implement new accurate positioning systems onboard future Space Platforms (Laser, Doris, altimeter,...).

APPENDIX 5

THE LAGEOS II PROJECT

Susanna Zerbini

1. INTRODUCTION

The LAGEOS II program, a cooperative effort between the National Aeronautics and Space Administration (NASA) of the United States and the Space Agency of Italy (ASI), was established in order to furthering international and national activities in earth sciences. The LAser GEOdynamics Satellite II (LAGEOS II) has been fabricated by AERITALIA for ASI and will be launched by NASA. The scientific objectives of the mission are focused toward research in solid earth geophysics, made possible through very precise satellite geodesy, with particular emphasis on the following areas:

- Regional Crustal Deformations and Plate Tectonics;
- Geodetic Reference Datum and Earth Orientation;
- Earth and Ocean Tides;
- Temporal Variations in the Geopotential;
- Satellite Orbital Perturbations.

The program initiated in 1982 when NASA and ASI established a joint study group charged with defining the optimum characteristics for both the spacecraft and the orbit, in order to ensure the scientific benefits expected from the availability of the two LAGEOS satellites. The study group recommended that

- LAGEOS II be the same physical size, mass and construction as LAGEOS I;
- with the exception of the inclination, the orbital characteristics of LAGEOS II be very similar to those of LAGEOS I.

In February 1988 a Research Announcement was simultaneously released by NASA and ASI to solicit basic and applied research proposals to perform scientific researches by using LAGEOS II and LAGEOS I laser ranging data.

2. LAGEOS SATELLITES CHARACTERISTICS

LAGEOS II, like LAGEOS I, is a passive satellite exclusively dedicated to laser ranging. The fabrication of the satellite has been completed, and in July 1988 the flight unit has been transferred to Goddard Space Flight Center (MD, USA) for the pre-launch laboratory tests. In Table 1 the characteristics of the LAGEOS satellites are described; as concerns LAGEOS II only the differences with respect to LAGEOS I are listed. The weight difference of about 2 kg, LAGEOS II being lighter than LAGEOS I, appears to be due to the slightly different density in the brass alloy used to construct the satellite core. This small weight difference will not affect the mission objectives. In Fig. 1 the LAGEOS II flight unit is shown.

Table 1 Orbit and Spacecraft Characteristics for the LAGEOS Satellites

	LAGEOS I	**LAGEOS II**
LAUNCH	May 4, 1976	August 15, 1991
SPACECRAFT	Spherical, 60 cm diameter 406.965 kg Brass core Aluminum shell 426 laser retroreflectors	2 kg lighter
ORBIT SEMIMAJOR AXIS INCLINATION ECCENTRICITY	12,265 km 109°.8 0.004	52°

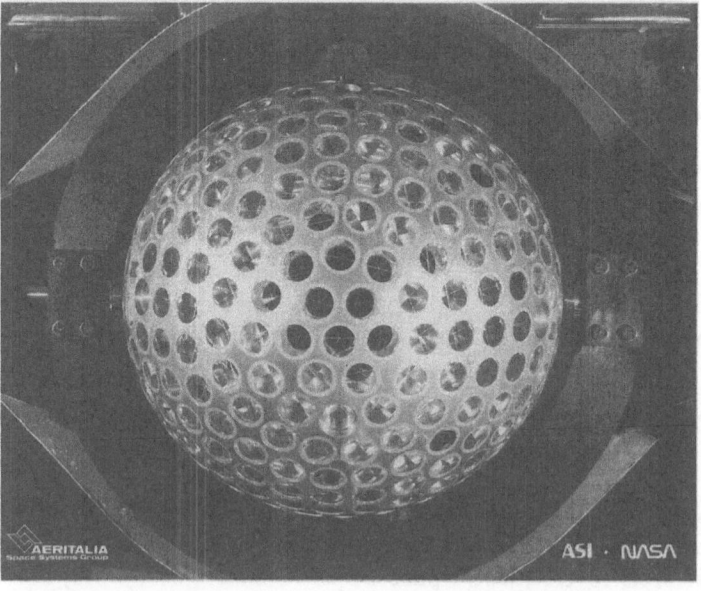

Fig. 1 The LAGEOS II satellite.

3. THE LAGEOS II MISSION

The satellite will be placed into the selected orbit through NASA's Space Shuttle (STS, Space Transportation System) in August 1991 from the Eastern Test Range. The LAGEOS II mission consists of the Italian Research Interim Stage (IRIS), a spinning solid perigee stage, and the apogee stage, a MAGE–1S class solid rocket motor. The LAGEOS II satellite and the apogee stage will be attached to the IRIS. This composite will be carried by the STS into low earth orbit (Space Shuttle parking orbit, I=28°.5). The mission profile is described in Fig. 2.

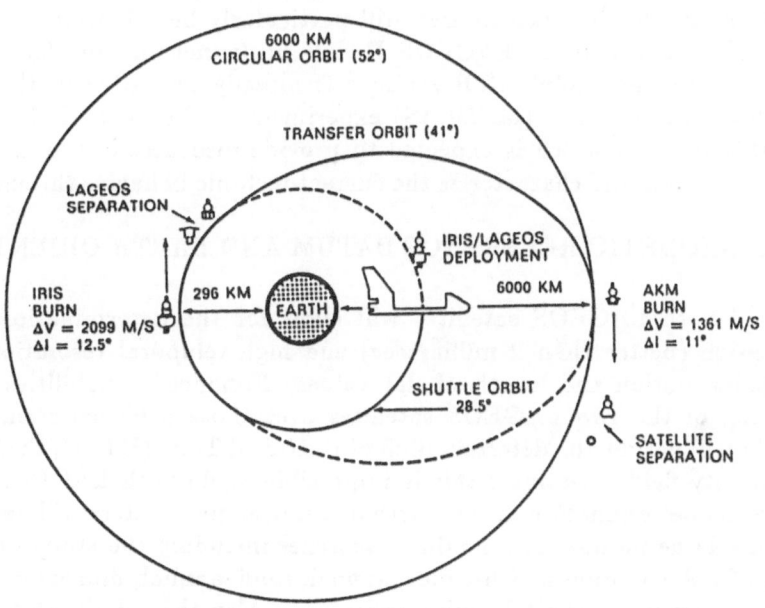

Fig. 2   LAGEOS II mission profile

After release from the Shuttle, at one of the nodal crossings the IRIS burn (both the IRIS and MAGE–1S maneuvers take place at nodal crossings) will inject the satellite and the apogee stage onto a transfer trajectory. On the opposite node the satellite will be inserted into a 6000 km, circular orbit with inclination of 52° by the apogee stage.

4. SCIENTIFIC OBJECTIVES

The scientific objectives of the mission are centered on further developing international and national scientific programs in solid earth geophysics by means of very precise satellite geodesy. The data which are expected to be acquired from both LAGEOS satellites will greatly contribute to the development of researches in different areas.

4.1 REGIONAL CRUSTAL DEFORMATION AND PLATE TECTONICS

The availability of LAGEOS II will significantly improve our capability of measuring global tectonic motions and will contribute by as much as a factor of 2 in the detection of regional crustal motions and deformations. There are several earthquake–prone areas at mid-latitudes where Satellite Laser Ranging (SLR) systems are or will be available, among them, the Mediterranean area will particularly benefit from the studies which will be undertaken soon as LAGEOS II will be launched. In this region a dense network of fixed and mobile SLR stations is already operative in the framework of the MEDiterranean LASer (MEDLAS) experiment of the WEGENER initiative (see Wilson, this volume) which is expected to provide measures of the kinematics of the plates and platelets and characterize the current tectonic behavior throughout this area.

4.2 GEODETIC REFERENCE DATUM AND EARTH ORIENTATION

Tracking of both LAGEOS satellites will allow for the determination and study of high–precision (better than 2 milliarcsec) and high temporal resolution (better than 2 days) polar motion and length–of–day values. Enhanced capabilities resulting from the tracking of the two LAGEOS satellites over those achieved from tracking only LAGEOS I will permit the determination of Universal Time (UT–1), thereby separating it from gravity field variations; this is impossible to do with LAGEOS I alone. The high precision determination of the Earth orientation parameters will have a significant impact on a large number of scientific researches including the study of the frequency structure of polar motion at Chandler, annual, semi–annual, diurnal etc., periods, the study of the nature of the Chandler excitation. Also the relationship between polar motion, earthquakes, and mass displacements in the Earth, as well as the yielding of the Earth with movements of the rotation axis can be investigated. The impact of atmospheric movements and mass transport on the orientation of the Earth and the core–mantle coupling can be studied.

4.3 EARTH AND OCEAN TIDES

The orbital inclinations of the two LAGEOS satellites are different enough to allow for enhanced capabilities for the recovery of Earth and ocean tidal coefficients at low frequency, for numerous ocean tide constituents and the frequency–dependent Love numbers. Improved understanding of the Earth and of crustal dynamics could be achieved through the study of core/mantle resonances, the computation of the Earth's

Q at intermediate frequencies and the study of tidal dissipation.

4.4 TEMPORAL VARIATIONS IN THE GEOPOTENTIAL

The orbital inclinations of the two LAGEOS satellites differ sufficiently for their contributions to the study of the geopotential to be complementary. Continuous monitoring of the evolution of the orbits of the satellites will permit the observation of changes in the geopotential. This may lead into a better understanding of numerous areas in geophysics including the Earth's rheology through observations of post-glacial responses, studies of mantle convection, observations of mass transports and observations of the rate of drift of the Earth's mean figure axis.

4.5 SATELLITE ORBITAL PERTURBATIONS

Studies can be performed on conservative and non-conservative forces (direct solar and earth reflected radiation pressure, air drag....) acting on the LAGEOS satellites. The different orbital inclinations of the two satellites might help in clarifying the nature of the unexplained along-track acceleration of LAGEOS I.

APPENDIX 6

THE WEGENER PROGRAMME

Peter Wilson

1. INTRODUCTION

As most participants will recognize by now, WEGENER is an acronym standing for the Working-group of European Geo-scientists for the Eastablishment of Networks for Earthquake Research. The idea for such a working group was first presented at the 45th Meeting of the Journees Luxembourgeoise de Geodynamique in December 1980 (Wilson, 1980) and the group met for the first time in March 1981 to formulate the outlines of a series of coordinated responses to the Announcement of Opportunity (AO) OSTA 80-2 on Crustal Dynamics and Earthquake Research, issued by the National Aeronautics and Space Administration (NASA) in August 1980 (NASA, 1980). The prime objective of the group is to initiate and to coordinate inter-disciplinary research activities directed towards the investigation of earthquake processes and the associated geodynamic phenomena and, in a later phase, to colate the results for joint interpretation.

Under this concept, complete projects can be carried out by groups of cooperating scientists or institutions under a common umbrella. Projects which have emerged to date cover the investigation of regional kinematics using satellite laser ranging (SLR) and the global positioning system (GPS) techniques and the establishment of regional networks of point determinations of absolute gravity.

For the purpose of conducting observations and analyses in connection with these activities agreements have been reached with the host countries which define the objectives of the research, establish permission for visiting investigators to conduct the observations and encourage the participation of the activities and the data analysis. All data taken under these agreements is available to the host countries and to any of the participating scientists submitting a request. The plans, ongoing activities and results are presented and discussed at the semi-annual meetings of the Crustal Dynamics Project, at workshops and at international scientific meetings. At this juncture it is appropriate to emphasize the strong international participation in this cooperative effort in which more than 30 European, African, Asian and North American institutions from 14 countries are directly involved. Publications have appeared in various journals. The work of the group to date has been concentrated in the Central and Eastern Mediterranean, with ties to established permanent reference stations, particularly in Western Europe.

2. WEGENER-MEDLAS

The first of the foregoing projects to become established was the WEGENER MEDiterranean LASer-ranging Project-WEGENER-MEDLAS. This Project was formally launched in 1985. The original objectives for WEGENER–MEDLAS (Table 1) were presented in proposals submitted to NASA for European participation in the Crustal Dynamics Project (CDP) and included the investigation of kinematic changes associated with major tectonic features in Switzerland, Southern Italy, Greece and Turkey as they are related to the Eurasian and the African tectonic plates.

Table 1   The Objectives of the WEGENER–MEDLAS Project

To estimate the rates of motion (extension, shortening, rotation) across some major tectonic features in the region, including:

– the North Anatolian fault,
– the East Anatolian fault,
– the Aegean basin,
– the Hellenic arc,
– The Adriatic promontory of the African Plate,
– the Peloponnes,
– the Calabrian arc,
– the Tyrrhenian basin;

to provide a fiducial network for further densification using other measurement techniques;

to improve existing and develop new models for computing:

– satellite orbits,
– station positions,
– baseline lengths and baseline variations,
– regional kinematic models,
– regional improvements in the gravity field;

to provide new information for geodynamic interpretation.

Other objectives covered the establishment of a network of fiducial reference stations for geodynamics which would later be densified by other techniques and the development and improvement of computational models. To test the feasibility of the techniques proposed it was desirable from the start to obtain first kinematic trends as soon as possible. The proposal therefore implied concentration on baselines across some of the major tectonic features of the region, which were suspected of demonstrating the largest annual rates of changes. However, delays encountered in the course of early negotiations, forced some changes in the original priorities. Though, in principle, the overall objectives have remained the same, circumstances have led to more frequent re–occupations of stations around the Aegean, where motions as large as 6 cm per year are considered likely. As a result, this area has assumed added significance for the early years of the Project and the first kinematic trends computed should be available for that area.

The tectonics of the Central and Eastern Mediterranean areas are extremely complex. The area is located along the collision zone of the plate boundaries between the continental paltes identified by Africa and Arabia to the south and Eurasia to the north (Fig. 1). The Mediterranean and Black Seas are seen as remnants of the ancient Tethys Ocean and cover significant portions of the boundary zone. The region displays the highest level of tectonic activity in Europe. The North and East Anatolian faults, the Hellenic arc, the Adriatic promontory of the African plate and the Calabrian arc are dominant tectonic features of the region and both the Aegean and the Tyrrhenian basins, with their characteristic volcanicity, have been identified as back–arc basins in different stages of development. All of these features localize bands of high seismicity in which a number of seismic gaps have been postulated. Areas exhibiting totally different tectonic characteristics are separated by small distances and even lie adjacent to each other (Fig. 1), with the result that kinematic modelling is very difficult. From the outset of the WEGENER–MEDLAS Project it has therefore been a primary objective to obtain direct evidence of the magnitudes and directions of the motions taking place. As it is neither economically or logistically feasible nor geodetically desirable to determine high density networks of extreme accuracy covering extensive areas in a single operation, it was decided to establish first a coarse fiducial reference network to obtain first estimates of the motions across the major tectonic features before moving in with other techniques to perform selective densification. In this way it is also hoped that additional information on the interaction between the various forces at work across the region over shorter time spans can lead to a better understanding of the tectonics of the region.

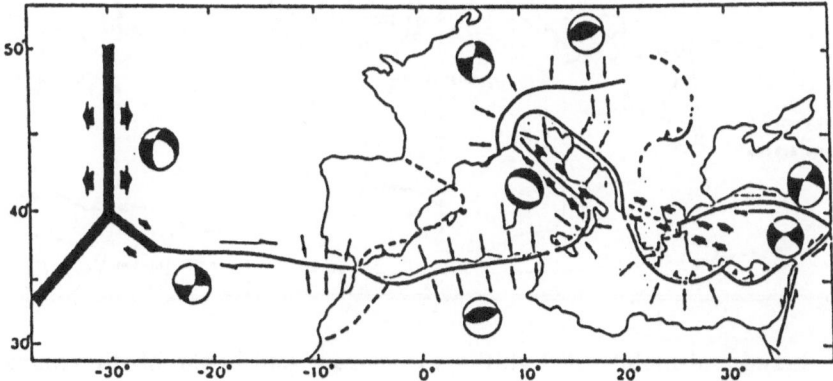

Fig. 1 Focal mechanisms and characteristic motions in the vicinity of the Afro–Eurasian plate boundary (after Udias, 1982).

Some 16 temporary stations have been occupied by mobile laser ranging systems in the region of the Central and Eastern Mediterranean since the commencement of field observations in 1985 (Fig. 2).

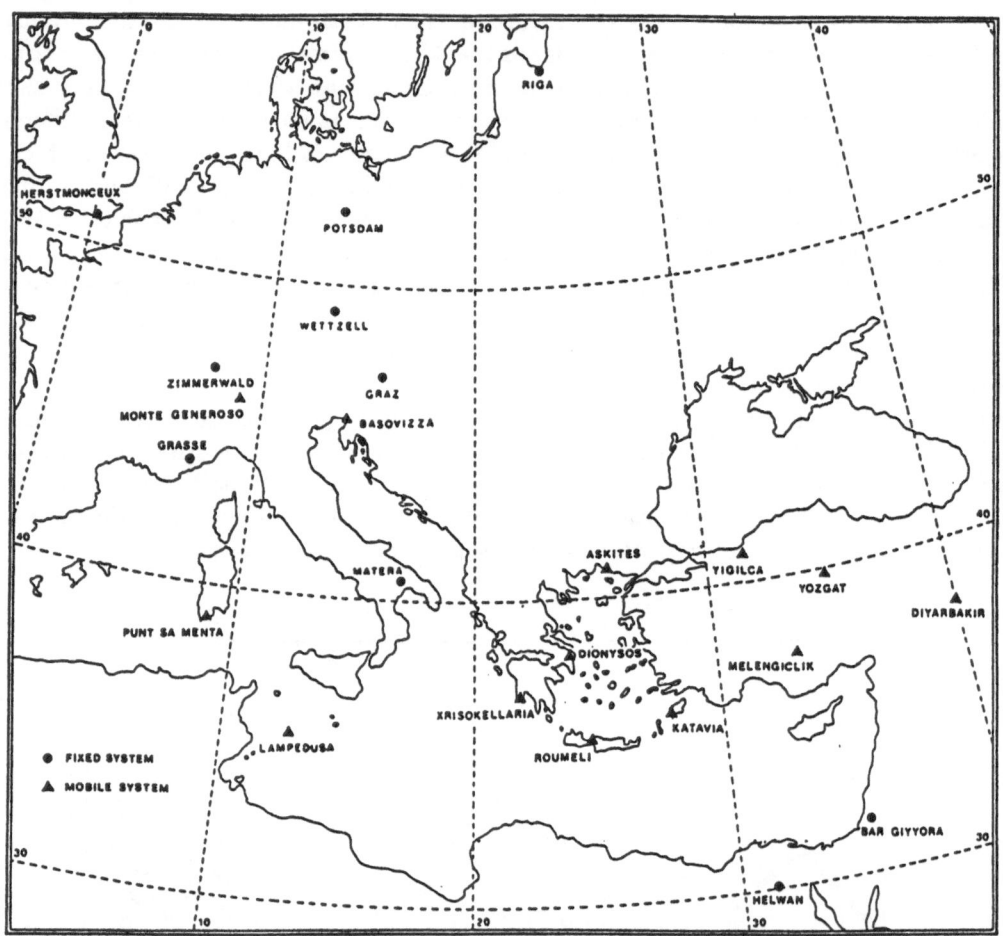

Fig. 2   Fixed and mobile SLR stations in Europe and in the Mediterranean.

Three systems have so far been available to conduct the observations. The Modular Transportable Laser Ranging System MTLRS-1 (operated by the IFAG from Germany) has been operational in the Mediterranean in 1986 and 1987 and MTLRS-2 (operated by the Delft University of Technology from the Netherlands) was in the field there in 1985, 1986, 1987 and in 1988. These systems were joined in 1987 by the Transportable Laser Ranging System TLRS-1 (operated for NASA from the U.S.A.). Through this cooperation six of temporary stations have so far been occupied more than once. The use of three mobile systems in the same regional project demonstrated the feasibility of obtaining large numbers of quasi-simultaneous passes (as many as 8 of the 10 stations occupied simultaneously in Central Europe and the Mediterranean were tracked one pass) and opened the way to the consideration of different modelling techniques, which until then appeared to be totally impracticable. It is anticipated that first kinematic trends along baselines across the Aegean Sea will be identified from the analyses of the 1987 data, the results of which will be presented at the CDP meeting in Munich in October. Further repeat observations will be performed every two years, with the core of repeat observations concentrating on the network observed in 1987. The next round of repeat observations is due to commence in March 1989 and will again run through to December.

The first observations for WEGENER-MEDLAS were conducted in Switzerland and Sardinia in 1985. Since that time 1328 passes of LAGEOS have been observed by the three mobile systems operating in the Mediterranean area. This result compares favourably with about 3000 passes collected by the ten permanent European stations submitting data to the Project over a similar period. A full break down of the data collected at all stations in the region is given in Table 2.

These activities have also drawn the attention of Eastern European investigators and discussions are being held on the feasibility of extending the investigations to the Balkans and the areas around the Black Sea in cooperation with the planned IDEAL Project (Georgiev et al., 1985). This interest is being followed up, and so far data has been submitted from stations of the Interkosmos network located at Helwan and Riga (Table 2). Further Eastern European stations will be added as evidence is presented that the data quality matches that of the Mediterranean networks.

Initial results from dynamic solutions computed with the 1985 and 1986 observation series were presented at CDP and WEGENER-MEDLAS meetings held in 1987 at Bologna and at Goddard Space Flight Center (GSFC). They showed good overall consistency, with formal standard errors of about 10 mm and 25 mm for the regional solutions computed from smaller data sets of quasi-simultaneous passes at the Royal Greenwich Observatory (Sinclair, 1987). These last results initiated a discussion on the data requirements, time and sky coverage needs for obtaining reliable station positions and baselines with a 1 cm error bar. As a consequence, a set of test data from the 1987 data set has been defined and distributed to the established analysis centres.

Table 2   Passes Observed by the European Laser Ranging Systems Participating in WEGENER–MEDLAS

Site number	Location	Passes obtained '85	'86	'87	'88	Projected occ. '89
Temporary sites						
7510	Askites (GR)		52	131		*
7512	Kattavia (GR)		56	61		*
7515	Dyonisos (GR)		49	155		*
7517	Roumeli (GR)		72	104		*
7520	Karitsa (GR)		46			*
7525	Hrisokellaria (GR)		23	57		
7540	Matera M1 (I)		33			
7541	Matera M2 (I)		62			
7544	Lampedusa (I)		87			*
7545	Punta sa Menta (I)	35			58	*
7546	Medicina (I)				12	
7550	Basovizza (I)		29			*
7575	Diyarbakir (I)			31		*
7580	Melengiclik (TR)			44		*
7585	Yozgat (TR)			36		*
7587	Yigilca (TR)			47		*
7590	Monte Generoso (CH)		48			
Permanent sites						
1181	Postdam (GDR)					
1884	Riga (SU)					
7530	Bar Giyyora (IS)		13	75	3	
7810	Zimmerwald (CH)		24	69	26	
7831	Helwan (ET)					
7834	Wettzell (D)		93	140	29	
7835	Grasse (F)		22	110	177	
7839	Graz (A)		88	114	34	
7840	Herstmonceux (GB)		402	243	224	
7939	Matera (I)		198	259	110	

Computations are now being made to attempt to get a better estimate of the data requirement for the mobile systems.   The results of these analyses should also be available for presentation at the CDP meeting in Munich in October.

3. NETWORK DENSIFICATION WITH GPS

Network densification for the Central and Eastern Mediterranean has been visualized at two levels. The first level aims at establishing large area densification to establish a 50–100 km spacing between points spanning the length and breadth of the major tectonic features (North Anatolian fault, Hellenic arc etc.). The second level envisages selective densification – 10 km spacing or less – over more restricted areas.

To date first level densification is being conducted in three areas, viz. the Calabrian arc and Tyrrhenian basin region of Italy (Fig. 3), the back arc basin of the Hellenic arc and trench (Fig. 4) and selected parts of Anatolia (Fig. 5).

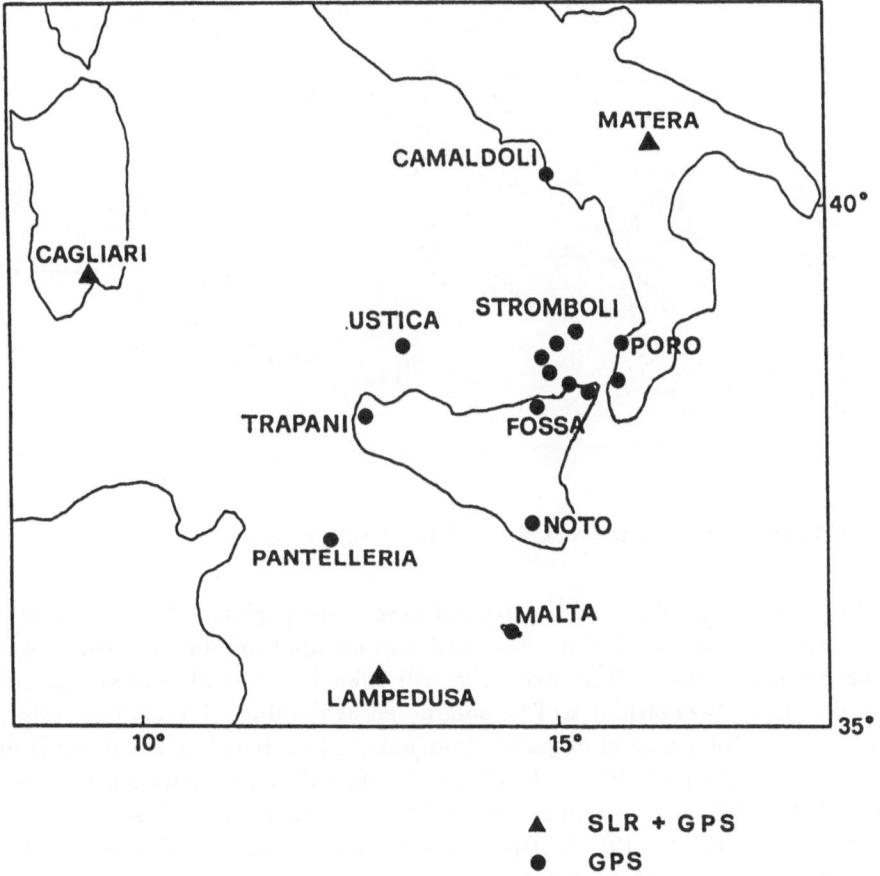

Fig. 3   GPS densification network in the area of the Tyrrhenian Basin.

Fig. 4   GPS densification network around the Cretan Sea.

Whereas observations in the Italian area were performed for the first time in 1986 and are being repeated this year, first site occupations in the other regions will only take place this year. The networks will then be expanded next year (Figs. 6 and 7) and repeat observations will be conducted at regular intervals thereafter. These projects are again the result of extensive international cooperation involving Italian and German groups in Italy (Baldi et al., 1988), Greek, U.S. and German groups in Greece (Kastens, 1986, 1987; Wilson and Seeger, 1987) and Turkish and U.S. groups in Turkey (Toksöz et and Reilinger, 1987). Up to ten dual frequency GPS receivers are being fielded simultaneously in these projects.

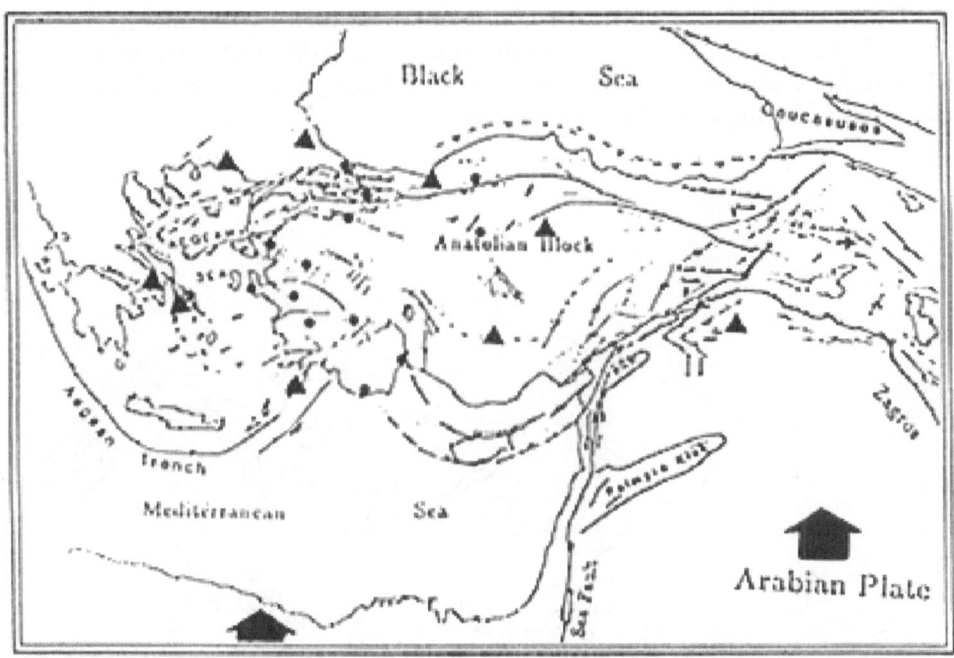

Fig. 5   Anticipated Phase 1 GPS densification network in W. Anatolia.

Two other projects are also being prepared at the second level of densification – one in central Greece (Cross, 1988) and the other in south western Turkey (Foulger, 1987). These cover areas of about 100 km diameter with something of the order of 80 to 100 points in each network. In the Greek network it is planned to use both dual and single frequency instruments, but in Turkey only dual frequency receivers are envisaged.

Each of these projects is addressing specific aspects of ongoing deformation as anticipated in the original objectives of WEGENER. As an example, evidence has been presented by different investigators for both linear and radial expansion across the Aegean and the current measurements in the southern area of this region, which are being conducted under the leadership of Kim Kastens and George Veis, have been designed to investigate the real nature of the expansion taking place there.

Using the (deforming) frame of the SLR stations in the Mediterranean as a fiducial reference, the first level of densification using GPS will serve to provide more detailed analyses of the patterns of deformation taking place and to localize areas of special interest for high density investigations.

Fig. 6   Proposed laser/GPS network for regional studies of the Hellenic Arc and the Aegean Basin. △ Established laser sites, • projected GPS sites, dotted lines: densification network boundary (high density).

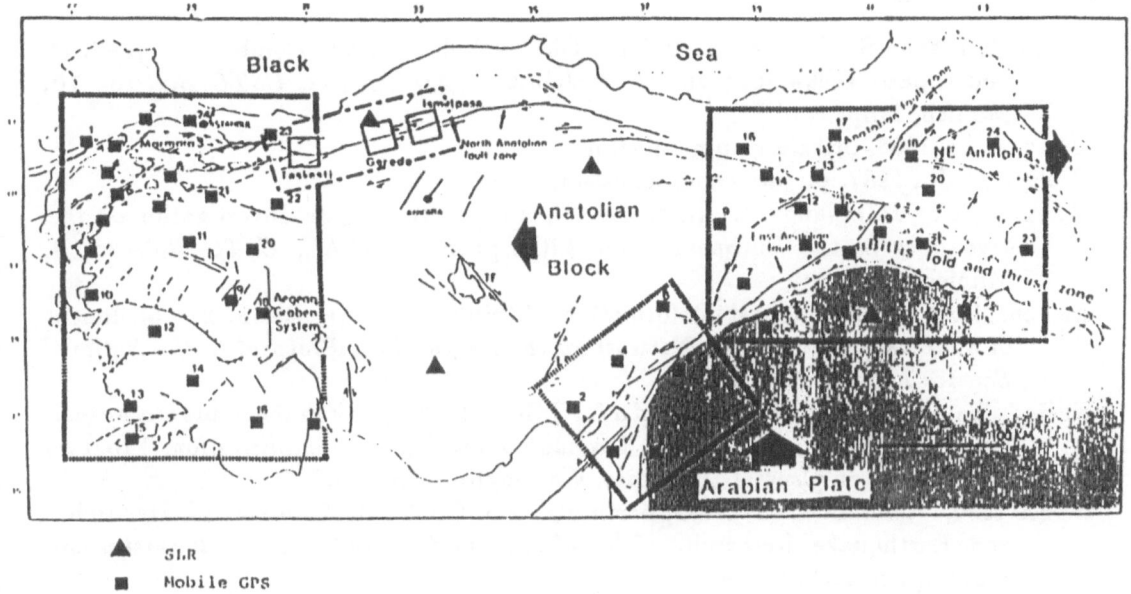

Fig. 7   GPS densification networks envisaged in Turkey.

4. CONCLUSIONS

The WEGENER concept was presented in an attempt to facilitate large scale multi-disciplinary investigations covering a whole region embracing different nations with different political backgrounds. The successes achieved to date serve to underline the urgency of the problems being addressed to the nations involved, which are not in a position to finance extensive investigations applying state of the art technology alone. The processes under examination necessitate repeated observations over long time periods. As a result, it is extremely important to obtain first observations as early as possible, though this may incur added expense. The WEGENER–MEDLAS Project established the necessary framework for follow-up activities and set standards for international cooperation which can be emulated in other regions of the globe confronted with similar problems.

REFERENCES

Baldi, P., Zerbini S., Drewes H., Reigber Ch., Achilli V., 1988, Combined Terrestrial and Space Techniques in the Calabrian Arc Project, *CSTG Bulletin*, **10**, Munich, 115.

Cross, P. A., 1988, private communication.

Foulger, G. R., 1987, private communication.

Georgiev, N., Totomanov I., Hadjiiski Al., 1985, Project for Investigation on the Dynamics of the European-Asian Lithosphere - IDEAL, *CSTG Bulletin*, **8**, Munich, 175.

Kastens, K. A., 1986, Internal Deformation Within a Back Arc Basin: Establishment of a GPS-based Geodetic Network in the Aegean Sea, *Proposal to the National Science Foundation.*

Kastens, K. A., 1987, Establishment of a GPS Geodetic Network in the Southern Aegean Sea, *Proposal to the National Aeronautics and Space Administration*, Program for Research in Crustal Dynamics, Washington, D.C., U.S.A.

NASA 1980, Announcement of Opportunity NO. OSTA 80-2 on Crustal Dynamics and Earthquake Research, *National Aeronautics and Space Administration*, Washington D.C., U.S.A.

Sinclair, A. T., 1987, The Determination of Coordinates and Motions of SLR Stations by Short-Arc Methods, *paper presented at the Crustal Dynamics Principle Investigators Fall Meeting*, GSFC, Greenbelt, Md., U.S.A.

Toksöz, M. N., Reilinger R., 1987, Global Positioning System Measurements of Faulting and Regional Deformation in Turkey, *Proposal to the Earthquake Research Institute*, Ankara, Turkey.

Udias, A. 1982, Seismicity and seismotectonic stress field in the Alpine - Mediterranean Region; Alpine - Mediterranean Geodynamics, *American Geophysical Union - Geodynamics Series*, **7**, 75.

Wilson, P., 1980, Proposal for an European Response to the NASA Announcement of Opportunity OSTA-2 on Crustal Dynamics and Earthquake Research *Report to the 45th Journees Luxembourgeoises* Walferdange, Luxemburg.

Wilson, P., Seeger H., 1987, A proposal for Conducting GPS Observations Along the Hellenic Arc and Across the Aegean Sea, *Proposal to the Hellenic Geodetic Commission*, Athens, Greece.

PANEL MEMBERSHIP

ENGINEERING

Co-chairmen:
F. BARLIER, CERGA/GRGS, Grasse, France
C. GOAD, The Ohio State Univ., Columbus, OH, USA

Panel members:
O. COLOMBO, E&G/WASC. Inc., Lanham, MD, USA
R. EANES, CSR–Univ. of Texas at Austin,TX, USA
S. KLOSKO, EG&G/WASC. Inc., Lanham, MD, USA
H. SEEGER, IFAG, Frankfurt am Main, FRG
S.H. YE, Shanghay Observatory, Shanghay, P.R. of China
T. YOSHINO, Kashima Space Research Center, Japan

REMOTE AND EARTH-BASED INSTRUMENTATION

Co-chairmen:
W. MELBOURNE, JPL, Pasadena, CA, USA
Ch. REIGBER, DGFI, Munich, FRG

Panel members:
T. CLARK, NASA GSFC, Greenbelt MD, USA
B. GREENE, Electro Optic Systems PTY LTD, Queanbeyan, Australia
Ph. HARTL, University of Stuttgart, FRG
Y.L. KOKURIN, Lebedev Physical Inst., Moscow, USSR
H.J. PAIK, University of Maryland, College Park, MD, USA
M. PEARLMAN, Harvard Smithsonian CFA, Cambridge, MA, USA
D. SONNABEND, Jet Propulsion Laboratory, Pasadena CA, USA
T. VARGHESE, Bendix Field Eng. Corp., Greenbelt MD, USA
E. VERMAAT, Delft Univ. of Technology, Delft, The Netherlands
P. WILSON, IFAG, Frankfurt am Main, FRG
L. YOUNG, Jet Propulsion Laboratory, Pasadena CA, USA
T. YUNCK, Jet Propulsion Laboratory, Pasadena CA, USA

SOLID EARTH PHYSICS SHORT-TERM

Co-chairmen:
I.I. MUELLER, The Ohio State Univ., Columbus OH, USA
J. ZSCHAU, University of Kiel, FRG

Panel members:
B.B. BAGHOS, Space Research Department, Helwan, Egypt
J. DICKEY, Jet Propulsion Laboratory, Pasadena, USA
S. DICKMAN, State University of NY, Binghamton, USA
T. HERRING, Harvard, Smithsonian CFA, Cambridge, MA, USA
R. O'CONNELL, Harvard University, Cambridge, MA, USA
R. REILINGER, MIT, Cambridge, USA
D.E. SMYLIE, University of York, Ontario, Canada
R. SABADINI, University of Bologna, Italy

SOLID EARTH PHYSICS LONG-TERM

Co-chairmen:
A. CAZENAVE, CNES/GRGS, Toulouse, France
D. TURCOTTE, Cornell University, Ithaca, NY, USA

Panel members:
A.J. ANDERSON, University of Uppsala, Sweden
B. HAGER, California Inst. of Technology, Pasadena, CA, USA
K. KASTENS, Lamont Doherty Geological Obs., Palisades, NY, USA
E. MANTOVANI, University of Siena, Italy
M. McNUTT, MIT, Cambridge, MA, USA
B. MINSTER, Scripps Institute of Oceanography, La Jolla, CA, USA
R. RAPP, Ohio State University, Columbus, OH, USA
J. RUNDLE, Sandia National Labs., Albuquerque, NM, USA

OCEAN PHYSICS

Co-chairmen:
M. LEFEBVRE, CNES/GRGS, Toulouse, France
S. WILSON, NASA Headquarters, Washington, D.C., USA

Panel members
E. HARRISON, NOAA/PMEL, Seattle, WA, USA
S. HOURY, CNES/GRGS, Toulouse, France
C.K. TAI, Scripps Institute of Oceanography, La Jolla, CA, USA
V. ZLOTNICKI, Jet Propulsion Laboratory, Pasadena, CA, USA

INTERACTION WITH OTHER DISCIPLINES

Co-chairmen:
F. GILBERT, Scripps Inst. of Oceanogr., La Jolla CA, USA
H. GRASSL, GKSS-Forschungszentrum, Geesthacht, FRG

Panel members:
K. BURKE, Lunar and Planetary Inst., Houston, TX, USA
I. FEJES, Inst. of Geodesy, Cart.and Remote Sensing, Budapest, Hungary
R. HIDE, Geophysical Fluid Dynamic Lab., Bracknell Berks., England
K. LAMBECK, The Australian National Univ., Canberra, Australia
J. MELOSH, University of Arizona, Tucson, AZ, Usa
A. MORELLI, Istituto Nazionale di Geofisica, Rome, Italy
D. SMITH, NASA GSFC, Greenbelt, MD, USA
G. VISCONTI, University of l'Aquila, Italy
C. YODER, Jet Propulsion Laboratory, Pasadena, CA, USA

EDUCATION IN GEODESY

Co-chairmen:
B. SCHUTZ, CSR-University of Texas at Austin, USA
G. VEIS, NTU Athens, Greece

Panel members:
A.A. ASHOUR, University of Cairo, Giza, Egypt
M. CAPUTO, University of Rome, Italy
D. CHRISTODOULIDIS, Jet Propulsion Laboratory, Pasadena, CA, USA
CH. HARRISON, University of Miami, FL, USA
J. RAIS, National Com. of Geodesy and Geophys., Jakarta, Indonesia
S. TATEVIAN, USSR Academy of Sciences, Moscow, USSR

AGENCIES

ESA – S. HIEBER, ESA H.Q., Paris, France
GLAVCOSMOS – V.A. GRITSAY and A.N. ZAHAROV, Moscow, USSR
NASA GEODYNAMICS – E. FLINN, NASA H. Q., Washington, D.C., USA
NASA OCEAN – S. WILSON, NASA H. Q., Washington, D.C., USA
NASA GEOLOGY – M. BALTUCK, NASA H. Q., Washington, D.C., USA
PSN – G. SYLOS-LABINI, Piano Spaziale Nazionale, Rome, Italy

MISSIONS

ACRE – R. BEARD, U.S. Naval Research Lab., Washington, D.C., USA
ARISTOTELES – B. PFEIFFER, ESA–ESTEC, Noordwijk, The Netherlands
CDP – R. COATES, NASA GSFC, Greenbelt, MD, USA
ERS-1 – B. PFEIFFER, ESA–ESTEC,Noordwijk, The Netherlands
EOS – S. COHEN, NASA GSFC, Greenbelt, MD, USA
GP-B – F. EVERITT, Standford University, CA, USA
GLRS – S. COHEN, NASA GSFC, Greenbelt, MD, USA
LAGEOS-II – S. ZERBINI, University of Bologna, Italy
STELLA – M. LEFEBVRE, CNES/GRGS, Toulouse, France
TOPEX/POSEIDON – M. LEFEBVRE, CNES/GRGS, Toulouse, France
WEGENER – P. WILSON, IFAG, Frankfurt am Main, FR

LIST OF PARTICIPANTS

JOSÈ ACHACHE
Inst. de Physique du Globe de Paris
4, Place Jussieu
Paris, Cedex 05
France

DARIO ALBARELLO
Istituto Nazionale di Geofisica
Dip. Fisica – Settore Geofisica
Viale Berti Pichat, 8
40127 Bologna – Italy

ATTIEH AL-GHAMDI
National Observatory Project
King Abdulaziz City for Sci. and Tech.
Riyadh
Saudi Arabia

FAHAD AL-SAEED
National Observatory Project
King Abdulaziz City for Sci. and Tech.
Riyadh
Saudi Arabia

ALLEN JOEL ANDERSON *
Planetary Geodesy and Geophysics
The University of Uppsala
Hallby
75590 Uppsala – Sweden

ATTIA A. ASHOUR ⊘
Department of Mathematics
Cairo University
Gizam
Egypt

DANIELE BABBUCCI
Dipartimento di Scienze della Terra
Università di Siena
Via Banchi di Sotto 55
Siena – Italy

BALEEGH BISHARA BAGHOS ★
Nat. Res. Inst. of Astronomy and
Geophysics – Helwan Observatory
Helwan
Egypt

MIRIAM BALTUCK ◁
NASA Headquarters – Code EEL
600 Independence Av., SW
Washington, D.C. 20546
U. S. A.

ALDO BANNI
Stazione Astronomica
Università di Cagliari
Via Ospedale 72
09100 Cagliari – Italy

FRANCOIS BARLIER ◇
CERGA/GRGS
Avenue Copernic
06130 Grasse
France

RONALD L. BEARD △
U.S. Naval Research Laboratory
Space Application Branch
4555 Overlook Av., S.W.
Washington, D.C.– U.S.A.

GIUSEPPE BIANCO
Piano Spaziale Nazionale
Centro di Geodesia Spaziale
Casella Postale 155
75100 Matera – Italy

GIUSEPPE BIRARDI ⊘
Dipartimento Idraulica, Tras. e Strade
Università di Roma
Via Eudossiana, 18
Rome – Italy

ENZO BOSCHI
Presidente
Istituto Nazionale di Geofisica
Via di Villa Ricotti 42
00161 Rome – Italy

JOHN BOSWORTH
NASA GSFC
Code 601
Greenbelt, MD 20771-0001
U.S.A.

KEVIN BURKE ⊗
Lunar and Planetary Institute
3303 NASA Road One
Houston, TX 77058
U.S.A.

MICHELE CAPUTO ⊘
Dipartimento di Fisica
Università di Roma
Piazzale Aldo Moro
Rome – Italy

ANNY CAZENAVE *
GRGS – CNES
18 Avenue Edouard Belin
31400 Toulouse
France

ALBERTO CENCI
Telespazio SpA
Via Bergamini 50
00158 Rome
Italy

PENGFEI CHENG
The Res. Inst. of Surv. and Mapping
16, Beitapinglu,The Western Suburbs
Beijing
The Peoples Republic of China

DEMOSTHENES CHRISTODOULIDIS ⊘
JPL/CALTECH
MS 238-640
4800 Oak Grove Drive
Pasadena, CA 91109 – U.S.A.

THOMAS A. CLARK ○
NASA GSFC
Code 621
Greenbelt, MD 20771-0001
U.S.A.

ROBERT J. COATES △
Crustal Dynamics Project
Code 601
NASA GSFC
Greenbelt, MD 20771-0001 – U.S.A.

STEVEN COHEN △
NASA GSFC
Code 621
Greenbelt, MD 20771-0001
U. S. A.

OSCAR COLOMBO ◇
EG&G/WASC. Inc.
5000 Philadelphia Way
Suite J – Bldg 16
Lanham, MD 20706 – U.S.A.

ANTONELLA DALL'OGLIO
Istituto Nazionale di Geofisica
Dip. Fisica – Settore di Geofisica
Viale Berti Pichat 8
40127 Bologna – Italy

JEAN O. DICKEY ⋆
JPL/CALTECH
MS 238-332
4800 Oak Grove Drive
Pasadena, CA 91109 – U.S.A.

STEVEN R. DICKMAN ⋆
Geology Department
State University of New York
Binghamton, New York 13901
U.S.A.

TIMOTHY H. DIXON
JPL/CALTECH
MS 264-802
4800 Oak Grove Drive
Pasadena, CA 91109–U.S.A.

RICHARD EANES ◇
Center for Space Research
The University of Texas at Austin
Austin, TX 78712
U. S. A.

C.W. FRANCIS EVERITT △
Hansen Lab. of Physics, GP-B
Standford University
Standford, CA 94305-4085
U.S.A.

ISTVAN FEJES ⊗
Fomi Satellite Geodetic Observatory
PO Box 546
1373 Budapest
Hungary

THOMAS L. FISCHETTI
Tech. Management Consultants, Inc.
2609 Village Lane
Silver Spring, MD 20906
U.S.A.

LUCE FLEITOUT
Laboratoire de Geophysique
Ecole Normale Superieure
75231 Paris
France

EDWARD FLINN ◁
NASA Headquarters Code EEG
Chief Geodynamics Branch
600 Independence Av., S.W.
Washington D.C. 20546– U.S.A.

JEAN GAIGNEBET
CNES/CERGA
Avenue Copernic
06130 Grasse
France

FREEMAN GILBERT ⊗
Scripps Inst. of Oceanography
Univ. of California, San Diego
IGPPA A-025
La Jolla, CA 92093, U.S.A.

CLYDE GOAD ◇
Department of Geodetic Science
The Ohio State University
Columbus, OH 43210-1247
U.S.A.

STANISLAV GORGOLEWSKI
Torun Radio Astronomy Obs.
Ul Chopina 12/16
87–100 Torun
Poland

HARTMUT GRASSL ⊗
GKSS-Forschungszentrum Geestacht
Max Planck Strasse 1
2054 Geesthacht
Federal Rep. of Germany

B. A. GREENE ○
Electro Optic Systems PTY LTD
55A Monaro St.
Queanbeyan, NSW 2620
PO Box 201, Wanniassa A.C.T. 2903
Canberra – Australia

VADIM A. GRITSAY ◁
GLAVCOSMOS USSR
9 Krasnoproletarskaya St.
103030 Moscow
U.S.S.R.

BRADFORD H. HAGER *
Seismological Laboratory
California Inst. of Technology
1201 East California Boulevard
Pasadena, CA 91125 – U.S.A.

CHRIS G.A. HARRISON ⊘
School of Marine,Atmospheric Sci.
University of Miami
4600 Rickenbacker Causeway
Miami, FL 33149 – U.S.A.

ED HARRISON •
NOAA/PMEL
7600 Sand Point Way, N.E.
Seattle, WA 98115
U.S.A.

PH. HARTL ○
Institute of Navigation
University Stuttgart
7000 Stuttgart
Keplerstrasse 11
Federal Rep. of Germany

THOMAS A. HERRING ⋆
Smithsonian Center for Astrophysics
Harvard University–MS 43
60 Garden Street
Cambridge, MA 02138 – U.S.A.

RAYMOND HIDE ⊗
Geophysical Fluid Dynamics Lab
Meteorological Office (METEO)
Bracknell, Berks RG12 2SZ
U. S. A.
England

JAMES R. HEIRTZLER
Geophys. Branch, Code 622
Lab for Terrestrial Physics
NASA GSFC
Greenbelt MD 20771-0001 – U.S.A.

SIGFRIED HIEBER ◁
EUROPEAN SPACE AGENCY
8-10 Rue Mario Nikis
75738 Paris Cedex
France

SABINE HOURY •
CNES/GRGS
18, Avenue Edouard Belin
31055 Toulouse Cedex
France

KIM KASTENS *
Lamont – Doherty
Geological Observatory
Palisades, NY 10964
U.S.A.

WILLIAM M. KAULA
Department of Earth and Space Sci.
University of California
3806 Geology Building
Los Angeles, CA 9002
U.S.A.

STEVEN KLOSKO ◇
EG&G/WASC, Inc.
5000 Philadelphia Way
Suite J – Bldg 16
Lanham, MD 20706 – U.S.A.

YURI L. KOKURIN ○
P.N. Lebedev Physical Institute
Academy of Science USSR
Leninsky pr. 53
117924 Moscow – U.S.S.R.

KURT LAMBECK ⊗
Research School of Earth Sciences
The Australian National University
GPO Box 4
Canberra 2601
Australia

MICHEL LEFEBVRE ●
CNES/GRGS
18, Avenue Edouard Belin
31055 Toulouse Cedex
France

HELMUT LENHARDT
Institute of Physical Geodesy
Technical University Darmstadt
Petersenstrasse 13
6100 Darmstadt
Federal Rep. of Germany

GIANCARLO LUCARELLI
Istituto Universitario Navale
Istituto di Navigazione Napoli
Via Acton 38
80133 Napoli – Italy

ENZO MANTOVANI *
Dip. di Scienze della Terra
Università Siena
Via Banchi di Sotto, 55
Siena – Italy

MARIA MARSELLA
Dip. Fisica – Settore Geofisica
Università di Bologna
Viale Berti Pichat, 8
40127 Bologna – Italy

MARCIA McNUTT *
Massachusetts Inst. of Technology
54-826 MIT
Cambridge, MA 02139
U.S.A.

WILLIAM G. MELBOURNE ○
JPL/CALTECH
MS 238-540
4800 Oak Grove Drive
Pasadena, CA 91109
U.S.A.

JAY MELOSH ⊗
Lunar and Planetary Laboratory
University of Arizona
Tucson, AZ 85721
U. S. A.

JEAN-BERNARD MINSTER *
Scripps Inst. of Oceanography
Univ. of California, San Diego
IGPP A-025
La Jolla, CA 92093
U.S.A.

FRANCO MONZANI
LABEN
Strada Padana Superiore 290
20090 Vimodrone MI
Italy

ANDREA MORELLI ⊗
Istituto Nazionale di Geofisica
Via di Villa Ricotti 42
00161 Rome
Italy

MARCO MUCCIARELLI
I S M E S
Bergamo
Italy

IVAN I. MUELLER ★
Department of Geodetic Science
The Ohio State University
1958 Neil Avenue
Columbus, OH 43210-1247
U. S. A.

RICHARD 0' CONNELL ★
Geological Sciences Laboratory
Harvard University
Cambridge, MA 02938
U. S. A.

HO JUNG PAIK o
Department of Physics and Astronomy
University of Maryland
College Park, MD 20742
U. S. A.

HAIM B. PAPO
Technion Israel Inst. of Technology
Department of Civil Engineering
Technion City
Haifa 32000
Israel

MICHAEL R. PEARLMAN o
Smithsonian Institution
Astrophysical Observatory
60 Garden Street
Cambridge MA 02138
U.S.A.

BARTOLOMEO PERNICE
Centro di Geodesia Spaziale
Casella Postale 155
75100 Matera
Italy

B. PFEIFFER △
Earth Obs. Programmes Dept.
ESA – ESTEC
Nordwijk
The Netherlands

DOMENICO PICCA
Dipartimento di Fisica
Università di Bari
Via G. Amendola 173
70126 Bari
Italy

JACUB RAIS ⊘
Nat. Com. Geodesy and Geophys.
c/o Bakosurtanal, P.O. Box 3546
Jakarta
Indonesia

RICHARD RAPP ★
Department of Geodetic Science
The Ohio State University
1958 Neil Avenue
Columbus, OH 43210-1247–U.S.A.

CHRISTOPH REIGBER o
Deutsches Geodätisches
Forschungsinstitut
Marstallplatz 8
München 22
Federal Rep. of Germany

ROBERT REILINGER ★
MIT – Earth Resources Laboratory
Dept. of Earth, Atm. Plan. Sciences
42, Carleton Street
Cambridge, MA 02142
U.S.A.

GEORGE W. ROSBOROUGH
Col.Center for Astronautical Res.
University of Colorado / Box 429
Boulder, CO 80309-0440
U. S. A.

JOHN RUNDLE *
Sandia National Laboratory
Division 6231
Albuquerque, NM 87185
U. S. A.

ROBERTO SABADINI *
Dip. Fisica – Settore Geofisica
Università di Bologna
Viale Berti Pichat 8
40127 Bologna – Italy

DAVID SANDWELL
Center for Space Research
The University of Texas at Austin
Austin, TX 78712
U. S. A.

GIANNINA SANNA
Dipartimento di Ingegneria Strutturale
Università di Cagliari
Piazza d'Armi
09100 Cagliari–Italy

RAFFAELE SANTAMARIA
Istituto Universitario Navale
Via Acton 38
80133 Napoli
Italy

BOB E. SCHUTZ ⊘
Center for Space Research
The University of Texas at Austin
WRW 402D
Austin, TX 78712 – U.S.A.

HERMANN SEEGER ◇
I F A G
Richard Strauss Allee 11
6000 Frankfurt a.M. 70
Federal Rep. of Germany

AVI SHAPIRA
Seismological Division
Inst. Petroleum Res. and Geophys.
P.O. Box 1717
58117 Holon – Israel

CHE-KWA SHUM
Center for Space Research
The University of Texas at Austin
Austin, TX 78712
U. S. A.

DAVID E. SMITH ⊗
NASA GSFC
Code 620
Greenbelt, MD 20771-0001
U.S.A.

D.E. SMYLIE *
Dept. Earth and Atmospheric Science
York University
4700 Keele Street
North York, Ontario–Canada M3J 1P3

DAVID SONNABEND ○
JPL/CALTECH
MS 301-125J
4800 Oak Grove Drive
Pasadena, CA 91109 – U.S.A.

GIOVANNI SYLOS-LABINI ◁
Piano Spaziale Nazionale
Viale Regina Margherita 202
00198 Rome
Italy

CHANG-KOU TAI ●
Scripps Inst. of Oceanography
University of California, San Diego
A-030
La Jolla, CA 92093 – U.S.A.

SURIA TATEVIAN ⊘
Astronomical Council USSR
48 Pjatniskaja St.
109017 Moscow
U.S.S.R.

PAOLO TOMASI
Istituto di Radioastronomia
C.N.R.
Via Irnerio 46
40100 Bologna – Italy

DONALD L. TURCOTTE *
Department of Geological Sciences
Cornell University
4122 Snee Hall
Ithaca, NY 14818 – U.S.A.

FRANCO VARESIO
AERITALIA
Gruppo Sistemi Spaziali
Corso Marche 41
10146 Torino – Italy

THOMAS K. VARGHESE ○
Bendix Field Engineering Corp.
10210 Greenbelt Road, Suite 700
Seabrook, MD 20706
U.S.A.

GEORGE VEIS ⊘
National Tech. Univ. of Athens
9, Heroon Politecniou Street
157 Zographos – Athens
Greece

ERIK VERMAAT ○
Observatory for Satellite Geodesy
Delft University of Technology
P.O. BOX 581
7300 AN Apeldoorn–The Netherlands

GRAZIA VERRONE
Dipartimento di Fisica
Università di Bari
Via Amendola 173
70125 Bari – Italy

GUIDO VISCONTI ⊗
Dipartimento di Fisica
Università dell'Aquila
Piazza Annunziata 1
67100 L'Aquila – Italy

ERNESTO VITTONE
AERITALIA
Gruppo Sistemi Spaziali
Corso Marche 41
10146 Torino – Italy

RANDOLPH WARE
Univ. NAVSTAR Consortium
University of Colorado
CIRES/499
Boulder, CO 80309 – U.S.A.

PETER WILSON ○△
I F A G
Richard Strauss Allee 11
Frankfurt a.M. 70
Federal Rep. of Germany

STANLEY WILSON ●
NASA Headquarters, Code EEC
600 Independence Av, SW
Washington, D.C. 20546
U. S. A.

SHU-HUA YE ◇
Shangai Observatory
Shangai
China

CHARLES YODER ⊗
JPL/CALTECH
MS 183-501
4800 Oak Grove Drive
Pasadena, CA 91109 – U.S.A.

TAIZOH YOSHINO ◇
Kashima Space Res. Center
Communications Res. Lab.
Hirai 893-1, Kashima-machi
Ibaraki-ken 314 – Japan

LARRY YOUNG ○
JPL/CALTECH
MS 238-600
4800 Oak Grove Drive
Pasadena, CA 91109
U.S.A.

THOMAS P. YUNCK ○
JPL/CALTECH
MS 238-640
4800 Oak Grove Drive
Pasadena, CA 91109
U.S.A.

ALEKSANDR N. ZAHAROV ◁
GLAVCOSMOS USSR
9, Krasnoproletarskaya Street
103030 Moscow
U. S. S. R.

SUSANNA ZERBINI △
Dip. Fisica – Settore Geofisica
Università di Bologna
Viale Berti Pichat 8
40127 Bologna – Italy

VICTOR ZLOTNICKI •
JPL/CALTECH
MS 300-323
4800 Oak Grove Drive
Pasadena, CA 91109
U.S.A.

JOCHEN ZSCHAU ⋆
Institut für Geophysik
Neue Universität
Olshausenstrasse
2300 Kiel
Fedearal Rep. of Germany

◇ ENGINEERING
○ REMOTE AND EARTH–BASED INSTRUMENTATION
⋆ SOLID EARTH PHYSICS SHORT–TERM
∗ SOLID EARTH PHYSICS LONG–TERM
• OCEAN PHYSICS
⊗ INTERACTION WITH OTHER DISCIPLINES
⊘ EDUCATION IN GEODESY
◁ AGENCIES
△ MISSIONS

STUDENTS

ALESSANDRA AVVENUTI   FEDERICO MATTIOLI
MAURO BALZANI   ANDREA TALLARICO
LIVIO CASAVECCHIA
all at:
Dip. Fisica – Settore Geofisica
Viale Berti Pichat 8
40127 Bologna – Italy

VINCENZO DONNARUMMA   CRISTINA TAMBURELLI
Both at:
Dipartimento di Scienze della Terra
Università di Siena
Via Banchi di Sotto, 55
Siena – Italy